□

SECRETS OF THE HOARY DEEP
———————————————————————

RICCARDO GIACCONI

SECRETS OF THE HOARY DEEP

A PERSONAL HISTORY OF MODERN ASTRONOMY

□

THE JOHNS HOPKINS UNIVERSITY PRESS

Baltimore

© 2008 The Johns Hopkins University Press
All rights reserved. Published 2008
Printed in the United States of America on acid-free paper
2 4 6 8 9 7 5 3 1

The Johns Hopkins University Press
2715 North Charles Street
Baltimore, Maryland 21218-4363
www.press.jhu.edu

Library of Congress Cataloging-in-Publication Data

Giacconi, Riccardo.
Secrets of the hoary deep : a personal history of modern astronomy /
    Riccardo Giacconi.
    p.   cm.
Includes bibliographical references and index.
ISBN-13: 978-0-8018-8809-0 (hardcover : alk. paper)
ISBN-10: 0-8018-8809-3 (hardcover : alk. paper)
1. X-ray astronomy—History—20th century.   2. Astronomy—History—
    20th century.   I. Title.
QB472.G53 2008
520.92—dc22   2007044287

A catalog record for this book is available from the British Library.

The last printed pages of this book are an extension of this copyright page.

*Special discounts are available for bulk purchases of this book. For more information,
please contact Special Sales at 410-516-6936 or specialsales@press.jhu.edu.*

The Johns Hopkins University Press uses environmentally friendly book materials, including
recycled text paper that is composed of at least 30 percent post-consumer waste, whenever
possible. All of our book papers are acid-free, and our jackets and covers are printed on paper
with recycled content.

In memory of my son, Marc Antonio Giacconi,
and dedicated to my wife, Mirella,
to our daughters, Anna and Guia,
and to my grandchildren, Alexandra and Colburn

Before their eyes in sudden view appear
The secrets of the hoary deep, a dark
Illimitable ocean without bound,
Without dimensions, where length, breath, and height,
And time and place are lost; where eldest Night
And Chaos, ancestors of Nature, hold
Eternal anarchy, amidst the noise
Of endless wars, and by confusion stand.
For hot, cold, moist and dry, four champions fierce
Strive here for mastery, and to battle bring
Their embryon atoms.

FROM MILTON'S *PARADISE LOST*

□

Salviati: . . . Pero', signor Simplicio, venite pure con le ragioni e con le dimostra-
zioni, vostre o di Aristotele, e non con testi e nude autorita', perche' i discorsi nostri
hanno a essere intorno al mondo sensibile, e non sopra un mondo di carta.
[However Mr. Simplicio, do come forth with reasons and demonstrations, your
own or Aristotle's, but not with texts or naked authority, because our discourses
should be about the real world, not about a world of paper.]

FROM GALILEO GALILEI'S *DIALOGO SOPRA I MASSIMI SISTEMI DEL MONDO*

□

Astrogation and military history he absorbed like water; abstract mathematics
was more difficult, but whenever he was given a problem that involved
patterns in space and time, he found that his intuition was more reliable
than his calculation—he often saw at once a solution that he
could only prove after minutes or hours of manipulating numbers.

FROM ORSON SCOTT CARD'S *ENDER'S GAME*

Contents

Contents

Preface

This book is an account of the development of astronomy from 1959 to 2006 as told by one of the participants. It is intended not as an autobiography but rather as a narrative of my own understanding of the field in an intellectual sense and its development as I experienced it. Biographical notes are thrown in as necessary to explain the changes in my perspective.

I was very fortunate in my career as a scientist to be involved in some of the most exciting discoveries and projects of the past decades. I participated in the start of a new field of astronomical space research—observations in the x-ray region of the spectrum—which grew from a subdiscipline of interest to a handful of scientists to an important and unique tool to study the universe, of interest to all astronomers. It turned out that high-energy phenomena, explosions, high-energy particles, and million-Kelvin plasmas play a fundamental role in the formation and evolution of the cosmos. All of these processes and entities are copious emitters of x-rays, and most of the normal matter in the universe is in the form of the hot plasmas whose existence was discovered in x-rays. X-ray observations also permitted discovery of stellar-mass black holes and the process of energy generation through accretion onto collapsed objects, which powers active galactic nuclei and quasars.

In 1981, after 20 years in this field, I was appointed as the first director of the Space Telescope Science Institute, an independent institute that is responsible for the scientific utilization of the Hubble Space Telescope. The institute played a central role in making Hubble the great scientific and popular success that it became. The institute led the world of astronomy in the operation of extremely complex systems that produce copious amounts of data at extremely high rates. The development of end-to-end systems of data management at the institute—the handling of guest observers' proposal submissions and scheduling, receipt of the data and on-line calibration, data distribution, and archiving—became the standards adopted by all major astronomy observatories both in space and on the ground. The scientific results from Hubble had a major impact on all of astronomy. The institute placed great emphasis on supporting research based on Hubble data by astronomers worldwide, with fellowships and research programs, but also on sharing the fascinating new vistas that were opened up by Hubble with the public at large.

Family circumstances led me to return to Europe in 1993, where for six years I directed the European Southern Observatory, during which time we were constructing the largest array of optical telescopes in the world—the Very Large Telescope now operating on Cerro Paranal in the Atacama Desert of Chile. With this project, European astronomy set new standards of excellence for ground-based astronomy.

On my return to the United States in 1999, I became president of Associated Universities, Inc., a nonprofit university-based organization that operates the National Radio Astronomy Observatory. The observatory is constructing the world's largest array of millimeter and submillimeter radio antennas, called ALMA, on the Llano de Chajnantor in Chile, in cooperation with Europe and Japan.

I was fortunate to have so many opportunities to be involved with enterprises at the forefront of astronomy in all wavelengths. My work took place in Europe and in the United States in a variety of institutional settings: universities, national and international institutes, and private industry. The agencies that funded these efforts were also varied; they included the Department of Defense, NASA, the National Science Foundation, universities, and such international organizations as the European Southern Observatory. In being exposed to so many different styles of work I was able to form my own opinions on how well the different systems functioned for science. In this book I describe my experiences with candor.

The title of the book is intended to convey the awe with which I contemplate nature and to hint at the irrational impulses that drive the work of scientists much more than is generally realized. It is also a nod to the fact that my narrative takes a peek behind the curtains of sanitized reporting on how great projects come about. I apologize in advance to all the scientists, engineers, and managers who contributed to the work I describe but whom I am not able to recognize in detail. I also offer my apologies to those I may inadvertently offend with my rather blunt style of writing. I hope this book may be accepted as a contribution to the description of the sociology of science.

Throughout this book I have tried to recognize the many people who have contributed to the work that I am describing. Here I recognize the debt of gratitude I owe to the people who helped with the book itself. My most candid and helpful critics were Ethan Schreier, my colleague and friend for more than 30 years, and my wife, Mirella, who read the manuscript in draft form. Mirella helped with the editing and also directly contributed to the writing of the first chapter. Their efforts substantially improved the book.

I also thank Miriam Satin of Associated Universities, Inc.; Hans Hermann Heyer of the European Southern Observatory; and Pete Medicino of the Johns Hopkins University for helping me to assemble the photographic material and for guiding my confrontational efforts with typing and computer usage.

I recognize my debt of gratitude to the staff of the Johns Hopkins University Press—in particular, to Trevor Lipscombe, who encouraged me along the way, and Bronwyn Madeo, who helped me organize the figures in the volume. A special thanks to Cyd Westmoreland of Princeton Editorial Associates, my copyeditor, for improving the legibility of the text, correcting my dysfunctional spelling of proper names, and posing helpful questions.

Finally, I thank my wife, Mirella, not only for her critical reading of the manuscript but also for her constant support throughout my career.

□

SECRETS OF THE HOARY DEEP

———————————————

ONE

My Italian Roots

Though I was born in Genoa (on October 6, 1931), I have always considered myself a Milanese, because it is in Milan that I was raised and educated. Today, having swallowed all its suburbs, Milan is a sprawling, multinational metropolis of five million people, with all the attendant problems of traffic and pollution. It may have gained in wealth and power over the past 50 years, but sadly the price for those gains has been the loss of much of its character.

In the 1930s and up to the time I left Italy in 1956, Milan was a smallish city, about the size of Boston proper. It was a commercial and industrial center, busy working and making money, but also a civilized and very livable place. It never had the grace of Renaissance Florence or the fantastic opulence of Rome; what it did have once was an old-fashioned dignity that reflected the spirit and work ethic of its entrepreneurial bourgeoisie.

While Rome was the seat of the government and a huge bureaucratic beehive, Milan was the undisputed business capital, the Italian equivalent of New York. Not surprisingly, it was also the country's cultural Mecca. Its opera house, La Scala, attracted the most famous singers in the world; its theaters staged memorable performances, from Shakespeare to Bertolt Brecht and Jean Anouilh. All that, plus museums, concerts, and wonderful bookstores.

It also had excellent schools, and whatever culture I was able to bring to the New World I owe, at least in part, to them.

Good schools depend on good teachers, and my mother, Elsa Giacconi Canni, was an outstanding teacher. A professor of mathematics, she taught math, geometry, and physics in one of the city's best-known scientific lyceums. During her exceptionally long career, she wrote numerous textbooks that were adopted by the entire school system and for which she received great praise and recognition.

My father, a World War I veteran, was a self-made man—an accountant, a carpenter, a trade union leader, a tinkerer, and a thinker. As an outspoken anti-fascist and card-carrying socialist, he always had great difficulty in finding and keeping a position in fascist Italy. During his relatively short life, he worked at many jobs, always with little success. I remember him as a decent, thoughtful, straight-thinking man, one of the few who could see events and their consequences through the fog of propaganda. It was from him that I first heard the famous saying "The emperor has no clothes." And yet, because of his lack of success, he was considered ineffective and was resented by those members of the family who had learned to conform for the sake of quiet living. Both my aunts were staunch fascists, and my mother, though apolitical, had managed to obtain a card that certified her as one of the earliest members of the party. Her membership was a fiction, but during the war it may have helped all of us.

Of the two, there is no question that my mother had by far the greater impact on my life. She was a highly ambitious woman, willful and domineering, and I believe that the bitterness caused by her marital problems led her to place great emphasis not only on her own success but even more on mine. This attitude began fairly early and remained unchanged throughout her life. I was supposed to be special, and whenever I failed to achieve the highest possible marks, she would ask me what was wrong and complain that I had not lived up to her expectations. Over time, her expectations grew to include my winning the Nobel Prize during her lifetime.

In the fall of 1937, my parents decided to separate. First because of the breakup of the family and later because of the dislocations caused by the war, I spent much of my childhood being shunted from place to place. It is probably one of the reasons I never developed a true sense of home until my old age. For a long time houses were for me just stops along the way.

When my parents separated they enrolled me, at age 6, in the military boarding school San Celso, in Milan. All I remember of that period is that I had to wear a uniform and that I once ran into a brick column, either in a fit of rebelliousness or during violent play. I cracked my head, suffering a concussion and temporary blindness. Apparently, though, I had shown a certain precocity and willingness to learn, and my teachers decided that I should skip the second grade and go directly to third.

The following year I was sent to Genoa to live with maternal relatives, and there I started the third grade in the private school of the Marist Brothers, a Catholic monastic order. Everyone was nice to me, but I was homesick, unhappy, and totally unmotivated. Then began my behavioral and disciplinary problems, ranging from absenteeism to bloody fistfights, which were to plague me for years. That year I often skipped school and spent time fishing from the rocky seashore or wandering about the town. Still, I was bright enough to finish the year with the minimum passing grade, equivalent to a C, in all subjects.

However, the following year, when I went back to Milan to live with my mother, she decided that a C would not do for her son, and to my intense dismay she made me repeat the third grade in the local public school. Boredom did nothing to improve my character or performance. It was only because my teacher took me firmly in hand that I was able to complete elementary school; however, some of his approaches, such as having me sit at an isolated desk near his own, further contributed to my sense of separation from schoolmates.

About the same time I was drafted into Mussolini's youth organization, the so-called Figli della Lupa (Sons of the She-Wolf; legend has it that Romulus and Remus, the twin founders of Rome, were suckled by a female wolf), who sported a scaled-down version of the uniform of Mussolini's infamous Black Shirts. I remember singing patriotic songs and performing endless group calisthenics for the entertainment of the fascist authorities. Unfortunately I also started to absorb some of the fascist rhetoric that was fed to us every day via radio, newspapers, and school lessons. As a result, for the rest of my life I had to fight an embarrassing tendency to speak with oratorical flourishes.

The summer of 1939 was one of the happiest times of my young life. I was sent to spend my vacations in the village of Pigra, on a mountain overlooking Lake Como. And there I lived for three months in the rustic house of an old widow, who treated me not as a paying guest but as any other village boy. During the day, it was my job to mind the calves in their high mountain pastures. In the evening, we would sit around the fire with other village families shucking corn and telling stories. I was brought back to Milan just as World War II started, on September 9, 1939, with the German invasion of Poland.

Conflict and Inner Turmoil: Childhood during World War II

I have only patchy memories of the war years.

I remember the excitement that swept through Milan—the rejoicing, the flag waving—when Italy entered the war on the side of Germany on June 10, 1940. People at that time seemed to generally support Mussolini's policy, perhaps believing that Germany had already won. (A belief presumably shared by Mussolini himself, as he must have known that the country was totally unprepared for war.)

I also remember that, in the midst of all the celebrations, my father tried to explain to a brainwashed 9-year-old why the coming war was not a splendid adventure but a terrible tragedy for everyone involved, win or lose. It took me a few years to fully understand what he meant, but then I never forgot.

For the first two years, the war hardly touched us; the fighting was far away, and we had no relatives or close friends in the army. The only problems as far as we were concerned were the strict rationing of food and clothing and the blackout, which had turned the city streets into gloomy canyons.

This uneasy quiet came to an abrupt end in the fall of 1942, when the Allies started bombing industrialized northern Italy in a series of surprise daylight raids. I have a vivid memory of an afternoon in October when, running with my mother to a nearby bomb shelter, I saw a string of bombs falling from the open bomb bay of a warplane. When we emerged from the shelter later in the evening, the city was ablaze, and from a terrace on top of our house we could see a ring of fire all around us.

Although the subsequent bombardments were not perhaps as bad as the London Blitz, there was a general exodus from Milan. My mother stayed, but she decided that I should go to live with her two older sisters in Cremona, a quiet, safe town on the banks of the Po River. The trains being out of commission, my father bought me a bicycle. Traveling with some acquaintances, I was able to reach Cremona, 60 miles away, in one day—not bad for an 11-year-old kid. What I found there was a happy time even in a world gone crazy.

My cousin Nanni became a brother to me, and a friend, which was a wonderful thing for a lonely child. We were the same age and attended the same class at school. I was also lucky enough to find a young Italian and Latin teacher who cared about us and showed it. As that is the only kind of teaching I ever responded to, I worked hard for her, only to discover that I was doing so for myself as well. On the other hand, I got into trouble with the math teacher by proving to her in front of the whole class that I knew more about math than she did. It was true, but I now regret my behavior, which was arrogant and rude.

At that time I also discovered books. My aunts had assembled a pretty extensive collection of classic works, and they opened up for me a whole new world of delights. My mother was not interested in literature—the bookcase we had at home was full of textbooks and little else. She was not religious, either, and saw the world as a stark place where God had created geometric archetypes.

More books were to be found in the summer camp that my Aunt Giulia directed. The founder of the camp, an old doctor, had left a large library, which included not only the classics but also old and rare books of anatomy, medicine, psychology, chemistry, and physics. During the war, the library was looted by retreating German troops, and we were able to save only a few volumes.

I still have the *Capricci Medicali* by Lorenzo Fioravanti, published in Venice in 1568 with a dedication to Alfonso d'Este, Duca di Ferrara. The author states quite clearly that he wished to depart from the ancients—Hippocrates, Galen, and Avicenna—and rely on only his own judgment and experience. The book, however, is a strange mixture of alchemy, medicine, and magic. Recipes are given for the philosopher's stone and the elixir of long life, among other things. My discovery of such books instilled in me a lifelong fascination with this period of transition between magic and science.

Besides giving me the gift of books, Aunt Giulia taught me how to write. I was expected to write often to my mother, but I always agonized over those letters, because I thought I was supposed to use the style taught in school for patriotic compositions. Here is a small sample: "From the frozen steppes of Russia to the arid sands of the Sahara our glorious soldiers" and on and on. This pompous nonsense was typical of the fascist style, a jargon from a never-never land where reality and truth had no place. My aunt suggested that I should think of what I wanted to say and then say it as simply as possible—an illumination, in its way. That simple rule taught me not only how to write but also how to think. If Aunt Giulia had been Mussolini's teacher, she might have changed the course of history.

The political situation changed drastically during 1943. In July, following the Allied landings in southern Italy, the fascist regime was overthrown, to the immense relief of the Italian people, who by then were sick of war and felt they had been made fools of by a posturing buffoon. Statues, effigies, and symbols of fascism and its leader were torn down in an orgy of destruction. (I was reminded of these events when I saw on television Saddam Hussein's statue being toppled after the fall of Baghdad in April 2003.) In September the government of Pietro Badoglio signed an armistice with the Allies, whereupon the Germans stationed in Italy were abruptly transformed from friends to enemies, and the northern half of the country just as suddenly found itself occupied by enemy troops. An ugly situation quickly developed. Feeling betrayed and knowing that the war was lost, the Germans vented their rage and desperation not only against the Italian soldiers, and later on the partisans, but all too often against the civilian population as well. At the same time a phony fascist republic was established in northern Italy with Mussolini, by now a German puppet, as its nominal leader. Its army consisted of squads of die-hard young fascists, the so-called Fiamme Bianche (White Flames) of the Brigate Nere (Black Brigades), who turned out to be as vicious as their German equivalent, the SS.

In 1944, as the Allies were fighting their way northward, I went back to live with my mother in Milan, where living conditions had seriously deteriorated: food, fuel, and clothing were growing scarcer by the day. (I still remember a dinner consisting of a slice of polenta and an orange.) I was then a 13-year-old schoolboy, but as the only man in the family, it became my responsibility to help provide for my mother and myself. One of my tasks was to purchase and transport bags of sawdust, which we used as fuel for a special stove that supplemented the inadequate heating in our apartment. I also had to stand in long lines, starting at two or three o'clock in the morning, to buy any available clothing or fabric. In addition I was often shipped to the country to get bread and butter from some friendly farmers.

It was while spending my summer vacations on one of those farms that I found myself somewhat closer to actual danger. One night two of us kids were sent to patrol the fields to prevent people from stealing the crops. We were carrying an old shotgun, and on hearing some noises, my companion fired a shot into the air. The strangers, whoever they were, shot back at us several times. Needless to say, we took to our heels. Another time I had to watch a bunch of young toughs, in the Black Brigades uniform, rough up with sadistic pleasure some poor peasants on the pretext of looking for weapons, but really to extort money from them.

The war came to an end in April 1945. I witnessed the final vicious fight between the partisans who had entered the city and the few Germans still entrenched in their headquarters. Then the Germans were gone. That left Milan as a sort of no-man's land, and there followed a short period of lawlessness and reprisals: collaborators were shot and left lying in the streets; prostitutes were marched around town with their heads shaved. I saw Mussolini's body hanging upside down from the truss of a gasoline station. A few weeks earlier, the same gasoline station had been used by the Germans to hang hostages. (Cola di Rienzo, the People's Tribune, had ended the same way in Rome in 1354, a story well known to Mussolini, to the partisans, and to every schoolboy in Italy.)

The first Americans to arrive in Piazza del Duomo (Cathedral Square, in the heart of the city) were two newspapermen on a Harley-Davidson. My father, who had insisted on taking me there to greet them, applauded along with all the other Milanese thronging the square. He then remarked, "I am glad to see them arrive, and I will be even happier to see them leave."

These are my war stories. I have often wondered whether the war was a profound, traumatic experience for me; I am not sure but do not think so. The memories are there and are still quite vivid, but they never turned into nightmares. This lack of response is generally attributed to children's resilience. To me, such resilience means that when confronted with painful or frightening events, children bottle up their emotions and freeze to the point of alienation, which allows them to view a hostile world as if through the cold eye of a camera. This is the way Imre Kertesz, the 2002 Nobel Prize winner for literature, describes his childhood experiences in a German concentration camp in *Fateless;* such complete lack of emotion can only be explained as total alienation. Perhaps it is the only way to cope with the horrors of war.

What I do know is that the war made me grow up faster than might have happened otherwise. I have often felt that my contemporaries growing up in the United States or in other places where the war was not actually being fought had the privilege of a longer childhood.

After the War

The war was over. For the first time in years, we saw the city lights shining and every building illuminated. It was a deeply emotional, magical moment, and with it came a great lifting of the spirit, as if a heavy darkness had finally been banished from our lives. Free from war and fascism, people began to hope for a renewal of the nation, and though I was still an adolescent, I also felt a surge of hopefulness. No longer would we have to cheat and dissimulate to survive. Italy would cleanse itself of the fascist inheritance of corruption and lies. The political party founded by several partisan leaders, Giustizia e Libertà (Justice and Liberty), summed up in its name all our aspirations. We did get a new constitution, and the king who had abandoned us to the Germans in 1943 was removed by popular referendum. Pretty soon, however, the various political factions crystallized into three major parties (the Christian Democrats, Socialists, and Communists) plus a cluster of minor ones, all of them more intent on attaining power than on improving people's lives or legislating justly.

And so Italy became once more a corruption-ridden society based on patronage, and one now also subject to Mafia influences. Even until recently, much of its stagnating industry was still owned by either a few families or the government. The division of the spoils of the electoral process between political parties became capillary after the war, with positions in all government or quasi-government enterprises assigned by political appointment. As a result, managers of industrial complexes, banks, hospitals, television stations, and research and academic institutions were often chosen not for their competence or expertise but only to share patronage. While living in Italy, I was too young and naïve to see what was happening, but with time and distance, it became clear that our dream of postwar renewal had simply failed. This realization is still a bitter one for me.

There was another notable phenomenon of the postwar years. As my father had wished, the American troops did not stay too long; yet they lingered long enough to start the Americanization of Italy. American cigarettes and chewing gum, Hollywood movies and modern literature, jazz, boogie-woogie, and Gershwin came pouring in as an endless flow. And with the exception of chewing gum, which mercifully never "took," the Italian people soaked it all up with great gusto. I was not immune: I started smoking Chesterfields, became addicted to westerns, and read all of Hemingway's books.

A sobering by-product of Italy's Americanization was the rediscovery of history. The fascist regime had been editing the news for nearly a generation (since the early 1920s)—at first to put a politically correct slant on every national and international event, such as the war in Abyssinia, the Spanish civil war, or Hitler's ruthless expansions. Later, during World War II, the regime manipulated the news to paint a rosy picture of the situation, and when that became impossible, to give a sanitized version of it. We knew nothing of the Battle of Britain, Stalingrad, or D-Day. All we ever heard on the radio was that for some reason our troops "had withdrawn to pre-established positions."

But if we didn't know much about the course of the war in Europe, at least we were quite aware that it was being fought all around us, whereas we only had the vaguest idea of what was going on in the Pacific. I do not remember if we were told of the surprise Japanese attack on Pearl Harbor (being a "victory" for the Axis, it was probably well publicized), but it is a fact that until the end of the war I had never heard of Guadalcanal or the Battle of Midway.

Even before books could be written about it, American movies and documentaries gave us a clear view of the war and what had led up to it, with such a great wealth of details and with such a tremendous visual impact that I still enjoy watching them today, some 50 years later. The search for history is still one of my passions. I share it with my wife, and we derive from it many hours of quiet enjoyment.

Another and unexpected effect of those old movies was to produce a rather peculiar view of the United States. According to Hollywood, there was a fabled Far West where the U.S. "ca-val-ry" battled it out with the ferocious Redskins, and then there was a glittering East Coast (or West Coast), where people appeared to spend a great deal of their time in evening clothes. Apart from leaving a huge void in the middle of the country, this Hollywood panorama almost completely ignored the realities of life in big cities, as well as in Small Town, USA.

In addition, those movies portrayed America not only as the undisputed "land of the free" but also as the cradle of justice. It was almost a religious tenet that the good guy always triumphed in the end over the bad guy(s). The Far West positively thronged with silent, fearless, honorable gunslingers who redressed any wrongs they encountered in their travels. And since neither Gary Cooper nor John Wayne could do it all by themselves, they were aided by a legion of policemen and private detectives who could always be counted upon to nab Al Capone or retrieve the Maltese Falcon. I do not know of any other film industry that produced so many heroes.

In a more serious vein, the United States had just won a just war against two dictators in Europe and a treacherous militaristic clique in Asia—perhaps the last holy war in American history—and its citizens could surely speak of "justice for all" with true and simple pride. Although clearly naïve, this view of America was so attractive to a young man thirsting for justice and liberty that perhaps subconsciously it acted as an added inducement for me to leave Italy for the New World.

High School

In the fall of 1945, I started high school. At this stage of our education, we had to make a choice between the classical and the scientific high schools. As I had shown some aptitude for math and geometry, and possibly with a little nudge from my mother, I was enrolled in her scientific high school, though not in her class. The difference between the two curricula had to do primarily with the study of a foreign language in the science curriculum in place of ancient Greek, in addition to Latin, which was a requirement in both. We also studied architectural and geometric drawing rather than art history, and more emphasis was given to mathematics, physics, and chemistry.

Beause the school system lagged the changed political situation, the only foreign language I could take in 1945 was German. I had also attended a German kindergarten (run by German nuns), and I learned enough between the two curricula to retain some knowledge of the German language for the rest of my life. The following year I was able to switch to English, which I actually studied with greater pleasure from Luke Short's western novels than from grammar books. As a "finishing school" for colloquial English, a few years later I went to England, spent some time in London, and then enlisted as a trimmer on an 80-ton trawler, the *Brisbane* out of Grimsby, for a two-week fishing trip in the North Sea.

Although life was returning to normal, it was by no means easy for kids who had lived through six years of war to fall back tamely into the school routine. I for one had great difficulty settling down. To begin with, I felt that I was older and more mature than my companions, which most likely was not true. Furthermore we all had discovered that we had a lot of living and playing to do to make up for the lost years, and that brought about an almost frantic explosion of activities. Every holiday I could squeeze out of school I spent trekking, rock climbing, skiing, camping, bicycling, scuba diving, or spearfishing with my schoolmate Antonio Berla. This kind of kinetic reaction seems to be pretty normal in the aftermath of war.

It was not all play, however. In 1947 I joined a group of young Catholics to which a friend of mine, Mario Farina, belonged. He was older than I was, and set for me an example of puritanical life devoted to study, religion, and charitable work. (He eventually became a distinguished professor of chemistry at Milan's Polytechnic Institute.) From Mario and his group I absorbed the spirit of public service. One of our tasks was to teach the children of refugees in the poorest quarters of the city. We also collected those children to take them to summer camps financed by Catholic agencies.

I became completely disillusioned with the Church and its political organizations during the 1948 elections, when they interfered with the electoral process much more than I deemed permissible. Nevertheless, I continued to work with the children, and in my first year at the university, I also helped organize strikes by the Catholic Workers Union in factories outside the city.

Then studying for my university courses became serious business and my "social worker" period came to an end.

Altogether I do not remember high school as a happy time. I was so impatient to be done with it and get on with my life that sitting in a classroom became increasingly irksome. Eventually I found a way to cut it short. There was a law in Italy whereby a fourth-year high school student could skip the fifth and last year and go directly to the university, provided the student obtained the highest grades in all subjects and then passed, in the fall of the same year, all the exams required to graduate. It took a tremendous effort and considerable help to do it, but I managed to pass half the exams in June and half in October. In 1949, at age 18, I was on my way to college.

Now came the time to choose a field of study, or more precisely, a career. In Italy we do not have four-year colleges where young people are parked until they decide what they want to do when they grow up. As soon as they graduate from high school, Italian students must make up their minds and choose one of the various disciplines: ancient or modern languages, architecture, chemistry, engineering, law, medicine, physics, and so on.

I loved architecture, and still do, but I didn't think I had the creative ability to envisage new shapes and have the impact of, say, Walter Gropius or Le Corbusier. I was interested in philosophy, but in that field the only career open to me would have been teaching, and I was not willing to live the life of a teacher, especially a high school teacher, because I had found that world narrow and stifling. My mother was pushing me to take engineering, with the thought of achieving prestige, power, and wealth.

I finally settled on physics, the subject I had done worst in during the final exams. My reasoning was entirely down to earth. I thought I had a good chance to be reasonably successful in physics, and in particular, I knew by then that I had better physical intuition than my mother or any of the teachers I had met. I also thought that with nuclear power coming of age, there would always be jobs for a physicist. Finally, the subject seemed more fundamental, scholarly, and intellectually stimulating than engineering.

Which is to say that one can arrive at the right decision by the wrong route.

Sink or Swim: My University Years

As soon as I set foot in the school of physics at the University of Milan, I discovered that some of my fellow students had spent the summer studying university texts (or so they told me), so that they would be ahead of the pack and could ask intelligent questions during the lectures. Fortunately, I found another soul, Claudio Coceva, who was as ignorant as I was, and we started studying together. Interestingly enough, we were the only ones to complete the doctorate requirements in the prescribed four years, with the rest of the class struggling along behind us.

That first year I took courses in differential calculus, projective geometry, analytical geometry, physics, and chemistry. We had outstanding teachers for the first two subjects and a hopelessly incompetent fascist appointee for chemistry. The chairman of the department and physics professor, Giovanni Polvani, though competent and well respected, was a relic of the pre-Einstein era. I promptly decided to forgo his lectures altogether and to skip as many of the chemistry ones as possible. As a result, I never learned to like chemistry and ran into some difficulties in physics.

Throughout the four years of university, which is free in Italy, I was given a small grant to defray the cost of books, which was dependent on my obtaining a grade no lower than 24 out of a possible 30 in all subjects. Once, during a physics exam, I was asked to withdraw and take it again later, so that I would not spoil my record.

I thoroughly enjoyed the projective geometry course, but I soon found I had no patience with analytical geometry. And that also got me into trouble. During my final exam, I was given three problems in analytical geometry to solve in writing. I looked at them and saw that they could be easily and quickly solved using projective geometry. I therefore solved them in 15 minutes and left the examination hall. Naturally the professor was incensed at my arrogance and gave me the relatively low grade, for a correct solution, of 28 out of 30. What I should have done, of course, was to go through the lengthy algebraic solution and thus satisfy the examiners that I knew analytical geometry. I was disappointed at the time, particularly because geometry had always fascinated me.

I used to look at a problem in geometry and either solve it easily or be unable to do so at all. If I had not solved it immediately, then the next day the solution would come in a flash of intuition, in which the entire problem was clear in all its aspects. I was particularly excited by the lectures of Oscar Chisini, who inveighed against trigonometry as a mind destroyer. He wanted us to conceptualize conic surfaces, not measure them. Chisini believed that "geometry teaches you how to carry out the correct reasoning on the wrong picture." E. La Rocca states that "Chisini expressed a dynamic vision of science in which history and mistakes are to play a prominent role. He believed that theorems should be presented as raw minerals rather than polished gems."[1] These views, which Chisini communicated to us in his lectures, profoundly influenced my approach to science on an almost subliminal level. However, I must say that conceptualization always implied for me a three-dimensional visual representation, an approach that I believe is typical of synthetic thinking.

Between the solution of geometric quizzes while asleep and Chisini's stress on conceptualization, I became sympathetic to the idea that one could learn by intuitive jumps based on fragmentary or confused information. Later in life I was quite interested in such books as *The Origin of Consciousness in the Breakdown of the Bicameral Mind* by Julian Jaynes and even more so in *The Sleepwalkers* by Arthur Koestler, as a way of understanding my own thought process.

During the first year of university, students were given the opportunity to volunteer to join an active research group and act as unpaid assistants by doing both bibliographical research and laboratory work. I eagerly volunteered and after a while found myself fully occupied. I used to joke that I needed permission to go to lectures and prepare for my exams. This was to be my situation during the entire four years of my university training.

I was lucky to have as my mentors two people who differed strikingly in character and talents. One was Antonio Mura, one of the best physicists at the university; I discussed with him the then-current literature in cosmic rays and the meaning of the experiments he was conducting with Antonio Lovati, Carlo Succi, and Guido Tagliaferri. He was clearly the thinker in the group; he was a man with a limpid logical mind who unfortunately died very young, just around the time I was completing my thesis work and leaving for the United States. I used to visit him in a sanatorium near Genoa and talk with him until the end.

Succi, on the other hand, was a clever and burly veteran of World War II. He had been the captain of a communications battalion right up to September 1943, when he marched with his entire battalion across the Allied lines and surrendered: his was a highly practical mind. He was a talented instrumentalist who could carry out experiments but could not as easily conceive them. I learned a tremendous amount from him, from designing and building (with surplus U.S. Army electronic components) power supplies, amplifiers, and digital counting systems to designing and operating diffusion and expansion cloud chambers. Cloud chambers had been invented by C. T. R. Wilson in 1911 and had come into general use, particularly after their development by Patrick M. S. Blackett, to make visible the tracks of ionizing elementary particles traversing their volume and to study their properties.

My introduction to cloud chamber operation was somewhat typical of the educational system in which we lived. An experiment on mu mesons was being carried out in an unused railroad tunnel some distance from Milan, which provided the necessary shielding from cosmic rays. I was taken there by Lovati, shown the equipment, introduced to some nuns living in a nearby convent (where I would be a guest), and left alone to operate the cloud chamber for the next week. It was typical of our university training to receive such scant guidance. Students were expected to sink or swim on their own; perhaps this can explain how Italy continues to produce excellent scientists despite its decidedly suboptimal university system.

While I was receiving on-the-job training in experimental physics, I was less fortunate in theoretical physics. Professor Piero Caldirola was interested in the classical radius of the electron, a subject that had little to do with the research going on in Milan. Although I did well on the exams, I found his lectures incredibly boring; because attendance was required, I developed the knack of sleeping with my eyes open while sitting in the front row.

By far the most important influence on my university life was that of Giuseppe (Beppo) Occhialini. Beppo had been a devoted anti-fascist, had spent many years abroad, and had contributed to outstanding research at the forefront of physics. He was working with Blackett when their group discovered electron-positron pairs and with Cecil F. Powell when they jointly discovered the pion. In each case, for reasons not clear to me, Occhialini did not share in the Nobel Prizes that were awarded for these discoveries. However, his work was later recognized by the selection committee for the Wolf Prize.

His arrival at the University of Milan was like a breath of fresh air in the rather provincial atmosphere prevailing there. He was in touch with the latest news in physics, the pre-pre-preprint phase. Only those who have struggled without benefit of such knowledge, before today's communications revolution, can appreciate the enormous advantage such connections could give to young and old researchers alike in making timely contributions.

Beppo never gave a formal lecture that I know of while I was in Milan, leaving this task to his assistants Alberto Bonetti and Livio Scarsi. He had started a nuclear emulsion group for cosmic ray research, which used many fourth-year students to read the emulsion plates. I was never attracted to the work, which I found to be highly specialized and focused entirely on data analysis for research projects controlled by Beppo and his assistants. Beppo never insisted that I work on what he was doing; on the contrary, he encouraged me to work with cloud chambers, which, as he put it, "would teach me plumbing." I learned from him more by osmosis than through any formal training, in informal and frequent discussions, as we traveled to meetings or to visit other laboratories. Beppo had an odd way of conversing on at least two subjects at once, just as he often smoked two cigarettes simultaneously. I was able to follow him without effort, and I still find myself, at times, speaking to people as he used to speak to me.

I did not do my thesis work, which had already started prior to his arrival, with him. It consisted of the study of nuclear interactions of protons in the lead plates of the cloud chamber, to investigate the fireball model of nuclear interaction at high energies, which had been proposed by Enrico Fermi. Because protons with energies on the order of 80 Bev could not then be produced artificially, cosmic rays provided the only means to study such interactions.

Because the proton components of cosmic rays interact with the atmosphere and are lost in it, the chamber had to operate at high altitude to intercept as many particles as possible. The cloud chamber was installed at the Laboratory of Testa Grigia (3,500 m), at the foot of the Matterhorn (Figure 1.1). Life at the laboratory was rough: we lived in a Quonset hut, melted snow to obtain our water, and suffered episodes of scurvy brought on by vitamin deficiencies. I spent, on and off, two years there to obtain the eighty proton events that formed the basis of my thesis.[2]

The result was a confirmation of Fermi's prediction, but the strength of my conclusion was hampered by the relatively poor statistics, the result of the

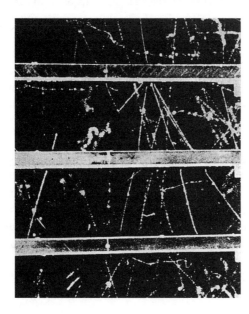

FIG. 1.1. *A photograph obtained with the 50-cm multiplate cloud chamber at the University of Milan. The photograph shows a nuclear interaction, which produces a pair of V particles, occurring in the top lead plate.*

limited number of events recorded. As I was doing the experiment, I was quite aware that our cloud chamber was not one of the largest in the world; with a bigger one, my rate of collection of events would have been much greater. Because we could not build or acquire a larger device in the short term, I dreamed of concentrating cosmic rays on my instrument by means of a magnetic funnel. Due to the extremely high energies of the particles I was studying, this magnetic focusing was also impossible; but that dream played an important role in guiding my thinking in the development of x-ray astronomy.

On finishing my thesis in 1954, with a modest score of 106 out of 110, I was immediately offered a position as lecturer in the Physics Department, to teach experimental physics. I was so overwhelmed by the honor that I immediately accepted without even asking about the salary. It turned out that the salary was the usual for my position, the equivalent of $100 per month, and it took about six months for the bureaucracy to send me my first check.

Apart from teaching my own courses, I was supposed to substitute for Professor Polvani when he was sick or not available. I normally would receive a call from him at 7:30 a.m. to teach the 9:00 lecture and pick up were he had left off; this of course implied that I was following his course, which I had never done, even as a student. My solution was to prepare a set of lectures on specific topics that I knew he would not teach, which were received with enthusiasm by the students. This ruse worked well, and I acquired the unmerited fame of being a good lecturer. Altogether the teaching load was quite heavy.

As to research, I was asked by Beppo to join him on a two-man committee to study the future of research in Italy using cloud chambers and diffusion chambers. My main contribution to the work was to act as a sounding board for the incessant stream of ideas he was producing. Many of them were impractical, some I could destroy by applying the second principle of thermodynamics, and others were outstanding but beyond our ability to realize with the available resources.

I was also given responsibility for the design and construction of a new and much larger cloud chamber of 120 × 120 cm, comparable to the chamber used by the group at the Massachusetts Institute of Technology (MIT).[3] While building this chamber, Lovati, Succi, Tagliaferri, Ettore Fiorini (then a graduate student), and I were considering how the instrument could be used. One possibility was to use it to study the interactions and decay of mu mesons. Such an experiment would have been of interest only if the energy of the mu mesons could be measured. I was shipped off to Manchester, England, to learn the techniques of the cosmic-ray group there. I was extremely impressed by the elegant simplicity of their magnetic spectrometer but not by their scientific program.

Another possibility was to continue the search for strange elementary particles, such as the lambda zero and the theta zero, which had been studied in both Europe and the United States. Unfortunately we had no clear and specific ideas of our own on what experiments to carry out, and we were looking abroad to find inspiration for our research. In my work with Occhialini on the Italian committee, I learned quite a bit about the prospects for cosmic-ray research with cloud chambers, and in particular with cloud chambers that had multiple plates but no magnetic field, and they seemed rather dim.

The rapid development of artificial particle accelerators (which could produce dense beams of particles) made the low fluxes available in cosmic rays uncompetitive in particle research. Many of the groups around the world were turning to other aspects of research on elementary particles. The multiplate chamber I had built with my colleagues was used by Fiorini and his colleagues for several years after I left Italy. At least until a few years ago, its skeleton was still exhibited in the lobby of the Physics Department at the University of Milan (Figure 1.2).

FIG. 1.2. *The 120 × 120-cm multiplate cloud chamber at the University of Milan. I led the group that built the device shortly before I left the university.*

The only hope for success seemed to be in the high-level research being carried out in the United States by R. W. Thompson at Indiana University using a cloud chamber with a magnetic field (Figure 1.3). The application of a magnetic field made it possible to measure the electric charge and energy of the decay products of the strange particles, and this approach had allowed Thompson to determine the most accurate value of the theta zero mass, still valid today.[4] This work was reported in the proceedings of the International Cosmic Ray Conference, held at Bagnerres de Bigorre, France, in 1953.

While the work reported by the MIT group led by Bruno Rossi, and that of many other groups, seemed to me quite similar to our own, Thompson's approach stood out as something quite different and worth learning about. Whereas we and the other groups were trying to squeeze what information we could from our multiplate chambers, Thompson had set himself a scientific goal, built the instrumentation, and developed the data analysis tools necessary to achieve it. That was the way in which I wanted to carry out my own research.

Occhialini visited Thompson during a trip to the United States and came back enthusiastic about his work. Among the things that impressed him was

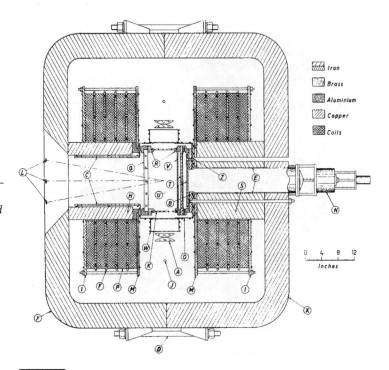

FIG. 1.3. *The magnetic-field cloud chamber at Indiana University. The application of a magnetic field alters the flight paths of decay products, allowing measurements of their charges and momenta.*

Thompson's dedication; he reported that the man worked so hard that he ate in a gasoline station, which turned out to be Beppo's description of a diner. Using an expression from his favorite Western, he told me: "Go west, young man." I applied for a Fulbright Fellowship to be spent initially at Indiana and possibly later at Princeton, where George Reynolds and his group were also carrying out cosmic-ray research. With Occhialini's and Thompson's support, I was granted the fellowship.

I must now introduce the most important person in my life: my wife, Mirella. Her father was a brilliant man who had obtained two degrees, one in engineering and one in law, at a very young age. Her mother belonged to a noble family dating back to the thirteenth century and known for their pride and fortitude. Mirella would need both pride and fortitude during our eventful life. An all-A student in her high school, the toughest in Italy, and at the top of her class at graduation, she dreamed of becoming an archaeologist, but a serious and lengthy illness prevented her from continuing her university studies—a great pity. She would have dearly loved an academic career. I had met her at sixteen, lost sight of her for a few years, and met her by chance again in the fall of 1954 on a streetcar in Milan. We have been together ever since.

In September 1956, I boarded the S.S. *Independence* at Genoa and sailed away to New York, like so many immigrants before me. Mirella followed me a few months later.

New World: The Fulbright Fellowship

——————————————————————

We sailed into New York Harbor on a fine day, which was fortunate, because I was able to see the Statue of Liberty and the Manhattan skyline as they should be seen—as millions of immigrants had seen them—from the sea. Arriving in New York by plane does not have the same emotional impact; the aerial view is actually rather banal, so I am glad I was able to sail up the Hudson just once before the great ocean liners were forced into early retirement by the advent of the jet.

From the dock I made my way to the Fulbright Foundation, where I was given a train ticket to Indianapolis, and from there I took a bus to the main campus of Indiana University at Bloomington. The whole trip lasted more than 15 long hours; it was my first inkling of just how vast the American continent was.

When I arrived at the Bloomington bus station, to my great surprise I found that the dean of the Physics Department was waiting for me, ready to help me with my luggage. It is difficult to convey how this simple courtesy affected me, except to remark that during my university years in Milan, not one of my senior colleagues had ever invited me to his home (except, of course, for Beppo Occhialini).

Bob Thompson was quite anxious to have me start and immediately installed me in a room at the Student Union. The room itself was quite modest and was entirely taken up by a bed and my steamer trunk, but Bob assured me that, given the pressure of work, I would not use it very much. We then went directly to the laboratory, where the larger magnetic cloud chamber was being built. The laboratory consisted of an empty warehouse outside the campus proper, which, from Bob's point of view, had the advantages of low cost, large size, and isolation. Bob was extremely enthusiastic about the new project and felt that with my experience in the construction of large cloud chambers, I would be of great help.

Some background on cloud chambers might be useful here. Their ability to reveal tracks of elementary particles is due to the phenomenon of supersaturation. A cloud chamber is filled with gas and water vapor at a given temperature, and the water vapor is in equilibrium with the gas at that temperature. A sudden expansion of the chamber volume results in too much vapor for that temperature, a state that is called supersaturation. Under these conditions, droplets of water condense on the ions created by a charged particle as it travels through the chamber volume. A photograph is obtained by a flash of light triggered within a fraction of a second of the expansion, so that the droplets have no time to disperse.

Cloud chambers were built with or without magnetic fields. Both the old one I had used in Italy and the new one I had built were chambers without magnetic fields. They had many plates across the path of an incident cosmic ray, so that we could observe the interaction of a particle with a nucleus in the material of the plates and study the type of interactions that occurred.

Magnetic field chambers were designed to detect the momentum and charge of the particles produced in interactions either in the walls of the chamber or entirely outside of it by measuring the curvature of their tracks. Once the decay of neutral particles in the cosmic beam was discovered, the magnetic cloud chamber permitted measurement of the mass of the decay products as well as that of the parent particle. Thompson's group was the most advanced in the world in the use of this technique. He and his colleagues had been the first to establish the existence of the θ^0 particle (which decays into $\pi^+ + \pi^-$) and in 1953 had obtained the most accurate measurement of its mass.

Plans for the New Chamber

In my first few weeks, I came to learn most aspects of the construction plans for the chamber, plans that Bob shared with me little by little. The overall plan was to increase the size of the chamber by a factor of two in linear dimensions, a factor of four in cross section, and a factor of eight in volume. This enlargement would increase by four the number of events that could be collected in a given time, improve statistics, and possibly allow for the study of rarer particles produced at higher energies. Bob also planned to improve the maximum detectable momentum. Its value for the old chamber at Indiana University was already the best in the world (5×10^{10} eV/c), yet he was hoping to improve it by a factor of two or more, which would allow a greater precision in the determination of particle masses.

It was clear that the technical challenge of building such a large chamber was the eightfold increase in the volume, in which a high magnetic field of 7,000 gauss had to be maintained with great stability and homogeneity. The magnet in turn required an eightfold increase in the weight of its iron yoke, from 11,000 to 88,000 pounds. The copper coils that circulated the current necessary to produce the field also had to increase from 6,500 to 52,000 pounds.

Many other aspects of the chamber had to be improved: the illumination system for photography; the field of view, depth of focus, and resolution of the camera; the temperature control system; and finally the speed of operations. The illumination system consisted of argon flash tubes, in which a sudden discharge of electrical energy stored at 1,000 V in a high-voltage capacitor bank would provide, with a 10-msec delay with respect to the expansion, the necessary luminous flux. This arrangement was no different from the system we had developed in Italy, but the high-energy capacitors were quite expensive.

The speed of operations could be increased if the chamber could be recycled in as short a time as possible after the expansion, so that it would be ready to record the next event without excessive delay. (The dead time in cloud chambers was typically 3–4 minutes, and it would have been interesting to decrease this time by factors of ten or more to make cloud chambers more useful as event detectors for particle accelerators.) In Milan we had started to study the effectiveness of supercompression (the process of rapidly increasing the pressure in the chamber after the photograph was taken). Although we had built a prototype, the technology we had tested seemed promising but not yet mature; perhaps it could be introduced later in Thompson's program.

What concerned me most as I started working on the project was the absence of an overall plan for how and when we would accomplish the goals described above. Bob was convinced that we should proceed to personally fabricate all parts of the chamber, including the magnets. His view seemed to be that only in this manner could we be assured of the quality necessary to achieve the required accuracies.

For a few months, our approach was to design something (for instance, the coils) and then locate the machine tools to fabricate it. Tools and machines could be obtained from army surplus depots and, so soon after World War II and the Korean War, they were abundant and of excellent quality. To illustrate the procedure, I describe my first task, which was to start the construction of the copper coils. They were made of strips of copper 4 inches wide and a quarter of an inch thick, to be wound to a final diameter of about 100 inches. To wind the coils, I built a mill made of steel I-beams. On the surface of the I-beam on which the coils would rest, I provided a cushion of hard wood (mahogany, I believe) that had to be milled to a flatness of one-sixteenth of an inch. An outside arm held the cutting tool precisely in position, and the wood was milled to the required precision over its entire surface. The material for the copper coils came in relatively short lengths, so that soldered joints were needed. This requirement implied grinding the ends of the coils to form wedges that could be butted together and making a fixture to hold the coils during soldering. To be fair, Bob was constructing this fixture personally.

The most disheartening aspect of the whole project came up in connection with the construction of the back plate of the chamber, which divides the sensitive volume (in which the particle track would be detected) from the expansion volume. Bob had decided that we ourselves needed to drill the required 3/8-inch-diameter holes, even though there were four thousand of them, because of the high precision required. This task would be hard for anybody to do manually, and today would be done under computer control.

In retrospect, I have often wondered whether Bob really believed that it was necessary for us to do all the work ourselves, or whether he simply did not have the funds to have it done by industry or research laboratories. I am afraid that he may have been in earnest about doing it himself, since by the time I met him he had become obsessed about the work and mistrustful of anybody not closely associated with him; his attitude may have had something to do with the perception that he had not received proper recognition from the scientific community for his discoveries and for the quality of his work.

Bob also appeared to have changed in a more fundamental way. He did not seem to have a clear idea of the science that he intended to do with the new instrument or whether, upon completion, it would still be competitive with the rapidly developing accelerators. This last aspect of our collaboration was of vital interest to me. After three years on the Fulbright Fellowship, I could not go back to Italy having completed yet another construction project but not a scrap of research; my purpose in coming to Indiana had been to learn, to become a better physicist, and to participate in active and significant research at the forefront of the field. I discussed my views with Bob, who, while unhappy to lose a valuable helper in carrying out his beloved project, understood my problem and tried his best to help me.

The Search for Anti-Λ^0

Thompson suggested that instead of building coils I should use the data he had obtained with the old magnetic-field cloud chamber to search for a new particle, the anti-Λ^0. Two of the particles predicted by P. A. M. Dirac's theory had already been found: the positron by Carl Anderson in 1932 and the anti-proton by Emilio Segrè and Owen Chamberlain in 1955. In 1952, Abraham Pais had discussed the hypothesis of associative production of V^0 particles, which made the existence of an anti-particle for the Λ^0 quite probable: it would decay into an anti-proton and a pion.

It seemed reasonable to me to start this project, even though I did not have a clear idea of either its difficulty or its chance of success. J. R. Burwell and R. W. Huggett had collaborated with Thompson on the θ^0 project and obtained their PhDs, but they were no longer working on the data analysis of the old results. Therefore, I had to proceed alone, my only point of reference being Thompson himself.

The work was quite challenging; for one thing, I suffered from the poor theoretical preparation I had received in Italy. Both Burwell and Huggett were far ahead of me, and I saw pretty clearly the difference between my doctoral degree and their PhDs. Thompson, on the other hand, insisted that I read all the relevant literature and made no allowances for any weaknesses. He himself taught experimental physics one year and theoretical physics the next. I remember with embarrassment that during a journal club meeting I was trying to report on Pais's work. Thompson's criticism was so devastating that I stopped and apologized for not fully understanding the subject. Bob later congratulated me on my courage in admitting ignorance, and clearly retained his good opinion of me. I took this episode as a sobering evaluation of my proficiency in the theory of elementary particles, a handicap that would be difficult for me to surmount.

There remained to actually carry out the project, and here I learned a great deal. The reduction of the data involved sorting out likely candidates by visually scanning the cloud chamber pictures. This chore was something I was familiar with from my work in Milan. Then followed the measurement of the momenta and the angles between particle trajectories. To quote Thompson:

"This required comparator measurement on the film and least square computation of the film curvature and angles. This was followed by stereoscopic reprojection and conversion of the film data to space."[1] To carry out these measurements I used a projected image on a screen rather than the microscope method used previously. It was much faster, and should a positive result be found, microscopic verification could still be performed on selected images. The x- and y-coordinates are measured on the screen, and the dispersion of the measurements is examined to spot anomalous values. For tracks whose curvature is sufficiently great, a rough value of the curvature can be immediately determined. This value can be measured as the reciprocal of kilometers; the probable error measured from no field curvature was 1/4.3 km, corresponding to a maximum detectable momentum of 5×10^{10} eV/c.

A highly detailed set of instructions had been worked out by Thompson, who ended his description of the recommended procedures with: "Although the measurements could possibly be entrusted to a capable technician, in the present work they are made by physicists concerned with the analysis of the event."[2]

There followed computations of the field curvatures and angles:

1. A least squares fit to the data was performed to obtain a curvature.

2. Least square curvature errors due to gas distortions were derived from tracks obtained with the magnetic field turned off.

3. A parabola fit to the data was performed, and the radius of curvature was defined as the inverse of the coefficient of the quadratic term in this fit.

4. The film angle was computed by least squares fits of parabolas to each track at the apex.

All these computations were done in a standard, prescribed manner, which was incorporated into printed forms for the observer to measure and record each piece of data and each calculation. (One should bear in mind that these calculations were done in the pre-computer era.) I had never seen such careful calibrations, painstaking error estimates, and methodical streamlining of procedure as those Thompson used.

Spatial reconstruction of the particle tracks was obtained by reprojection through the actual cameras of the original negatives of the event and by the use of sector shutters (used to blink the images from the two cameras, one against the other, in the plane containing the decay fragment track to reconstruct the curvature in three dimensions). After careful evaluation of potential error sources and rigorous justification of the formulas used for the derivation, a standard procedure was again developed for the reconstruction. The procedure itself was tested with dummy tracks and angles, and the statistical errors were determined. Thus, for each fragment of the V particle, the fragment's space momentum could be derived.

At the end of this process, the mass of the parent particle could be determined. It is almost always the case that the fragments are too relativistic to obtain a direct mass estimate from the measurements of ionization and momentum. Thompson developed a method of analysis based on a clever scheme that did not depend on knowledge of the nature of the fragments. He used a Q function defined as the kinetic energy of the decay products with the decaying particle at rest. It could be shown that on a three-dimensional representation with coordinates p (the momentum of the fragments), α (the inclination of the angle of the helix on which the particle travels, and p_y (the momentum of the fragments measured in a center-of-mass frame of reference), the data define a unique surface on which the events must lie (Figure 2.1). The mass of the parent particle could then be obtained from the sum of the assumed masses of the fragments and the value of Q.[3]

Thompson had found a Q value for the θ^0 (which decays into π^+ and π^-) of 213.9 ± 2.8 MeV—an astonishing precision at the time. He had found for the Λ^0 (which decays into a proton and π^-) a value of 36.9 MeV. I will not describe the tests to evaluate two-body versus three-body decay hypotheses, the tests to reject the possibility that the θ^0 would decay into a pion and a mu meson, and a host of other tests that were carried out in our analysis.

To search for the anti-Λ^0, I examined the sample of V^0 events (230), using the same criteria that Bob had used to find the Λ^0. He had found fifteen measurable events. I analyzed twice that number and found no evidence for the anti-particle. Mirella, who by this time had joined me in Bloomington, helped me to reduce and analyze the data and was as disappointed as I was about the outcome.

The Λ^0 particle was eventually discovered a year later (in 1958) by D. J. Prowse and M. Baldo-Ceolin, who used emulsions exposed to the beam of the Berkeley Bevatron with an intensity of 10^6 pi mesons/cm^2. Out of the myriad interactions produced by this beam, one anti-Λ^0 was found!

In the end, what I learned at Indiana University was the need to work carefully and hard, and how to analyze data competently. I learned about the careful planning that must go into the design and construction of instrumentation to facilitate the reduction and analysis of the data. The deepest impression that I carried with me for the rest of my scientific career was the

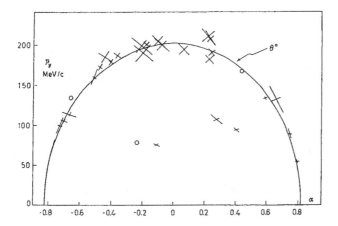

FIG. 2.1. *Q curve plot of the* θ^0 *data.*

importance of having clear scientific goals when starting a project. I also learned how unprepared I was to join in accelerator experiments, where the ability to fully understand the ongoing theoretical debates and to select a viable experiment were more important than experimental skill.

At the end of the Λ^0 search there was little to hold me in Indiana, and I accepted the invitation of George Reynolds to join him at the Princeton Cosmic Ray Laboratory. We left Bloomington with some regrets. The unfailing kindness that members of the faculty showed to both my wife and me was truly heartwarming and helped a great deal to soften the megrims of two young immigrants. We were also fortunate to be befriended by four exceptional people: Roger Newton and his wife, Ruth, of the Physics Department; Dick Bennet, a graduate student who became the state district attorney; and Jim McManus, a Korean War veteran who became a senior correspondent for NBC. Bob Thompson moved to the University of Chicago soon after I left. I believe he stopped publishing altogether after 1957. The big new chamber was never built.

I should add, however, that we did not meet with universal benevolence. When my wife and I looked for a house to rent, we came face to face, for the first time in our lives, with prejudice and hostility directed particularly against Italians. As it turned out, our future landlord was reluctant to rent us his bungalow on the grounds that Italians might be expected to splatter tomato sauce on the walls. My wife, fresh off the boat from Europe and carrying strong prejudices of her own, was so incensed by the ignorance of the "natives" that she had a hard time readjusting her perspective. She was experiencing a mild form of Kultur shock.

Princeton: A Stranger in a Strange Land

I arrived in Princeton, New Jersey, on a snowy day in February 1958. What immediately struck me, coming from rural Indiana (in a 1948 Chevy), was the sign in the Wine and Game Shop on Palmer Square that read: "Reserve Your Pheasant Now." In taking the Princeton position, I had to accept a 10 percent cut in salary, probably due to the honor they were bestowing on me by hiring me; thus the pheasant was a little out of my range. Mirella joined me a few days later with our daughter, and we were assigned an apartment in the housing set aside for young faculty on Prospect Avenue.

Reynolds's Cosmic Ray Laboratory was a large Quonset hut on the edge of the campus, where his group had prepared and executed experiments for a decade. While I was there, Jim Cronin and Val Fitch were working, in the space adjacent to mine, on the lifetimes of the K^+ and θ^0 mesons. In 1964, in their experiment at Brookhaven, they showed for the first time the violation of charge-parity (CP) symmetry in weak interactions, a discovery for which they were awarded the 1980 Nobel Prize in physics.

These were the most notable people—that I recall—working in the laboratory. I soon found that there wasn't much opportunity to interact with them or, for that matter, with other scientists in the Physics Department or the Institute for Advanced Studies, located near the campus of Princeton University. (I met Robert Oppenheimer for the first time a few years after I had left Princeton.) It was very disappointing. Furthermore, I soon discovered the arrogance that seemed to sprout from Princeton's soil along with the ivy.

Shortly after our arrival, we were invited to a soirée where Eugene Wigner was the guest of honor. We entered our hostess's living room to find the great man ensconced in an armchair with many of the invitees sitting on the floor at his feet (quaint but acceptable). Wigner then proceeded to discourse at length on various subjects of his choosing (with only a few nods and smiles from the audience). During his dissertation, Mirella and I stood quietly in the back; quietly, that is, until the subject of architecture came up. At this point Wigner delivered himself of a profound statement: "There is no such thing as architecture," said he dismissively. Having a strong love for and a solid background in both classical and modern architecture, I could not resist piping up and objecting that such a statement was at least debatable. Our hostess immediately took us aside and asked where we had come from. "Indiana," I answered. "Oh, well, Hoosiers," said she—a remark that both my wife and I found as offensive as it was preposterous. We did not receive many invitations from faculty members after that, nor did we seek them. Another junior couple advised us to refrain from such comments, and to prepare instead, as they did, suitable witticisms to be delivered at the next party.

As I started my research at the laboratory I found that work on cosmic rays using cloud chambers had been significantly cut back, and that Reynolds's group had largely dispersed. I came across an experiment on mu mesons being carried out by a graduate student who seemed to need a lot of help. It was aimed at the detection of mu mesons by means of their Cerenkov radiation in air and therefore at energies greater than 4 GeV. The detection of the extremely low-intensity flash produced by these mesons was quite challenging, but we achieved it. The experiment was set up in a 4-meter-long light-tight tube (Figure 2.2). A mirror of 35.5 cm radius focused the Cerenkov light emitted by a downward-moving particle onto a Dumont 6263 photomultiplier with a cathode diameter of 12.5 cm.

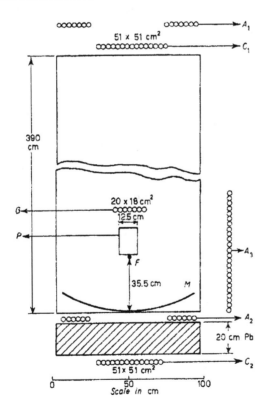

FIG. 2.2. *An air Cerenkov counter for detection of high-energy mu mesons.*

The muon beam was defined by the two Geiger counter assemblies C_1 and C_2, with A_1, A_2, A_3, and G in anti-coincidence (see the figure). The first two veto counters better define the beam, and the anti-coincidence G prevented the measurement of particles traversing the photomultiplier to avoid spurious counts. We simultaneously recorded $C_1 C_2 P$ coincidences (with G, A_1, A_2, and A_3 in anti-coincidence) and $C_1 C_2$ coincidences alone with the same anti-coincidence. The resolving time of the veto circuitry was 2×10^{-6} second. We measured and recorded the pulse height from the photomultiplier P.

We used two different photomultipliers for this experiment, with about 200 hours of running time for each. We recovered a counting rate of about $1,075 \pm 36$ counts/hour, in agreement with previous data. The differential pulse-height distributions of the two counters were in agreement with those that could be computed from simple theory. The fraction of muons emitting Cerenkov light in the phototubes was found to be 19.9 ± 2 percent in the first and 25.5 ± 3.5 percent in the second, close to the predicted value of 27 percent.

It is interesting to note that we were measuring, in the two photomultipliers, pulses as small as one to three photoelectrons produced by the Cerenkov radiation impinging on the cathode of the photomultiplier. It was good training for low-flux photon counting. The results were actually published in January 1958.[4] It is not clear to me in retrospect why this experiment had been started, as there did not appear to be any urgent problem in physics that was being investigated. I tried, therefore, to pursue interests of my own: one was the search for a particle 500 times the mass of the electron; another was the development of scintillation chambers for possible use at the planned Princeton-Pennsylvania particle accelerator, which was never built.

The search for particles of mass $500 m_e$ was stimulated primarily by the work of A. I. Alikhanian and his colleagues in the Soviet Union. Alikhanian had been among the first to find a component of cosmic rays with mass between that of the proton and the mu meson. Therefore, some credence was given to their claim that they had evidence for the existence of a $500 m_e$ particle with an abundance of one in a thousand mu mesons. Herbert Gursky, who had just completed his PhD thesis and who was to become a lifelong friend; Fred Hendel, an experienced cosmic-ray physicist from Austria; and I decided to undertake a search for this particle by means of a system of counters. We thought that the particle might decay into a neutral pion and a muon.

We put together an array of counters under a lead shield to reduce background noise. We used a Cerenkov counter to measure velocity and two scintillator counters to measure the energy and the energy loss per unit path length of the particles selected. The Cerenkov counter is based on the detection of the light emitted by a charged particle when it passes through a dielectric at a speed greater than the speed of light in that medium. A scintillation counter measures the ionizing radiation produced in a transparent crystal or in plastic (doped with anthracene), which fluoresce when struck by radiation. The intensity of the light produced is proportional to the ionizing energy deposited in the scintillator by a particle traversing its volume. The selection criteria were that the particle must be heavier than a muon, stop in the counter, and decay. The decay was required to occur between 10^{-1} and 5×10^{-6} second and be followed by another decay within 10^{-6} second of the first. Geiger counters surrounded the scintillation counters and provided veto anti-coincidence.

We found the ratio between $500m_e$ particles and muons to be at least a factor of fifty below that claimed by the Soviet group. I reported this result at a meeting of the American Physical Society in 1959.[5]

The more remarkable aspect of our experiment is that we worked at it with incredible dedication. We would get up, go to the lab, work until exhausted, and go home to eat and sleep. We left messages for one another so that any of us could proceed with the work night or day, as he happened to come in. We completed our work in just a few months, and we learned a great deal about how to work together and how to accomplish remarkable feats of engineering.

I have always thought that this undertaking showed just how desperate we were to find a new and worthwhile problem. In all probability, had we been smarter, we would not have wasted our time on it. It was a small consolation to find out later on that some eleven groups had pursued this chimera as late as 1961. One of the groups included Marcello Conversi and Carlo Rubbia, in collaboration with my colleagues in Milan and, I believe, using the cloud chamber I had built.

Technical Developments: Building Expertise and Gaining Insight

The experiment to search for the mass $500m_e$ particle required a great deal of relatively fast electronics; delay lines; and a combination of scintillators, Cerenkov counters, and Geiger counters. Coincidence and anti-coincidence techniques were utilized, as well as more sophisticated algorithms to eliminate spurious events, such as pulse-shape discrimination. I mention these facts because the familiarity I had acquired with all these techniques became extremely useful later on, when I applied them to x-ray astronomy.

One area of particular interest to me, aside from the search for mass $500m_e$ particles, was the construction of scintillator chambers for possible use with accelerators. The idea was to image a particle crossing a scintillator directly from its scintillation light. I built a plastic scintillator chamber composed of planes of plastic scintillator fibers, which I myself had fabricated with a machine of my own design. Each plane was oriented at 90 degrees to the preceding one. An imaging camera placed on two sides of the matrix of fibers revealed the passage and direction of a particle with high spatial and temporal resolution, in three dimensions.

The amount of light that could be collected at the end of each fiber was, however, quite small, and the image had to be intensified by a factor of a million to enable a camera to photograph the track. We worked with the RCA Laboratories in Princeton and used their prototype image intensifier, configured in a stack of three. Each intensifier had an accelerating voltage of about 1,000 V and achieved a gain of roughly one hundred; thus, we could obtain a total gain of one million with a cascade of three intensifiers.

In actual use for either cosmic-ray studies or particle accelerators, the intensifier had to be turned on or off on command. The work required a square-shaped pulse of 1,000 V with a precisely specified duration and rapid rise and fall times; to be useful, this pulse had to follow the trigger pulse less than 10 msec after the command was issued. This rapid response was to prevent too large a decay of the scintillation light before the event was recorded.

In Princeton we had made reasonable progress with our imaging system, and I was invited to MIT to run our system with a chamber of their design made of a high-Z crystal of cesium iodide. Herb Bridge was involved in the project, together with others in his group. After a few days of work, Herb asked me what prospects I had for my career, and he later suggested to Martin Annis, president of American Science and Engineering (AS&E), that I would be a good candidate for the company to hire.

At the time I was deeply concerned about my future and that of my family. I had visited Beppo Occhialini at CERN, but he was in one of his depressive periods and more agitated than ever. He did not encourage me to work at CERN or in Milan, and by then I did not find either prospect terribly attractive either. On the other hand, my visitor visa was to expire in October 1959. AS&E offered an exciting prospect in a new area of research (space physics), concrete help in changing my visa, exactly twice the salary I was earning at Princeton, and the prospect of steady employment for a few years. I jumped at the opportunity and never looked back.

THREE

Introducing X-Ray Astronomy

In September 1959, I started working at AS&E, a private research corporation founded in 1958 by a group of scientists and engineers primarily from MIT. Many of the stockholders were employees or consultants of the corporation. The chairman of the board, Bruno Rossi, was a professor at MIT. The president of the company, Martin Annis, had obtained his PhD at MIT (working on cosmic rays in the group led by Rossi), as had the vice-president in charge of the Geophysics Division, Jack Carpenter (in theoretical astrophysics). Yet another MIT professor, George Clark, who was working on gamma ray astronomy, was to play a major role as a consultant to the company. The corporation provided research services and products to the government in diverse areas, ranging from defense to medicine to education.

In 1959 AS&E had only twenty-seven employees and received its support mainly through research grants from the Department of Defense (DOD). Before founding AS&E, Annis had been the head of the Physical Science Division at Allied Research Associates and had been responsible for several original contributions on the effects of nuclear weapons against ICBMs (intercontinental ballistic missiles). This work continued at AS&E and was the basis for the development of most of the early passive instruments used to measure such effects near nuclear detonations.

I had been hired to lead the corporation's effort in space physics. One obstacle, however, was that, as a Fulbright fellow, I had a visitor visa that would soon expire. I also had not been cleared for classified research and thus could not become involved in DOD work. Eventually my visa was changed, and I was allowed to work on subjects classified as "confidential" and, later, on all aspects of the company's work. Yet these initial handicaps were providential in that they allowed me the freedom to work for several months on almost anything I wished. I started by solving some simple problems of gamma ray transmission in the atmosphere and thinking of particular experiments that could be carried out in space. George Clark suggested that it might be interesting to measure the ratio of alpha particles to protons in the radiation belts trapped by the earth's magnetic fields, and I designed a detector for this purpose. (This instrument was finally flown in 1963 but experienced an electronic failure with total loss of data.)

The focus of my attention changed dramatically in the fall of 1959 after a casual conversation at the home of Bruno Rossi, where I was introduced to my host by Martin Annis. Bruno was then a member of the Space Science Board, which had been established in June 1958 by the National Academy of Sciences (NAS). The creation of the board was part of the U.S. response to the launch of the first artificial satellite, *Sputnik,* by the Soviet Union on October 4, 1957. *Sputnik* was seen as a threat to the technological superiority of the United States in a field with significant military potential. The feat could not go unchallenged. The charge of the Space Science Board was to "survey the scientific problems, opportunities, and implications of man's advance into space."[1]

The board was divided into several committees that were to prepare reports on a variety of subjects. Of particular interest to astronomers serving on the committees was the possibility of studying celestial objects in all wavelengths of the electromagnetic spectrum unimpeded by the absorbing effects of the atmosphere. In an interim report by the committee, "Physics of Fields and Particles in Space," John A. Simpson suggested that the celestial sphere should be mapped in gamma rays and x-rays.

Lawrence Aller (a member of the Committee on Optical and Radio Astronomy) realized the more profound implications of x-ray astronomy. The study of high-energy ultraviolet emission at wavelengths shorter than 912 Å (or 13.6 eV, the ionization potential of hydrogen) had to overcome the absorption of the incoming radiation by the earth's atmosphere. Even then, the work would be severely impeded by absorption from hydrogen gas in interstellar space. Through its 21-cm emissions, this gas had been discovered by radio astronomers to exist in large quantities. Short ultraviolet radiation would not be observable except from the nearest stars, and interstellar space would become transparent again only at wavelengths shorter than 20 Å, which fall within the x-ray region of the spectrum. X-ray observations therefore provided a unique tool to study high-energy emissions in the universe. Leo Goldberg, chairman of the abovementioned astronomy committee, agreed with this point but noted the technical difficulties in pursuing x-ray astronomy. He reported that "instrumentation for x-ray optics is in a very rudimentary state, and image forming optics is non-existent. Attention to this problem is urgently required."[2]

Bruno Rossi was chairman of the board's Committee on Space Projects, which also discussed the possibility of x-ray astronomy. At the party in his home, Bruno took me aside to discuss the direction that space science could take at AS&E. He suggested that, in addition to particle experiments, I should consider x-ray astronomy. He did not mention the discussions at the Space Science Board or follow up his suggestion by sending me any of the board's reports. (Most of what I knew about the board's activities I learned in 1979, when Richard F. Hirsh sent me his PhD thesis, "Science, Technology, and Public Policy: The Case for X-Ray Astronomy.")[3] I did not see Bruno again until I had worked out the idea of an x-ray telescope. The situation I found myself in was not so different from the "sink or swim" approach typical of my university training. Nevertheless, I owe Bruno a great debt of gratitude for having made a suggestion that completely changed the course of my scientific career.

Coming Up to Speed on X-Rays

Because I was quite ignorant of x-ray properties, experimental techniques, and previous observational work, I decided to start by studying the subject on my own. I began by reading the Arthur Compton and Samuel Allison book *X Rays in Theory and Experiment,* published in 1935. This was a wonderful book that started with a historical introduction to the field, narrating the developments from Wilhelm Roentgen's discovery of "a new type of rays" to 1935. For more up-to-date information, I relied on Springer-Verlag's *Encyclopedia of Physics,* edited by S. Flugge.

X-rays are electromagnetic radiations having wavelengths from a fraction of an angstrom to tens of angstroms (10^{-8} cm), corresponding to quantum energies from a fraction of an electron volt to a hundred thousand electron volts. X-rays are produced most commonly in the laboratory by the use of high-vacuum tubes containing a filament as a thermoionic source of electrons. The electrons are accelerated by a positive potential and strike a metal target, from which x-rays are emitted in all directions. The radiation thus emitted is called "bremsstrahlung" (braking radiation) and has a continuous spectrum up to a maximum energy $h\upsilon < eV$, where h is Planck's constant, υ is the frequency of the radiation, e is the electron charge, and V is the potential across the tube. This relationship was established by William Duane and Franklin Hunt in 1915. The constant of proportionality is Planck's constant, which enters the equation for the photoelectric effect proposed by Albert Einstein in 1905. The Duane-Hunt equation simply says that the maximum energy of the x-rays is equal to the kinetic energy of the electrons that strike the target.

Hans-Joachim Kulenkampf studied the efficiency of this mechanism in the production of x-rays and found that it is very low. For example, a tube with a tungsten target (atomic number $Z = 74$) at 100 kV has an x-ray yield of 0.8 percent; a carbon target ($Z = 6$) at 10 kV yields even less, 0.007 percent. These efficiencies are so small that most of the energy is dissipated in heating the target and reappears in infrared radiation containing a thousand times more power than that in the x-ray domain.

This was a sobering first finding in trying to plan for celestial x-ray observations. If this vacuum-tube process was an appropriate model for the emission mechanism in stars, only extremely small stellar x-ray fluxes would be detected on Earth. But as it turned out, the thermal bremsstrahlung emission from gases at extremely high temperatures (ten million Kelvins) results in a spectrum in which the power emitted at long wavelengths (visible and infrared) is a thousand times smaller than that emitted at x-ray wavelengths.

In addition to continuous emission, an x-ray tube emits a characteristic line spectrum that results from excitation processes in which the incident electrons knock an electron from the inner shell of an atom in the target. When outer electrons make the quantized transition to fill the hole in the inner shell, radiation of a characteristic wavelength is emitted, resulting in an emission line in the spectrum. W. H. Bragg was the first to observe this phenomenon, using the spectrometer that bears his name. Henry Moseley began the systematic study of x-ray spectra of different elements in 1913, when he was 25 years old, and found a striking regularity. Each element exhibited a spectrum identical to that of other elements except for a regular shift of the wavelength scale that is characteristic of the element. The spectral measurements completely supported Niels Bohr's predictions based on the quantum theory of spectral lines that followed from his model of the atom, which he had presented in the same year. Apart from its significance for the theory of matter, Moseley's discovery opened up a new field of analysis of the composition of materials, both in the laboratory and in celestial objects. Called spectroscopy, this tool is now one of the most valuable in x-ray astronomy. (Moseley was to die in the ill-fated battle of Gallipoli in 1915, when he was only 27 years old.)

To analyze the atomic constituents of a material, one can use not only emission lines but also the sharp absorption edges, which were discovered by Louis de Broglie in 1914. They occur at wavelengths independent of the material's chemical composition. This property provides an important diagnostic tool for astronomical observations of interstellar gas.

Scattering of x-rays by electrons consists of a coherent part at the same wavelength as the incident radiation and an incoherent part (the Compton effect). The coherent scattering can be understood in classical terms, but explanation of the Compton effect requires quantum physics. In Arthur Compton's own words: "An x-ray photon is deflected through an angle φ by an electron, which in turn recoils at an angle ϑ, taking a part of the energy of the photon."[4] This process provided the first direct evidence that light quanta carry momentum in their direction of propagation, in addition to energy.

The inverse of this phenomenon is the so-called inverse Compton effect, in which a photon acquires energy in a collision with an electron of higher energy. It turns out that this effect is observed frequently in cosmic settings. For example, the interaction between the 3-K cosmic background radiation and electrons in cosmic plasmas results in a temperature shift of the radiation, which is observable in radio waves. This effect was first suggested by Rashid A. Sunyaev and Yakov B. Zeldovich and was named after them.

An important but less well-known effect has to do with the refraction and reflection properties of x-rays. According to Lorenz theory, it was expected that, for x-rays of sufficiently high energy, the index of refraction in a medium would be less than unity. This prediction was confirmed experimentally by Karl Stenstrom in 1919. The index of refraction being less than unity implies that, when x-rays strike a polished surface at sufficiently small angles, they are totally reflected. (We are all familiar with external reflection by mirrors and total internal reflection by visible light, a phenomenon that permits fiber optics of high efficiency.) The experimental work by Elmer Dersham and Marcel Schein measured the dependence of the x-ray reflection efficiency on the angle of incidence. The practical application of these findings has had extraordinary importance for x-ray astronomy. It has made possible the design and construction of x-ray telescopes, based on total-reflection grazing-incidence optics, with imaging capability comparable to that obtained in visible-light telescopes.

The classical work in x-ray science offered a strong theoretical foundation and a variety of experimental techniques that had been demonstrated in the laboratory and could be extended and applied to x-ray astronomy.

X-Ray Observations of the Sun

I started the second part of my study of the literature on x-rays by reading the article by S. L. Mandel'shtam and A. I. Efremov in the second volume of *Russian Literature of Satellites,* published by the Academy of Sciences of the USSR, using the translation provided by the International Physical Index of New York.[5] It was clear from their article that most of what we knew about x-ray emission from the sun was due to the pioneering work of Herbert Friedman and his group at the United States Naval Research Laboratory (NRL) in Washington, D.C. Friedman's first launch, which took place in 1948, consisted of a V-2 rocket. He had instrumented the rocket with a narrow-band Geiger counter to measure x-rays from the sun in the wavelength band of 1–8 Å. He obtained the first clear and indisputable evidence of solar x-ray emission. The use of photon counters, rather than film, for x-ray measurements was well established in the laboratory, but it had not previously been used in space; after its successful use by NRL, it became the standard approach in the field. I was happy to find that the instrument of choice for x-ray astronomy was the Geiger counter, as I was completely familiar with the device from my previous research on elementary particles.

Geiger counters consist of a cylinder at ground potential and a thin wire stretched along the cylinder axis at high positive voltage. The cylinder is sealed and filled with a noble gas and an organic quencher. Electrons produced by ionizing radiation in the gas are accelerated to the central wire and produce an electrical pulse, which can be counted, stored, and transmitted to the ground from a rocket-borne instrument. Friedman's counters had holes drilled in their sides that were covered with thin metallic or organic films to permit x-rays to enter the active volume of the counter. Table 3.1 summarizes the results of a particular flight (NRL-16) using the Aerobee rocket, which was more powerful than the V-2. When the detectors with aluminum, Mylar, and Glyptal windows faced the sun, they detected a signal that increased with increasing altitude of the rocket. The detector with a beryllium window, which could transmit only radiation of wavelengths less than

TABLE 3.1. *Rocket-based narrowband Geiger counter measurements with different filters*

	Filter		
	Aluminum	Mylar	Glyptal
Thickness (mg/cm^2)	1.60	0.73	0.18
Diameter of opening (mm)	3.00	0.25	0.13
Count rate extrapolated to the boundary of the earth's atmosphere (counts/cm^2 · sec)	4.5×10^4	2.8×10^6	4.9×10^7

8 Å, gave no signal. The detectors on this flight had small windows, typical of the state of the art at the time (see Table 3.1). In the x-ray band from 8 to 20 Å, the results given in the table correspond to an incident solar spectrum of x-rays with a temperature of about 10^6 K and a total flux of about 0.1 erg/cm^2 · sec. For all x-rays up to 100 Å, the energy flux was on the order of 1 erg/cm^2 · sec. This quantity is minute compared to the total solar energy flux of 6.25×10^{10} erg/cm^2 · sec.

For a decade, the NRL scientists continued to study the sun over an entire solar cycle, but perhaps their most ambitious experiment took place on October 12, 1958, when six Nike rockets were launched during a solar eclipse to study the distribution of x-ray sources on the sun. At totality, the extended corona still emitted 13 percent of the uneclipsed flux. The experiment also demonstrated the association of x-ray emissions with plages on the sun.

The Russian review by Mandel'shtam and Efremov also discussed the theoretical predictions by Cornelius De Jager and Gerhard Elwert. They interpreted the solar x-ray emission as coming from a hot corona and attempted a first computation of the expected spectrum, based on the assumption that the corona had a composition identical to that of the chromosphere. Although some qualitative agreement was found, the mechanisms for heating the corona to these high temperatures were not at all understood. The solar x-ray work by the NRL group proceeded with great success, but tentative searches by them for stellar x-ray emitters gave only negative results, with an upper limit of 2×10^{-8} erg/cm^2 · sec for 2-Å (6-keV) photons.

The conclusions from my review of the Soviet work regarding the feasibility of stellar x-ray astronomy were rather daunting. It was clear that one could continue x-ray studies of the sun in great detail, but if all stars emitted the same flux as the sun, their fluxes measured at Earth would be so attenuated by their large distance from us as to become almost undetectable. The light from a sunlike star as close as 1 light-year would be decreased by a factor of about ten billion and thus would produce a flux on top of the atmosphere a thousand times less than the detection limit set by Friedman. Detection of these weak fluxes would require an improvement in detection capability by many orders of magnitude.

First Light: Conceiving the X-Ray Telescope

At this point I found myself once again mired in a field in which the sparseness of the data that could be obtained would result in poor statistics, thus preventing fundamental new discoveries, as had happened in cosmic-ray research at the end of its productive years (1945–55). However, I remembered my dreams at the University of Milan of concentrating cosmic rays in my cloud chamber with magnetic funnels. And then geometry came to the rescue. While reading and rereading Flugge's *Encyclopedia,* I again found the description of attempts to use total external-reflection grazing-incidence optics for microscopy, which had not succeeded. It occurred to me that if I scaled up the size of the optical surfaces as appropriate for a telescope, many of the difficulties experienced in building optics for microscopy would disappear. As soon as this idea crystallized, I remembered my beloved projective geometry and understood that I could build a mirror that could concentrate x-rays by using the outer edges of a paraboloid, where the grazing-incidence conditions could be satisfied. I then roughly computed the gain in sensitivity that such a device could achieve, which turned out to be on the order of one thousand to one hundred thousand, depending on the angular resolution of the mirror and the detector (Figure 3.1). Thus, however hard the work might prove in the beginning, the potential for discoveries in x-ray astronomy was secure. This finding proved a powerful stimulus for me to continue research in the field; its impact can be compared to that on an astronomer making measurements in the visible with the naked eye who is promised, in the not-too-distant future, the sensitivity of a 30-inch telescope.

As soon as I hit on this idea, I brought it to Martin Annis, who suggested we call in Bruno Rossi. I had not seen Bruno since the party at his house, but he immediately grasped the concept and its advantages and suggested that one could use multiple surfaces nestled one within another to further improve the gain. After this discussion, I wrote up a paper and (at Martin's suggestion) asked Bruno to be a coauthor, in recognition of his suggestion to

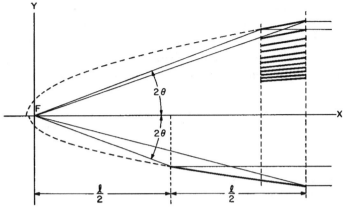

FIG. 3.1. *The first published conceptual diagram of a single-reflection x-ray telescope with single and nested mirrors. The incoming lines parallel to the x-axis represent x-rays incident at the grazing angle θ. When they strike the mirror (or mirrors), they are totally reflected and are focused at the focus F of the parabolic mirror (indicated by the dashed curve).*

study x-rays and his contribution of the nesting concept. "A 'Telescope' for Soft X Ray Astronomy" was published in the *Journal of Geophysical Research* in February 1960.[6] The paper was sent to the journal on December 7, 1959, about two months after I had started work in this new field.

Faint Echoes: Early Estimates of Stellar X-Ray Fluxes

In addition to the study of the literature and the article on the telescope, with George Clark I prepared a white paper titled "A Brief Review of Experimental and Theoretical Progress in X Ray Astronomy" (AS&E-TN-49) for inclusion in proposals for funding that AS&E intended to submit in 1960. We discussed some aspects of the work with Stan Olbert of MIT, particularly with regard to x-ray emission from magnetic bremsstrahlung of high-energy electrons in strong magnetic fields. While I provided most of the instrumentation ideas, George was much more cultured than I in subjects astronomical and also reflected some of the views of the gamma ray and plasma physics groups at MIT. The purpose of the paper was threefold: to investigate what astrophysical processes could produce x-rays, to discuss potential sources of x-rays outside the solar system, and to discuss possible experimental approaches that might provide sufficient sensitivity to achieve source detection.

In addition to the processes that were believed to give rise to solar x-rays (line emission from multiply ionized atoms, bremsstrahlung from hot coronal gases, and recombination emission), we considered the process of magnetic bremsstrahlung (or synchrotron emission) that, although not very important for the sun, might be significant for other celestial bodies. The main indication of this significance came from radio observations of strong magnetic fields in interstellar space and of the envelopes of novas and supernovas, which we suggested as a possible source of x-rays.

We considered a number of other types of stars that might be detectable in x-rays. Bright hot stars like Sirius would not be detectable from Earth if they had fluxes like the sun; however, because Sirius is a thousand times more luminous than the sun in visible light, we reasoned that it might also be a thousand times more luminous in x-rays. This luminosity would result in a flux of about 0.25 photons/cm^2 · sec at the top of the earth's atmosphere. Still, this flux corresponds to only 2×10^{-9} erg/cm^2 · sec—fifty times weaker than the upper limit of detectors such as those flown by NRL.

We considered flare stars, because they show sudden increases in their optical brightness and because, as in the case of the sun, such flares could be accompanied by x-ray bursts. George was particularly keen on peculiar A-stars, which are rapidly spinning, have large and variable magnetic fields, and show intense spectral lines for the heavy elements. Their rapidly changing fields might accelerate protons to high energies. Nuclear interactions of the protons would produce gamma rays and x-rays. In fact, William Kraushaar and George were collaborating at MIT on gamma ray astronomy, and these objects were one of their targets.

We considered supernova remnants, such as the Crab Nebula, which were known to emit both radio and visible radiation by the synchrotron process. Electrons of energy as high as 10^{12} eV were believed to be trapped in a 10^{-4}-gauss magnetic field. If higher-energy electrons existed and interacted with locally stronger magnetic fields, x-rays would also be produced. Herbert Friedman and other scientists had also suggested the Crab Nebula as a potential source of x-rays at about the same time as our white paper was published. On top of all the other uncertainties, the lifetime of such energetic electrons would be quite short, requiring an unknown mechanism of energy replenishment operating even today, hundreds of years after the supernova explosion.

Finally, we considered the moon as a potential source of x-rays. Clearly the moon should be a source of fluorescent x-rays produced on its surface by incident solar x-rays. Such an effect was computed to result in 0.4 photons/cm^2 · sec. This flux was below the threshold of previous detectors but not impossible to achieve. We also considered the possibility that energetic electrons in solar ejected plasmas could produce much greater x-ray intensities by bremsstrahlung. The estimate of fluxes of high-energy electrons in the solar ejecta was uncertain, at the time, by a factor of a thousand, and the x-ray fluxes could correspondingly range from nothing to 1,000 photons/cm$^2 \cdot$ sec.

The paper (AS&E-TN-49) was completed on January 15, 1960, and the authors included Riccardo Giacconi, George W. Clark, and Bruno B. Rossi.[7] We did not include the section on x-ray emission from the moon in the proposal for funding that we submitted to NASA. This section was of particular interest to the Air Force Cambridge Research Laboratories in Bedford, Massachusetts, and thus we included it in the proposal we submitted to them.

The World's First Conference on X-Ray Astronomy

On May 20, 1960, the first conference on x-ray astronomy was held at the Smithsonian Astrophysical Observatory in Cambridge, Massachusetts, and was attended by most of the scientists interested in the field. A notable exception was Herbert Friedman. The speakers included such experts in x-ray optics as A. V. Baez (who chaired the meeting), Art Berman, Paul Kirkpatrick, Leonard Rieser, and D. H. Tombulian; experimentalists, including Philip Fisher, H. E. Hinteregger, Jim E. Kupperian, and myself; theoreticians, such as Robert J. Davis, Stan Olbert, and Bruno Rossi; and technical experts, such as Robert Schneeberger.

We heard presentations on predicted x-ray fluxes from Davis and Olbert, which provoked many comments. I heard for the first time about the work of William Grasberger and Louis Henley ("Preliminary Report on Interplanetary X-Ray Radiation from Stellar Sources"), which Fisher seemed to admire. No new suggestions were presented beyond what was contained in our AS&E white paper. There were presentations by Baez, Kirkpatrick, and me on x-ray telescopes. Kirkpatrick and Baez mainly emphasized the difficulties of obtaining reflecting surfaces sufficiently smooth and precise to render useful images. I was more optimistic and mentioned briefly the visible-light models, based on total internal reflection, that we had already built. Replica techniques for the construction of telescopes were suggested by Kirkpatrick. Imaging detectors based on channeltron multipliers (such as the ones I had used in my research at Princeton) were discussed. Baez also considered the possible use of Fresnel-zone plates and of two sets of grazing-incidence mirror plates at right angles to each other. Finally, we were shown a pinhole camera picture of the sun in x-rays obtained by the Friedman group on April 19, 1960. The picture showed, with rather poor resolution, the glow of the emitting corona and brightening that coincided with a solar plage.

In rereading the proceedings, I noticed some remarks I made that do not sound bad even today:

> I would like to comment on the processes involved in x-ray emission. Until now we have examined the sun and its peculiar type of emission as if it were the norm for all other stars. However, the various processes by which x rays are produced in different amounts—collision bremsstrahlung, magnetic bremsstrahlung, Cerenkov radiation as well as K ionization—would lead to large variations in the x-ray flux to be expected from different stars. X-ray astronomy, in other words, is of broader interest than the astronomy of optical emission, and we would expect to see different objects in many cases than those identified only by their visible light. X-ray emission is much more closely related to electron temperatures; for example peculiar A-stars with a thin outer shell of high-Z materials would contain a great number of high-energy electrons, and despite their relatively low visible-light transmission, would be expected to be a high x-ray emission source. We would expect a large x-ray flux from the Crab Nebula, but for reasons different from what we would call its kinetic temperature. If novas represent the mechanism by which cosmic-ray energy is injected in the universe, then one can ask the question: Is there high local acceleration of cosmic rays in the region of the Crab Nebula for example? If so we would expect x-ray production that would overshadow all other mechanisms. And so my point is that x-ray astronomy is especially interesting because of these new mechanisms and you cannot simply extrapolate from ordinary temperature considerations.[8]

While I am a little embarrassed by the apparent brashness of my preaching to the audience, and also by a few barbarisms, I believe the view I expressed then is still valid today. Having had the opportunity to compare my work with that of other scientists in the field gave me renewed confidence and the determination to succeed in opening up this new observational window.

The First Celestial X-Ray Source: Discovering Sco X-1

The Challenges of Imaging

While writing the 1960 white paper, I had come across the work of Hans Wolter,[1] which described several possible optical designs based on two reflections from conic surfaces. The use of two reflections ensured that the Abbe sine condition for imaging would be satisfied. Ernst Abbe, while working with Carl Zeiss on the improvement of microscopes, had shown that this general condition (between the angles of the incoming and outgoing rays) had to be met by any optical system for it to produce sharp images of off-axis objects, as well as of those directly on the axis of the instrument. Thus, a paraboloid-hyperboloid mirror could yield undistorted images at the focus within a significant field of view. Wolter's designs were intended for microscopes, and the exceedingly small dimensions, coupled with the requirement for high-precision surfaces and an extra-smooth finish, made a practical application impossible.

Even though, as mentioned previously, I believed that the construction of telescopes would be easier, the technical difficulties of figuring high-precision mirrors with a surface roughness of less than 20 Å presented, at the time, a formidable challenge. It took my group from 1960 to 1968 to build the first really high-performance x-ray telescope for solar work and until 1978 to build one suitable for extra-solar x-ray astronomy. We were fortunate in obtaining NASA funding to "design, construct, and test a prototype x-ray telescope" as early as October 1960 (NAS-5-660). This contract was the first of a series that ultimately succeeded in developing the world's most advanced x-ray optics for astronomical research.

I was unwilling, however, to wait 10 years or more to start actual observations of the sky. While beginning the work on the development of x-ray telescope technology, I considered the problem of designing rocket-borne experiments with traditional counters. I soon realized that to make significant progress, one would need an instrument at least fifty times more sensitive than that used by Herbert Friedman at NRL, but I thought this increase could be achieved by improving both the detector system and its utilization in flight.

With my background in cosmic-ray research, I was well aware that Geiger counters were sensitive to the charged particles of cosmic rays. The cosmic-ray flux at Earth is about 1 particle/cm$^2 \cdot$ sec. Because the entire counter volume is sensitive to cosmic rays, a 10-cm-long counter of 2-cm diameter, such as that used by Friedman, would have a cross section of 20 cm^2 and would measure about 20 counts/sec. The size of Friedman's x-ray windows, on the other hand, were less than a fraction of a square centimeter, and the expected counting rate was less than 1 count/sec. Thus, the background count was higher than any of the expected fluxes from stellar x-ray sources.

This effect could easily be reduced by applying the anti-coincidence technique that was commonly used in elementary particle research. The technique consists of using a scintillation counter, which surrounds as much of the x-ray detector as feasible (without, of course, covering the entrance windows). Energetic cosmic-ray particles will traverse both the scintillation and Geiger counters and give a simultaneous signal in both, whereas x-rays will be stopped in the gas of the Geiger counter, which alone will detect them. By eliminating electronically timed coincident signals, the background can be reduced by 90 percent.

A second consideration had to do with the tiny field of view (3 degrees) used by Friedman in his searches. This field covers only one part in four thousand of the sky. If the detector's field of view were to sweep the sky uniformly during the flight, it would spend only one part in four thousand of the time on any one source. Even if the detector were made to sweep the same great circle in the sky during the entire flight, it would observe a possible source for only 1 part in 120 of the available time. Given that the duration at altitude of rocket flights was only 300 seconds, and the expected flux less than

1 count/cm^2 · sec, fewer than three counts would be accumulated during the entire flight, far too few for detection of x-ray sources. These disadvantages could be overcome by pointing the detector at a fixed point in the sky during the flight. However, without knowing where to point the instrument, only 0.025 percent of the sky would be explored with each flight.

To have the best chance of observing an unexpectedly bright x-ray source anywhere in the sky, my solution was to open up the field of view by a huge factor: 40 times in angle and 1,600 times in solid angle. Such an increase would allow the detectors to see a large fraction of the sky at any given time. Furthermore, if a source was observed, it would remain observable for a large portion of the available time at altitude (Figure 4.1).

A third and most important consideration had to do with the available detection area. We thought it would be possible to build pancake-shaped Geiger counters with many windows to allow more x-rays to penetrate the counter. By using many windows, we could keep them quite thin but increase the total detection area by at least a factor of ten (Figure 4.2).

These considerations convinced me that the NRL surveys could never have discovered x-ray sources as faint as those we expected to find, but that we could design a rocket experiment to search for celestial x-rays with a sensitivity of 0.1 photon/cm^2 · sec, some fifty times lower than the upper limit given by Friedman.

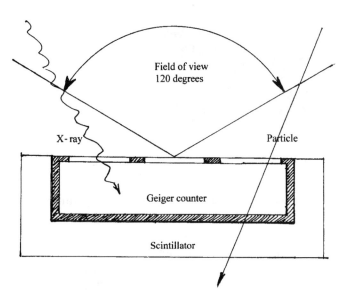

Field of view
120 degrees

X-ray

Particle

Geiger counter

Scintillator

FIG. 4.1. *Schematic of the 1962 detector showing the field of view and the anti-coincidence system used to reduce background caused by incoming cosmic-ray particles.*

FIG. 4.2. *The 1962 rocket payload. At the bottom can be seen the pancake-shaped x-ray counters and the housings for the anti-coincidence scintillators, light pipes, and photomultipliers.*

This overall approach to the design of the experiment owes much to the training I had received as a physicist in instrumentation design and development, but even more to the scientific methodology I had learned from Beppo Occhialini and R. W. Thompson. They taught me that one had to look carefully at a problem and then try to design an experiment with the maximum likelihood of obtaining critical results. This process often required not just a small improvement of the previous state of the art but a major jump in sensitivity by the use of new technologies.

This approach was imposed by the new circumstances one confronted when starting observations in a new domain, as in x-ray or gamma ray astronomy in space or neutrino astronomy from Earth, but it differed from the traditional methodology of astronomy. Ground-based optical astronomy, for instance, had incrementally improved its observational capabilities first by introducing charge-coupled device (CCD) detectors in place of film for recording faint objects and later by constructing larger telescopes. Much of the most interesting work (for instance, the study of the nature of radio quasars and of x-ray sources) did not immediately require higher sensitivity. This lack of urgency is reflected in the fact that it took 50 years to go from the 5-meter Palomar Telescope to the 10-meter Keck, an improvement of a factor of four. By contrast, the sensitivity of x-ray observations increased by a billion in less than 40 years. In the past decade, increased worldwide competition, better funding, and improved technology prompted a much more ambitious pace of development in optical astronomy, both in space and on the ground.

Construction of a New Breed of Instruments

There remained the problem of obtaining grant support for the research we planned. We approached NASA in the summer of 1960 with two proposals: one to develop an x-ray telescope and the other to scan the night sky for x-ray stars with rocket-borne detectors. We were fortunate that John Lindsay of the Goddard Space Flight Center (GSFC), Greenbelt, Maryland, was interested in the development of telescopes for solar research and, as mentioned at the beginning of this chapter, we started work on the telescope in October 1960 with NASA funding.

We were less fortunate in obtaining support for our rocket research. During our discussions with NASA officials, we were not told that the agency was already supporting Philip Fisher's group at Lockheed Missiles and Space Company and Malcolm Savedoff's at the University of Rochester, New York. Moreover, we were regaled with jocular remarks questioning why NASA should encourage the search for objects not known to exist.

However, one of the great strengths of the research system in the United States at the time was the opportunity to receive funding from a variety of agencies. This is to some extent still true, though with the usual politically correct concern for duplication of effort. So we immediately started discussions with scientists at the Air Force Cambridge Research Laboratories (AFCRL). AS&E had obtained support from AFCRL for several classified and unclassified research projects. (George Clark and I, for instance, had proposed an "electron-proton spectrometer" for space research in August 1960.) Although the Air Force had lost its bid for leadership in space to NASA in 1958, it still retained a great deal of interest in space exploration.

We were able to persuade John W. Salisbury, chief of the Lunar and Planetary Exploration Branch of AFCRL, to support a first flight of a 1-cm² x-ray detector on a Nike-Asp (a very small rocket) and later four flights on the much bigger Aerobee rockets. The entire funded program ran from January 11, 1961, to February 28, 1963.

Even on the first flight, which took place in 1960 (before we had been awarded the contract), the detector design followed the criteria I have discussed: large field of view, anti-coincidence shield, and as large a detection area as possible (1 cm²). By then Frank Paolini had joined my small group, which also included two great technicians, Al di Caprio and Tom Quinn. In June 1960, we journeyed to Eglin Air Force Base in Florida with our first payload. A few seconds after launch the rocket blew up.

We started again with a bigger payload, made possible by the use of an Aerobee rocket, which had a diameter of 30 cm. The most important improvement was the completion by Frank Paolini of the pancake counter with seven mica windows, which achieved an effective window area of 10 cm². Our payload consisted of three of these counters mounted at 120 degrees from one another around the azimuth of the rocket's axis and 55 degrees in elevation. Each counter was embedded in an anti-coincidence scintillator.

The windows in two of the counters were made from 0.2-mil-thick mica, the thinnest film that could withstand the pressure changes and vibrations during launch, and that in the remaining counter was of 1.0-mil mica. The different thicknesses resulted in different x-ray transmission bands, and the ratio of fluxes in the two different counters gave us a measurement of the x-ray spectrum. The counters peaked in sensitivity between 2 and 8 Å.

For each counter we used a digital electronic processor that would veto signals when a signal from the Geiger counter occurred within 10 msec of one from the scintillation counter, thus rejecting cosmic-ray particles. To distiguish x-rays from cosmic rays, we were obliged to count the x-rays as individual quanta and to know each x-ray photon's time of arrival with great accuracy. We decided to preserve these timing data and devised a clever system to transmit the information to the ground, even though the rocket had only an analog telemetry system: we designed the circuitry so that each jump in the analog voltage corresponded to a single x-ray quantum.

This approach was different from the analog reading of averaged counting rates used by Friedman's NRL group in their work and was more appropriate for counting individual photons in low fluxes. It also was more suitable for retaining accurate information on the time of arrival of photons from variable or pulsating sources. This capability was to prove essential in future x-ray observations.

We used an optical aspect system (similar to that previously used by J. E. Kupperian Jr. and R. W. Kreplin) that allowed correlation of the direction of arrival of the x-rays with the azimuthal orientation of the detectors with respect to visible objects in the night sky.

More than a year had passed since our first unsuccessful attempt with the Nike-Asp. By this time, Herbert Gursky had joined the AS&E group and the three of us (Frank, Herb, and I) attempted another launch with the new equipment on October 1, 1961, from the White Sands Missile Range in New Mexico. This time the rocket flew beautifully, but the doors to the detectors (which were supposed to jettison once the rocket achieved altitude) stuck to their frames and refused to open. No data were obtained, and the payload was of course lost when it crashed to the ground. We could build a new payload in less than a year, so we rescheduled a new launch for June 1962.

Discovery Launch: June 18, 1962

The launch took place at one minute before midnight at White Sands, New Mexico, when the moon was one day past full and 35 degrees above the horizon. The rocket reached an altitude of 225 km and was above 80 km for 350 sec. All three doors opened properly, but one of the 0.2-mil mica window counters did not function.

The rocket spun on its axis every half second, and the axis was pointed at the zenith (Figure 4.3). As the rocket ascended through the atmosphere, the counting rate in the two good counters rose rapidly. A large peak in the counts was immediately apparent to Herb, who was anxiously monitoring the strip-chart recorder. The peak appeared to be in the general direction of the moon but did not coincide with it. The x-ray flux was much larger than expected from the moon or, for that matter, any other celestial source (Figure 4.4).

What had we found? We knew immediately that the data were sensational —if indeed they were valid—but before announcing to the world that we had

FIG. 4.3. *Rocket scan geometry.*

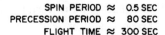

SPIN PERIOD ≈ 0.5 SEC
PRECESSION PERIOD ≈ 80 SEC
FLIGHT TIME ≈ 300 SEC

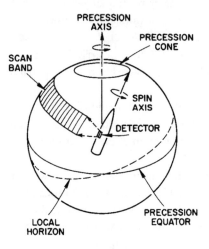

FIG. 4.4. *Results obtained during 300 seconds of the June 18, 1962, rocket flight from two of the counters, folded over different rotations.*

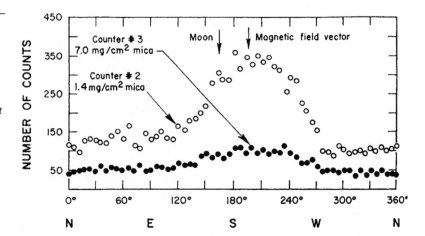

discovered something new, we had to make absolutely sure that the observed effect wasn't something old, or trivial, or downright wrong. For one thing, we could not pinpoint the source of the radiation, because the wide field that had made the discovery possible was not suited to accurately measure positions in the sky. That did not help us validate our findings. We agonized over the results for three long months.

By careful analysis, we were able to exclude all simple explanations for the observed flux. It could not be due to ultraviolet radiation, because our counters were coated with lamp-black to prevent sensitivity to ultraviolet (and, in fact, we did not observe any of the well-known ultraviolet sources). Nor could it be caused by particles spiraling along the earth's magnetic field, whose direction roughly coincided with the moon's position. Our understanding of the characteristics of the trapped radiation in the Van Allen Belts was crucial in excluding that possibility on the basis of our own data. We were also able to dismiss the possibility that we were dealing with auroral effects. Last, but not least in this case, we were sure that our detectors had functioned properly.

The first conclusion was inescapable. *We had found a new object, an individual celestial source with an x-ray flux at Earth of 5 photons/cm² · sec at wavelength 3 Å.*

Similar considerations led us to conclude that we had also measured an isotropic x-ray background. Lack of correlation with the ecliptic plane or the plane of the galaxy was taken as evidence that the background was extragalactic in origin.

The second conclusion, therefore, was that *we had observed a diffused, isotropic x-ray background of 1.7 photons/cm² · sec · steradian of extragalactic origin.* This finding was soon to have important cosmological implications.

Bruno Rossi had not participated in the experimental work leading to the discovery, but he contributed greatly to the writing of the paper reporting our results. I was told by George Clark that he went so far as to personally guarantee the integrity of the work to the editor of the journal *Physical Review.* The paper by Riccardo Giacconi, Herbert Gursky, Frank Paolini, and Bruno B. Rossi, "Evidence for X Rays from Sources outside the Solar System" was soon published in *Physical Review Letters.*[2] We ended our letter with a carefully worded conclusion (patterned after that of Victor Hess when, in 1912, he announced the discovery of cosmic rays): "We believe the data can best be explained by identifying the bulk of the radiation as soft x rays from sources outside the solar system."

The flux of x-rays from the moon, which we had estimated as between 0.1 and 1.0 photon/cm² · sec, was not detected until much later by the German x-ray satellite ROSAT. In our flight, the flux from the moon was completely swamped by the star of the show, Sco X-1. We named the source after the constellation Scorpius, in which we found it. X stood for x-ray, and 1 marked it as the first x-ray object discovered in Scorpius.

The discovery of Sco X-1 proved to be as exciting as it was unexpected. It was clear that, for this celestial source, the ratio of x-ray to visible-light energy was vastly different from that in the sun. While in the sun the x-ray emission constitutes less than 10^{-6} of the visible-light emission, in Sco X-1 the x-rays dominate by a factor of a thousand over visible light. This result by itself was astounding, because no x-ray source in laboratories on Earth could produce x-rays with that efficiency. Therefore, some new physical process and/or unknown conditions had to be involved.

Furthermore, the x-ray luminosity of the source had to be immense to produce such large fluxes over interstellar distances; it had to be at least a thousand times more luminous than the luminosity of the sun integrated over all wavelengths. This prodigious output was another strong indication that we were dealing with some new type of celestial object.

After the discovery of Sco X-1, x-ray astronomy would never be the same. The field's rapid progress in the following years had an impact on all of astronomy.

The Underpinnings of Discovery: Institutional Support

Before describing what happened next in the field, I comment on some aspects of the working environment that enabled my group at AS&E to vigorously pursue this new opportunity. I sometimes wonder whether x-ray astronomy could have developed as quickly had it started in a different kind of institution. The advantages we had at AS&E compared to working in an academic institution were great:

> no distraction from research and the management of research;
> no limits on expansion of personnel and facilities (provided funds could be
>   obtained);
> ability to carry out several development programs simultaneously;
> ability to hire and fire at will (although few people were ever fired); and
> ability to take risks and make management investments in the areas of capital,
>   equipment, and seed money.

Management errors and technical incompetence could quickly be remedied; new skills could be acquired rapidly through expansion of the workforce. Because AS&E was interested in growth not only in scientific but also in commercial and defense-related research, rapid expansion occurred from

1959 to 1970, and thus technical competence could be drawn from a constantly expanding pool.

For instance, I was hired in September 1959 to start the science program. I was able to use consultants of the caliber of George Clark and Stan Olbert and to rely on existing shops to build prototypes of the telescopes. As soon as I obtained the first NASA contract to develop an x-ray telescope, I was able to design and acquire a vacuum facility for testing x-ray instruments. The confidence and support of Martin Annis, the president of AS&E, was a constant source of strength.

I was able to have assigned to my project an electronic and a mechanical technician, as well as three PhD physicists. (Norman Harmon had joined my group in 1960 in addition to Frank Paolini.) By creating a division of Space Research and Systems, of which I became chief, I was able to exercise direct technical and management control over the project and the staff assigned to it.

Significant for our development was an opportunity that presented itself in late 1961. As a response to the atmospheric testing of nuclear weapons by the Soviets, President John F. Kennedy decided to carry out a test program for high-altitude bursts of nuclear weapons, as a bargaining chip to be used in negotiating a test ban treaty. Rapid response was essential. In October 1961, AS&E was asked to carry out several projects for AFCRL, to be completed by March 1962; money was no object, but time was of the essence. Martin discussed this possibility with me and asked whether I would be willing to take part in the effort. Moved by confidence in Kennedy's judgment, a feeling of gratitude and love for my adopted country, and loyalty to AS&E, I committed my group to this work.

The tests were carried out from Johnston Island, a mile-long sandbar 500 miles southwest of Oahu, Hawaii. I spent that summer shuttling back and forth between Cambridge, Hawaii, and the test island, while simultaneously working on the experiments designed to detect the effects of the nuclear explosion and on the x-ray astronomy sounding rockets.

As it turned out, from the fall of 1961 to the summer of 1962, my group expanded from five or eight individuals to seventy or eighty. During those months, we regularly worked 72 hours a week (Monday through Saturday), 12 hours a day—a pace difficult to sustain for long. We prepared instruments to measure electrons, x-rays, and gamma rays produced by bursts in the atmosphere from nuclear weapons. We designed, built, tested, integrated into vehicles, and launched nineteen rocket payloads from Johnston Island. We built and integrated six satellite payloads launched from Vandenberg Air Force Base in California. We furnished a debris-measuring instrument for an airplane in the record time of one week. In 95 percent of the cases, our pay-loads worked properly and the experiments were successful.

Completely separate from the above frenzy, our unclassified program continued at its own fast pace. We managed to launch four rockets from Kiruna, Sweden, to search for noctilucent clouds (of interest to geologists at AFCRL) and the stellar x-ray discovery rocket from White Sands (of great interest to us).

These efforts built up a loyal, gung-ho crew, technically skilled and of high morale. Our management capabilities were immeasurably strengthened and demonstrated; we had acquired a reputation for technical competence and a "can-do" attitude.

Throughout this period and in the following months, we greatly strengthened our system, electrical, and mechanical engineering support capability. We developed the habit of giving each PhD scientist who was carrying out an experiment a crew of three engineers and one or two technicians. With this kind of support, experiments could quickly be carried out; they would normally succeed, and therefore a relatively small group could accomplish a great deal. Laboratory space kept expanding from one 10,000-square-foot former milk truck garage (owned by MIT) to six or eight facilities.

I believe that this background of strong and flexible institutional support explains, at least in part, how we were able to quickly pursue and expand our research endeavors after the initial discovery flight of June 18, 1962.

Reactions to the Discovery of Sco X-1

The paper reporting the results of our work was published in the December 1, 1962, issue of *Physical Review Letters*. I had first presented our results at a symposium on x-ray analysis techniques held in late August at Stanford University in California. Although Herb Friedman, who was at the meeting, introduced himself and congratulated me on our discovery, some astronomers were skeptical about the validity of our extraordinary results.

We decided to use our last two Aerobee rockets to confirm our findings. By flying at different times of the year, we could determine whether Sco X-1 was fixed in celestial coordinates rather than in solar or terrestrial ones. We used the same detection systems as in the first launch, but we included beryllium windows, which are more transparent to x-rays. The first of the two launches occurred on October 12, 1962, and the second on June 10, 1963. During the October flight, the bright source Sco X-1 was below the horizon and should not have been observable. This is in fact what happened: no Sco X-1 was seen, but we observed a new source in the constellation Cygnus (Cyg X-1) and one near the Crab Nebula. During the flight of June 1963, we were again able to see Sco X-1 and thus reconfirm its existence. We also observed a new source east of that bright object. The existence of a diffuse background was confirmed in both flights. We reported these results in a paper titled "Further Evidence for the Existence of Galactic X Rays," which appeared in September 1963.[3]

These investigations satisfactorily verified for us the existence of celestial x-ray sources. However, such an important discovery needed independent confirmation to be fully accepted by the scientific community. Soon the work of the NRL group provided important and clear confirmation of our measurements.

Friedman had immediately recognized the profound implications for astronomy of the existence of very luminous x-ray sources, which outshined the ultraviolet sources he had been studying by orders of magnitude. He and his group had accumulated years of experience in carrying out space experiments and were in an excellent position to explore the new field.

As it happened, Stuart Bowyer, a graduate student in physics, joined the NRL group in late 1962. He had worked previously during summer recesses at NRL and was quite familiar with the technology. He had come up with a new idea for how to construct large-area x-ray detectors. NRL had already planned the launch of an Aerobee rocket equipped with ultraviolet detectors for April 1963. By removing some of the ultraviolet instrumentation, Bowyer's x-ray counter could be flown instead, thus giving NRL a chance to promptly follow up on the AS&E discovery. The only question remaining was whether Stu could do it in the short time available before launch.

The counter consisted of an ensemble of small, stubby, proportional counters packed together in a bundle. (Proportional counters are identical to Geiger counters except for being operated at a lower voltage. The pulse produced by an x-ray photon is proportional to the energy of the photon.) The entrance window was at the end rather than on the side of the counters and was covered with a thin layer of beryllium. It had a total sensitive area of 65 cm^2, about twice the total area of three AS&E counters combined. Although the device lacked the anti-coincidence feature, the background was little affected, because (as demonstrated in the AS&E observations) the x-ray isotropic sky emission was larger than the cosmic-ray component. The field of view was defined by a metal honeycomb to about 10 degrees. Bowyer completed the detector in time for the launch on April 29, 1963, and the NRL group was able to scan the region near the galactic center where we had located our source (Figure 4.5).[4]

The relatively narrow field of view of the detector permitted much greater precision in determining the location of Sco X-1 than we had been able to obtain on the discovery launch. The NRL data clearly demonstrated that the galactic center was not the origin of the x-ray emission. The NRL group also detected radiation from the direction of the Crab Nebula and confirmed the existence of an isotropic background.

Fisher's group at Lockheed, which had received NASA support as early as February 1961, had been plagued with technical and bureaucratic delays. They did not succeed in launching until three months after the successful

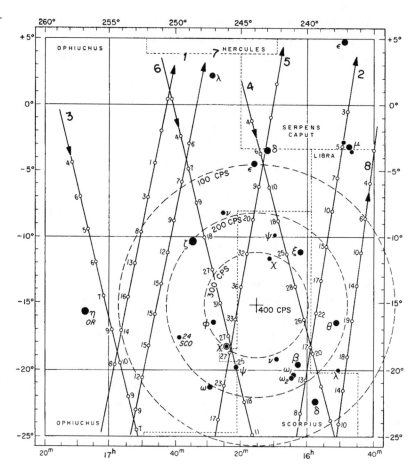

FIG. 4.5. *Results obtained during the NRL scans of the Sco X-1 source on the April 29, 1963, rocket flight. The tracks with arrows represent the trace of the center of the detector's field of view as it scans the sky, plotted in right ascension and declination. The numbers along the tracks are the counts accumulated in 0.1 sec. The circles represent lines of equal counts estimated from the known angular response of the detector (the rate is in counts per second), if the bright source in the field is centered on the cross labeled 400 cps.*

AS&E flight. A rocket flight in September 1962 obtained very little data, and early submissions to the *Astrophysical Journal* were rejected, because the journal editor thought that the results could be due to random noise fluctuations rather than to discrete sources.[5] In a new flight in March 1963, Fisher and Arthur J. Meyerott obtained data that led them to conclude that they had observed supergiant stars, nebulae, and a radio source.[6] Although this paper was published, the results were received with great skepticism. In a letter to the editor of the *Astrophysical Journal,* Bowyer showed a direct contradiction between the NRL and Lockheed data in regions where they overlapped.[7] Bowyer suggested that the sources claimed by Fisher and Meyerott could be explained by random background fluctuations above an improperly chosen low-average background value. In their response, Fisher and Meyerott agreed that "the published results and interpretations of the Aerobee 4.69 data were inconclusive."[8] The results derived from the 4.70 rocket were described as "tentative and suggestive only" because of their low statistical significance.

The discrepancies between the Lockheed results and those of the AS&E and NRL groups caused some confusion in the research community and made initial acceptance of our results a little more difficult. Nevertheless, Fisher retained NASA's confidence, and the sponsorship of his group continued throughout the 1960s.

We at AS&E, on the other hand, were facing a serious problem. Notwithstanding our success, AFCRL was not inclined to continue a program of pure astronomical research that had no immediate relevance to the exploration of the moon. They therefore told us that their funding would end in 1963. We could only go back to NASA and try again to win support.

FIVE

A Blueprint for X-Ray Astronomy: AS&E Proposals to NASA—
Surging Interest in X-Ray Astronomy in the 1960s—
The Nature of the Beast: The Engines of Galactic X-Ray Sources—
The NRL Crab Nebula Experiment—The Optical Identification of Sco X-1

Plans and Progress in X-Ray Astronomy

A Blueprint for X-Ray Astronomy: AS&E Proposals to NASA

One of the difficult aspects of building up and retaining our technical staff at AS&E was the lack of any safety net to carry on the research when government support was substantially reduced. Thus it was imperative that we plan in advance for programs that had financial as well as logical continuity.

We were successful in late 1963 in obtaining further support (beyond the October 1960 contract) from the Solar Research Group at the Goddard Space Flight Center; as I mentioned earlier, John Lindsay, the head of that group, was keenly interested in the development of x-ray telescopes for solar research. We were awarded funds to develop a solar x-ray telescope for an initial flight on a rocket in June 1963, and later, the pointed section of the Orbiting Solar Observatory (OSO) IV. We also were assigned a space on the wheel section of OSO IV for stellar x-ray observations. Work on the OSO instruments started in September 1963.

It was not, however, until June 1964 that we were funded for rocket research on x-ray stars from NASA's astronomy section at headquarters. We had submitted a proposal to that office as early as May 13, 1963, but had received no response for a few months. We decided that perhaps we had made a mistake in our approach to NASA by requesting support in a piecemeal fashion. To be taken seriously, perhaps what we needed was a multiyear, comprehensive plan for x-ray astronomy research. On September 25, 1963, we submitted to NASA a document written by me in collaboration with Herb Gursky, with contributions from several other people on the AS&E staff. It was titled "An Experimental Program of Extra-solar X Ray Astronomy."[1] This document presented a clear vision of the direction the field should take and would profoundly affect the development of x-ray astronomy for four decades.

We first summarized the results of our own investigations as well as those of NRL, noting that the existence of nonsolar x-ray sources was now well established. Three separate discrete sources were known in Scorpius, the Crab Nebula, and Cygnus.

We also reviewed theoretical efforts to explain these findings. S. Strom and K. Strom had shown in 1961 that interstellar and intergalactic gas was expected to be quite transparent to x-rays of wavelengths less 20 Å, so that we were assured that celestial x-rays would reach Earth without being absorbed by these giant clouds in outer space. Robert J. Gould and Geoffrey Burbidge had suggested in 1963 that the Scorpius x-ray source could be identified with the center of our own galaxy and the isotropic x-ray background with the combined emission from external galaxies. While the first suggestion was soon proven to be untenable in view of the NRL results, the second suggestion (on the nature of the background) was validated in 2001, although with a different production mechanism than the one they had proposed. The possibility of detecting x-rays produced by synchrotron emission from cosmic-ray electrons spiraling in the magnetic field in the halo of our galaxy had been studied by Vitaly L. Ginzburg and Sergei I. Syrovatskii. James E. Felton and Philip Morrison had investigated the possible contribution from Compton scattering of starlight against the same electrons. In both cases, the computed intensities were small compared to the observed background. Finally, we reported Fred Hoyle's calculation of the x-ray flux predicted from his "hot universe" model of continuous creation. It turned out that this model predicted a thousand times more x-rays than were actually observed. Burbidge considered this discrepancy as definitive evidence against the model. Clearly the field of x-ray astronomy was of great interest to some of the best astrophysicists in the world and was well worth pursuing.

We then proposed a set of themes for future observations:

all-sky surveys with increased angular resolution and sensitivity to further study
  of individual sources and the diffuse x-ray background;
higher-resolution studies of the structure of individual sources;
increased spectral resolution for both discrete and diffuse sources; and
study of the detailed properties of the x-ray emission, such as secular changes
  and polarization.

Considering the instrumentation needed to carry out this research, we described photoelectric x-ray detectors as potential successors to Geiger counters and summarized the technical development of x-ray telescopes, which was under way for solar astronomy.

We then outlined the seven-phase program illustrated in Figure 5.1:

I. We repeated our proposal for a rocket-based program of investigations.

II. We reiterated our justification for the OSO IV wheel experiment.

III. We proposed a high-resolution x-ray survey by a scanning satellite. We discussed in detail the unsuitability of existing spacecraft, such as OSO and the Polar Orbiting Geophysical Observatory (POGO), to carry out our proposed work. The need was for a slowly spinning satellite with a controllable spin rate, because a slow spin would permit us to observe a given source for a longer time. Each x-ray photon could also be tagged with time information directly in the telemetry stream without onboard storage. While we were ultimately forced to bin our data in 0.1-sec intervals, the relatively high time resolution would prove to be one of the most important features of the satellite that resulted from this proposal (Uhuru).

IV. We briefly mentioned a program of astronaut-assisted x-ray observations using the manned orbiting spacecraft of the Gemini program, which we intended to propose to the Johnson Space Flight Center.

V. We proposed a simple first experiment as a piggyback on the Orbiting Astronomical Observatory (OAO) to study rapid variability from x-ray sources.

VI. We proposed a prime experiment on OAO, consisting of a high-resolution x-ray telescope identical to the one we had proposed jointly with Goddard for the Advanced Orbiting Solar Observatory (AOSO). This solar telescope ultimately flew on Skylab in 1973 with spectacular results.

VII. Finally, we proposed the flight of a 1.2-m (4-foot) telescope with a focal length of 30 feet, to be launched with a standard Agena fairing and a Thor or Atlas booster (Figure 5.2). The telescope size would permit imaging of discrete sources or the granularity of the background on the scale of 1 arcmin. This was the first proposal for what became one of NASA's "Great Observatories": Chandra, launched in 1999 and still operating today.

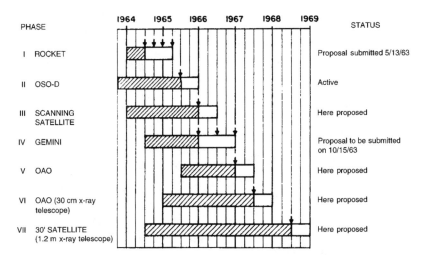

FIG. 5.1. *Comprehensive plan for x-ray astronomy missions presented by AS&E to NASA in September 1963. This strategic plan helped determine the path taken by x-ray astronomy for nearly four decades.*

We did not for a moment dream of convincing NASA to embrace the entire program, but we wanted NASA officials to know that we had a long-term, rational plan that extended over many years. The plan was based on step-by-step development of instrumentation in which we had the utmost confidence. Our experience and success in DOD classified research made us fairly sure of the feasibility of the proposed schedule. We were actually hoping that NASA would support the rocket program of x-ray studies we had so successfully started under AFCRL sponsorship. Beyond that, we were hoping to kindle some NASA interest in our future plans.

I was amazed and delighted when, after an oral presentation of this program to Nancy Roman at NASA headquarters, we were not only promised support for the rocket program but also encouraged to submit an unsolicited proposal for the scanning satellite (phase III of the proposed program) to be reviewed by a NASA committee of astronomers chaired by her. The proposal, "An X-Ray Explorer to Survey Galactic and Extragalactic Sources," submitted to NASA on April 8, 1964, was endorsed by the committee, and the spacecraft was ultimately launched on December 12, 1970. It was named Uhuru—meaning "freedom" in Swahili—in honor of its launch from Kenya on that nation's independence day.

Not all programs discussed in the 1963 AS&E plan came to fruition or were as successful as Uhuru and Chandra, but this plan was conceptually far ahead of the thinking in the research community at the time. Its execution provided unique observational data for many fields of astronomy.

FIG. 5.2. *Schematic of a 1.2-m x-ray orbiting telescope mission.*

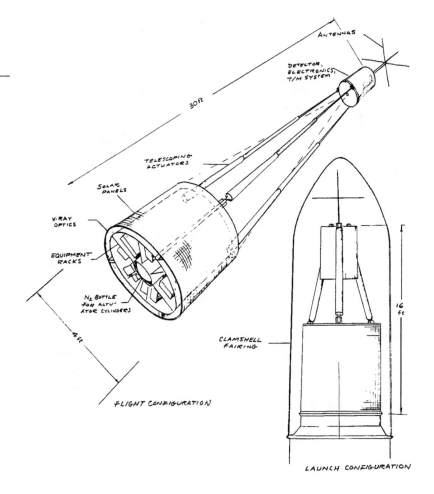

The time delay between conception and execution in the 1960s seemed interminable to us young scientists (I was 32 at the time the AS&E plan was drafted), but in retrospect, it was sufficiently short that a scientist could see ideas come to fruition during his or her scientific career and could train a new generation of astronomers. Delays today are much longer. It is doubtful whether the current proponents of Constellation X (the next great observatory for x-ray astronomy) will still be active scientists when it is launched 20 or 30 years hence. This delay will produce an observational gap in the field greater than the 20 years experienced by those of us waiting for Chandra after Einstein (HEAO 2) ceased operations.

Surging Interest in X-Ray Astronomy in the 1960s

Soon after the initial discovery, many groups joined AS&E, Lockheed, and NRL in searching the sky for x-ray sources and in studying their properties. George Clark and William Kraushaar of MIT first detected gamma rays from NASA's Explorer XI satellite as early as 1961. Contrary to theoretical predictions by Morrison, the gamma ray rate was extremely low, and during the first six months of satellite operations they detected only twenty-two celestial gamma ray photons. X-ray astronomy seemed much more promising in terms of scientific returns, and both Clark (who remained at MIT) and Kraushaar (who created a new group for high-energy astronomy at the University of Wisconsin) shifted their interest to x-rays. They carried out a number of balloon and rocket experiments during the 1960s.

Laurence Peterson's group, at the University of California at San Diego, specialized in the measurement of spectra of x-ray sources in the energy range of 20–250 keV. Groups at Rice University in Texas, Goddard, and the universities of Adelaide and Tasmania in Australia, which had all started to study gamma radiation with balloon-borne instruments, soon entered the field of x-ray rocket research. The government weapons laboratories also became engaged in gamma ray and x-ray research. The group at Los Alamos Scientific Laboratories was led by Hans Mark and that at Lawrence Radiation Laboratory (LRL) by Fred Seward. These groups had ample access to rocket carriers and carried out significant measurements mostly as opportunities for piggybacking arose.

Groups interested in x-ray astronomy became active in the 1960s in many foreign countries, including Australia, Canada, India, Italy, Japan, the Netherlands, and the United Kingdom. Many of these groups contributed to the experimental and theoretical progress in the field with a large number of new observations and discoveries.

By the late 1960s the number of known x-ray sources had increased to about forty. While most of the sources appeared to be stellar systems in our galaxy, one source coincided with the position of a well-studied radio emitter in the Virgo Cluster of galaxies. A claim by NRL to have discovered emission from 3C273—a nearby, powerful quasar—did not stand up because of poor statistics due to the scant observational data. There was evidence that some of the sources (in particular, Cyg X-1 and Cen X-4) showed wide variations in their intensities. Although this evidence rested on data obtained by various groups using different instruments and therefore was difficult to cross-calibrate, the effect appeared to be real. Spectral measurements were obtained on some of the strongest sources, including Sco X-1, Crab, Cyg X-1, Cyg X-2, and Cyg X-3. These measurements excluded blackbody spectra and favored exponential or power-law spectra, whose significance is discussed later in this chapter.

I will not attempt to offer an account of these discoveries. To do justice to the contributions of all the investigators involved (whose number had increased, in the United States alone, to more than a hundred by 1970), would take a book in itself. I describe in detail only two experiments, one by NRL and one by AS&E, which I consider representative of the high level of sophistication that had been reached.

The Nature of the Beast: The Engines of Galactic X-Ray Sources

Both experiments attempted to address the fundamental question of x-ray astrophysics. The character of the emission process was becoming clearer through spectral measurements, but what was the power source of these extraordinarily intense x-ray emissions?

A number of theoretical models were being debated throughout this period. They mainly involved the possibility that the energy source was connected with the end states of stellar evolution. In particular, some theorists considered the possibility that neutron stars (remnants of some supernovas) could provide the required energy. The x-ray emission could occur, for instance, through the expenditure of the kinetic energy of rotation stored in the neutron star at the time of its formation during a supernova explosion. Or perhaps the emissions were fueled by energy acquired by gas flowing from a companion star and accelerating as it fell into the deep gravitational potential of the neutron star. In either case, the energy source would have only a few degrees of freedom and could most effectively provide x-ray emission through particle acceleration to high energies or gas heating to high temperatures. These processes are highly inefficient in normal stars.

The existence of neutron stars was first suggested by George Gamow on theoretical grounds after the discovery of the neutron in 1932. A star made up of a degenerate neutron gas could be shown, by quantum theory, to be a possible next phase in stellar evolution after a star had become a white dwarf. The star would have a mass similar to that of the sun but a diameter as small as 10 km. Its density could be as large as 10^{15} times that of the sun.

In 1934 Walter Baade and Fritz Zwicky suggested that neutron stars were the result of the collapse of a normal star in a supernova explosion. In 1939 Robert Oppenheimer and George Volkoff presented their calculations on the collapse of massive neutron cores and continued gravitational contraction in two papers that are still regarded as the cornerstones of relativistic astrophysics and gravitational physics. Alistair G. W. Cameron extended and refined these computations in the early 1960s. Hong-Yee Chiu further computed models of formation of neutron stars and their surface properties and concluded that neutrino heating from the core would heat the surface of the neutron star, believed to be composed of iron, to the extremely high temperature of 10^6 K. He concluded that even a thousand years after the formation of the star in the supernova explosion that gave rise to the Crab Nebula, the neutron star would still radiate as a blackbody at a temperature of a million Kelvins. An early preprint of this work was circulated in 1963, and it was published in 1965.

The NRL Crab Nebula Experiment

The NRL scientists were well aware of these theoretical speculations. They had become persuaded since their first experiment that the radiation from the Crab could not be due to synchrotron emission, although this mechanism produced both radio and visible light emissions from the nebula. Their conviction was motivated by the large energy requirements, but even more by the fact that the high-energy electrons (10^{14} eV) required to produce x-rays by synchrotron radiation had a lifetime of only a few years when spiraling along the lines of the nebula's strong magnetic field. Such a brief lifetime would require some mechanism in the nebula to continually re-accelerate the particles, and they could not account for such a process. Thus Chiu's suggestion of blackbody radiation appeared attractive. Friedman asked Don Morton of Princeton to review Chiu's computations, and Don came to the same conclusion as Chiu, namely, that a neutron star could produce the x-ray radiation by thermal emission. The x-ray source would therefore appear as a point source in the center of the Crab Nebula.

In 1964 Friedman devised a clever way to test this hypothesis. He had used a lunar occultation to study the structure of x-ray sources on the solar disk. The same technique could be used to study the structure of the Crab x-ray source. He learned from NRL radio astronomers that the moon would eclipse the Crab in July 1964 and every 9 or 10 years thereafter. Friedman thought that if a neutron star at the center of the nebula was the x-ray source, then the radiation ought to disappear suddenly during the occultation, whereas if the source was the entire nebula, its disappearance would be gradual.

The NRL group faced substantial technical challenges to develop the rocket instrumentation in time for this unique opportunity. They had to launch the rocket so that the x-ray detectors would be at altitude precisely when the moon occulted the target, requiring a precision in launch time of about 100 seconds. This kind of requirement had never before been imposed on Aerobee rocket launches, because it was not needed, and the countdown procedure for these liquid-fuel rockets was lengthy and of somewhat uncertain duration. NRL further required directional control (which at the time was not fully developed) that would keep the axis of the rocket during flight steadily and precisely pointed (within a few degrees) at the Crab. The two Geiger counters used were again the brainchild of Stu Bowyer; they had thin Mylar windows with an effective area of about 100 cm^2 each.[2]

Everything worked perfectly. Figures 5.3 and 5.4 summarize the results, which were very disappointing to Friedman and his associates. The angular size of the source, which could be determined from the slope of the counting rate, turned out to be 1 arcmin, corresponding to a diameter of 1 light-

FIG. 5.3. *Count rates observed by the NRL counters during the July 1964 Crab Nebula occultation experiment.*

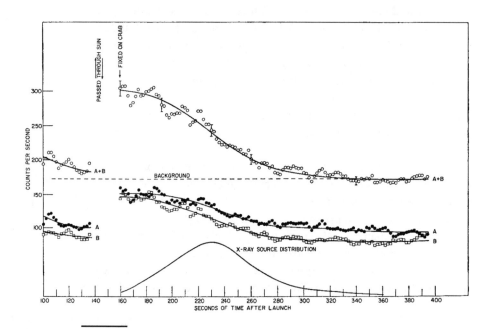

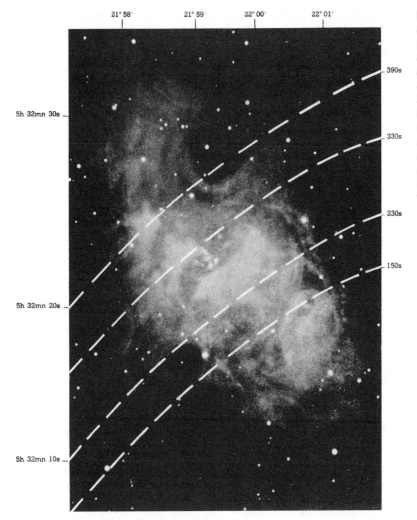

FIG. 5.4. *Positions of the limb of the moon (indicated by dashed lines) at different times during the Crab Nebula occultation experiment. The times, given in seconds after the start of observations, are shown for each limb position.*

year at the distance of the Crab. This large size completely excluded the possibility that the bulk of the flux came from a pointlike neutron star.

The possibility that the Crab emitted x-rays by blackbody radiation was shown to be untenable by Clark, who shortly thereafter measured an x-ray spectrum for radiation energies of more than 15 keV from the nebula. This radiation could not be produced by a thermal source unless the star surface was even hotter than allowed by theory. The AS&E group also showed in their August and October rocket flights that the spectrum of Sco X-1 was not that of a blackbody. So the hypothesis of thermal emission from neutron stars as the explanation of the x-ray emissions from galactic sources was abandoned.

Ironically, we now know that the pulsar in the Crab Nebula is a neutron star and that it powers both the acceleration of high-energy electrons and the shocks that heat the entire nebula. Had their experimental sensitivity and/or the time resolution been better, Friedman's group could have discovered the first pulsar.

The Optical Identification of Sco X-1

The AS&E scientists had decided that, while building Uhuru, they would focus their efforts in rocket research not on broad surveys but on the detailed study of individual intense sources (such as Sco X-1) to measure their spectra, angular sizes, and precise positions in the sky. In an AS&E proposal to NASA in October 1965, I stated:

> In past programs large regions of the sky have been explored to search for x-ray sources. A number of sources have been detected; location, angular sizes, and spectral data have been obtained and published. The experiments which are currently planned and which are proposed here aim at a more detailed study of small regions of the sky with particular emphasis to the precise determination of the characteristics of single sources.[3]

The description of the program that led to the identification of Sco X-1 with an optical counterpart gives a good idea of the methodology we followed. We started by utilizing scintillators in addition to Geiger counters in our rocket payloads to extend our sensitivity to higher-energy photons. We measured the spectrum of Sco X-1 from 1 to 25 keV during the August and October 1964 AS&E flights. The spectrum was clearly not that of a blackbody, but rather that of emissions from synchrotron radiation or thermal bremsstrahlung.[4] (A year later the LRL group again measured the Sco X-1 spectrum with greater precision from a rocket-borne instrument and definitely established the nature of the emission as thermal bremsstrahlung.)

Bruno Rossi was the first to propose that x-ray sources could emit radiation through thermal bremsstrahlung from a hot, optically thin plasma, and he emphasized the extreme efficiency of the process in producing x-ray radiation. (He discussed this point in some detail at the Solvay Institute's Thirteenth Physics Conference, held in Brussels in September 1964.)

In fact, the best fit to the AS&E and LRL measurements was an exponential spectrum, as would result from thermal bremsstrahlung from optically thin plasma at a temperature of 5×10^7 K. It is a characteristic of this spectrum that the power per unit frequency increases (or at least remains constant) with decreasing frequency.

If it did indeed have an exponential spectrum, then Sco X-1 ought to produce visible light with a brightness equivalent to that of a 13th magnitude star and therefore should be easily observable in the visible. An alternative possibility was that Sco X-1 was an extended nebula: if much larger than 1 or 2 arcmin, it could have low surface brightness and thus escape optical detection; if smaller, it ought to be visible. Studying the region near Sco X-1 in the visible, we could not see any nebula and therefore we concluded that the optical counterpart was either diffuse or one of the many faint stars appearing in the region.

We set out in 1964 with our MIT colleagues to measure the angular size of Sco X-1 in the x-ray region, using a new type of collimator invented by Minoru Oda, a Japanese visiting professor at MIT. The difficulty we had to overcome was that, with mechanical collimators, one could achieve angular resolution only by narrowing the field of view, which implied that we could scan only a tiny region of the sky during each experiment. As we did not know the position of our source accurately, we would most likely miss it altogether. Oda suggested the simple but elegant idea that a collimator could be formed with two sets of parallel wires placed at a distance from each other, as shown in Figure 5.5. The field of view of such a device consists of a set of narrow bands in the sky. As a source passed through this field, a strong modulation of the signal would occur only if the source was pointlike rather than diffuse. The advantage of the device was its ability to measure angular sizes with high accuracy over a wide field. Its disadvantage was that it could not measure position, because the source could be on any of the many bands in the sky defined by the collimator.

We piggybacked Oda's modulation collimators on the two AS&E rockets launched in 1964 and observed the predicted modulation, indicating a source size of less than 7 arcmin. The position could only be measured with a precision of half a degree, still too coarse to permit identification.[5]

We decided to try again, this time using the entire payload of a rocket to carry out the observation with increased accuracy. Herbert Gursky conceived a new payload consisting of two modulation collimators placed side by side along the axis of the rocket (Figure 5.6). In this manner, they could be made longer with wider wire separation and greater angular resolution. The rocket

FIG. 5.5. *Schematic representation of the angular response of a four-grid modulation collimator.*

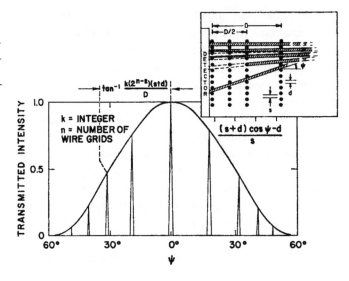

FIG. 5.6. *Layout of the instrumentation for the AS&E March 1966 rocket.*

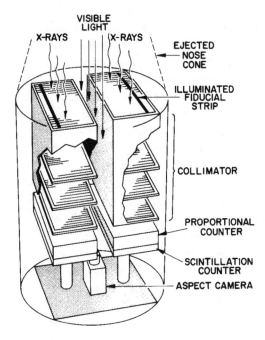

axis would be pointed to Sco X-1 during the flight (utilizing the same pointing control used by NRL for their Crab experiment) and would slowly scan
across the source. As the source crossed the transmission bands of the collimator, the signal would be strongly modulated. The width of the peaks determines the angular size of the source. The time of appearance of the peaks
and knowledge of the positions in the sky of the transmission bands yield a
set of lines on which the source must lie. The lines define equally probable
positions of the source and although quite narrow, they are separated by
5 arcmin. To resolve the resulting ambiguity in position, Herb came up with
the idea of slightly changing the separation of the grids in the two collimators, by about 5 percent. This variation permitted the use of a vernier technique to substantially reduce the ambiguity.

The collimators were built mainly by the MIT group (Oda, Hale Bradt,
Gordon P. Garmire, and Gianfranco Spada). The AS&E group built the proportional counters with beryllium windows and a visible-light camera to determine accurately the direction of the transmission bands in the sky. They
also integrated the entire payload under Herb's direction.

The launch, in March 1966, was completely successful, and the results are
shown in Figure 5.7.[6] Analysis of the results yielded an angular size of less than
20 arcsec and two equally probable locations in the sky, as shown in Figure 5.8.
It was clear that the optical counterpart of Sco X-1 would appear as a 13th magnitude star in one of the two boxes shown in the figure. The area to be searched
was quite small, less than a thousandth of a square degree. The possible location was more than half a degree away from previously given positions. During the flight, we also observed the Crab Nebula as a calibration source, and
the x-ray source was found to coincide with the center of the nebula.

There now remained the problem of identifying the optical object. None
of us at AS&E had any background in optical astronomy, but I had met Alan
Sandage at the 1965 summer school of the Italian Physical Society in Varenna,
on Lake Como. In one of my lectures on x-ray astronomy I had explained the
rationale for what we intended to do, and during breaks in the program
(which we often spent rowing a small boat around the lake), Alan and I discussed the need for optical follow-up on the x-ray observations. Alan offered
to undertake this search, and I agreed that I would give him the position of
Sco X-1 as soon as we had it. Little did I know at the time that, by a stroke of
luck, I had stumbled across one of the great observational astronomers of
our generation. I communicated the positional information to him and his
colleagues at the Mt. Wilson and Palomar observatories as soon as we completed our data analysis. Oda, apparently unaware of my agreement with

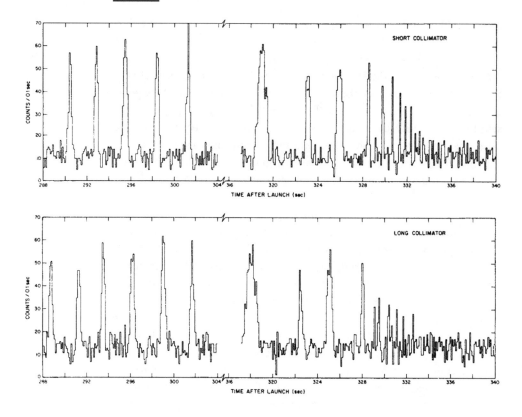

FIG. 5.7. *Results obtained during the AS&E-MIT rocketborne experiment of March 1966. Shown are the x-ray counts registered by the two detectors in the energy range of 1–20 keV during the time that Sco X-1 was in the field of view of the collimators.*

Sandage, communicated the same information to the staff of the Tokyo Observatory. The search was immediately successful at Tokyo on June 17–18, 1966, and within a week at Palomar. The star was a blue object at right ascension 17 h 17 min 4.3 sec and declination –15 degrees 31 min 13 sec.[7]

Having found the optical counterpart of Sco X-1, we had clearly established the existence of stellar objects whose x-ray luminosity was a thousand times greater than their optical luminosity. The close agreement between the intensity of the computed visible-light emission (based on the x-ray spectrum alone) and that observed made it quite attractive to consider, in this case, a single emission process for both x-ray and visible light.

Detailed optical spectra of Sco X-1 were obtained by Sandage; they resembled the spectra of an old nova and permitted a direct estimate of 250 parsecs for its distance. The total emitted power could now be computed: in the range of 1–12 keV, Sco X-1 emitted 10^{36} erg/sec, a thousand times the luminosity of the sun at all frequencies.

Robert B. Kraft had shown that old novas were members of binary star systems. Thus, the similarity of the spectra led to considerable interest and discussions of the possibility that x-ray sources could be in binaries. Rossi had announced (with our permission) the results of our work at the Thirty-First International Astronomical Union Symposium held at Noordwijk, the Netherlands, in August 1966. The eminent astronomers Geoffrey Burbidge, Vitaly L. Ginzburg, Kevin H. Prendergast, and Josef S. Shklowsky were present and joined others in animated discussions. Burbidge published a detailed account of these discussions a year later.[8] Shklowsky proposed at about the same time that Sco X-1 was a binary system containing a neutron star.

Naturally, we were interested in testing the hypothesis that x-ray sources were old novas, and on the next rocket flight (October 1966), we scanned the Cygnus constellation to measure the position of other sources with sufficient precision to permit an optical identification. Optical astronomers were able to identify a 14th magnitude blue star as the counterpart of Cyg X-2 (a Sco X-1 look-alike), but nothing was found for Cyg X-1. Moreover, we detected no x-ray emission at the known locations of old novas. Because there was no direct evidence of the binary nature of the sources from either optical or x-ray observations, the binary model was abandoned for a time.

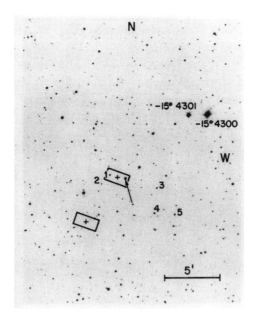

FIG. 5.8. *Sky survey showing the first successful optical identification of a pointlike celestial x-ray source. The boxes show the probable locations of Sco X-1, as determined by calculations based on the results of the AS&E-MIT observations of March 1966; the arrow marks the actual position.*

The discovery by Antony Hewish and Jocelyn Bell of radio pulsations from the Crab's pulsar, NPO 0531, occurred in 1967. Soon thereafter the NRL, MIT, Goddard, and Saclay groups observed the pulsations in the x-ray domain using rocket-borne instruments; NRL was the first group to report them. The MIT group, led by Hale Bradt, also observed the pulsations at almost the same time. The time-averaged flux of the pulsed component is about 8 percent of the total Crab intensity in the energy range of 1–10 keV.

The most widely accepted explanation of the pulsar emission was that proposed by Thomas Gold. He suggested that the radio pulsations were due to beaming from a rapidly rotating neutron star. By analogy, it became fashionable to study the possibility that all x-ray sources might emit radiation through a combination of synchrotron emission from the pulsar and thermal bremsstrahlung from the nebula surrounding it. No evidence was found in the available x-ray data to support this view, and a full understanding of the emission process had to await the launch of Uhuru in 1970.

Both the NRL and AS&E experiments failed, in a sense, to achieve their most ambitious objectives, namely, to definitively establish the nature of the celestial x-ray sources being studied. However, they are good examples of the systematic approach adopted by x-ray astronomers in their research and of the high degree of sophistication that had been achieved in the observations and in their interpretations.

Furthermore, the close connection between the mysteries unveiled by x-ray observations and the novel objects being studied in high-energy relativistic astrophysics made research in this field exciting and rewarding. For me, it provided the opportunity to do what I had always wanted: to work on challenging and worthwhile problems in physics to which I could actually contribute technical and scientific competence and creative ideas.

The First Orbiting X-Ray Observatory: Uhuru

A Slow Start

Although the concept for the first orbiting x-ray observatory was already included in the strategic blueprint submitted to NASA in 1963 (described in Chapter 5), the formal proposal, "An X-Ray Explorer to Survey Galactic and Extragalactic Sources," was submitted to NASA on April 8, 1964.[1] The proposal was approved shortly thereafter by a NASA Reviewing Committee of Astronomers. The proposed launch date was December 1965, 18 months after the start of the contract that we expected to receive in June 1964. We were jubilant and most anxious to start work. But conflicting institutional interests came into play, so that NASA delayed the program and its launch by five years.

We proposed to build both the experiment and the spacecraft. Though the spacecraft we needed was not complicated or expensive, it was different from any other in the NASA program; we proposed a design almost identical to that of the Television and Infrared Observational Satellite (TIROS). The new requirement was that it should rotate slowly in a controlled manner; this feature was uniquely suited for our research, but it meant that the craft would not fit the mold of OAO, OSO, POGO, or any other series of NASA spacecraft. Although I did not realize it at the time, I was entering into a philosophical conflict with NASA that was to last throughout my scientific career.

The conflict had its origin at the formation of NASA, when a decision had to be made on the selection process for the scientific research to be undertaken by the agency. Should the responsibility to select the science be entrusted to the Space Science Board of the National Academy of Sciences? Or should NASA itself make the scientific decisions? John Naugle (one of the best science administrators NASA ever had) described this early struggle in the book *First among Equals,* a 1991 publication of NASA's History Office. He outlined how NASA had wrestled the power of making science decisions from the National Academy of Sciences, and how the system had subsequently worked. When he sent me a complimentary copy of the book, I congratulated him on his scholarly approach but asked whether he still thought that NASA's victory had ultimately been in the best interest of science.

Making NASA responsible for the scientific program favored a system standardized at the top (the spacecraft level) rather than at the bottom (the component level). For ease of management, NASA preferred to plan a small number of missions, which its centers could manage. For these missions, a NASA contractor would build identical spacecraft that would be flown in a series. The scientific community was then asked to submit instrument proposals that would fit within the pre-established design characteristics of the series.

While worrying about this uniformity at the top, NASA never paid attention to the standardization of electronic components common to all spacecraft, such as batteries, solar cells, power systems, transmission systems, aspect systems, and recorders. If standardization at this level had been pursued, it would have allowed the assembly of custom-designed spacecraft from off-the-shelf items, with little need for specially built subsystems.

It is difficult to fathom why this standarization of electronic components was not institutionalized at NASA. But an essential element of the decision was that the diversity implied by this approach entailed decentralized control of the process of design and integration; it would have rested in the hands of the experimenters themselves, with concomitant loss of control and power by the NASA bureaucracy. No matter how attractive the concept might be for the proposed X-Ray Explorer, the AS&E scientists could not be allowed to have responsibility for the spacecraft and its integration. A NASA center willing to manage the program had to be found, but because it consisted of only a single spacecraft, no NASA center was interested. It was not until the X-Ray Explorer became the prototype for a new series of small astronomy satellites that it became acceptable.

My own opinion was that NASA was there to enable scientists to carry out their work and not to establish scientific direction (a task better left to the National Academy) or shoehorn scientists into pre-established satellite designs. Clearly, the philosophical disagreement I had with NASA ran deep. For my part, I began tilting at windmills, trying to explain to the authorities at NASA the unique features of the X-Ray Explorer and to convince them that we were entirely competent to proceed. But all the while, NASA used bureaucracy's simplest and most effective weapon: to simply do nothing. The five-year delay that resulted could have cost the United States its lead in x-ray astronomy.

While immersed in this political conflict, we at AS&E sought every other opportunity to continue our research. We were dismayed to learn that in February 1965 the Lockheed group, headed by Philip Fisher, had been invited by NASA to fly a piggybacked x-ray detector on the OAO, due to be launched in 1966. We felt wronged; having waited since early 1964 to receive a NASA green light on our (approved) Explorer, we were now going to be scooped by another group flying a detector almost identical to ours. I asked NASA to allow our group to compete for this opportunity, with an appropriate peer review process, but the agency turned a deaf ear to my request. Ultimately, it did not matter, because the OAO I carrying the Lockheed x-ray detector experienced a total power failure soon after launch on April 8, 1966.

It was only after this failure that NASA started in earnest a negotiation with AS&E to first design (October 1966) and then construct (November 1967) an x-ray experiment for the Small Astronomy Satellites (SAS). This proposal was for a new program of small satellites, which ultimately proved highly successful. I believe that this new start owed much to the personal efforts of John Naugle in encouraging young scientists and new and exploratory research. During his tenure as NASA's associate administrator for science, the foundations were laid for many of NASA's most successful scientific programs.

Even though the new series of SAS seemed a good idea, it still needed a NASA center willing to administer the program. Goddard was chosen, and Marjory Townsend was appointed as the NASA project manager. An electrical engineer who had worked at NRL on basic and applied sonar research, and later at Goddard, where she had been involved in a number of satellite programs, she was the first woman to become a project manager for a NASA program.

AS&E received the contract to build the satellite's instrumentation, aspect system, and processing electronics. Once again, NASA broke new ground by giving us a contract on a fixed-price basis. (I believe this award was the first such contract for scientific space research; previously such projects had been funded on a level-of-effort, or cost-reimbursement, basis.) The total sum allotted to us was $5.131 million. As wages and prices experienced one of their fastest growth spurts in U.S. history between 1964 and 1970, this type of contract placed a burden on all of AS&E and on me in particular to maintain a great deal of discipline. It also to some extent influenced the design of the experiment by forcing us to emphasize simplicity and reliability rather than advanced technology.

To manufacture the spacecraft, NASA chose the Applied Physics Laboratory (APL) in Laurel, Maryland, a huge not-for-profit division of the Johns Hopkins University with more than three thousand employees and a yearly budget of $600 million. Until then, APL had worked on military satellites—this venture into space science was new to them. For the SAS program, they relied on the experience they had accrued in designing and building Transit, the first satellite navigation system. It is interesting to read the recent statement by Eric Hoffman of APL, then a team member and now the chief engineer of APL's Space Department: "We were so far ahead of everybody else in showing the world the utility of small satellites . . . [that] years went by before others followed up . . . and rediscovered the faster, better, cheaper formula for space missions."[2] I only wish that NASA had understood it quite so clearly early on.

Design Concept

The overall concept of Uhuru was first hammered out during long discussions and consultations among AS&E's technical and engineering staff. Before joining AS&E in 1963, Don Friklund, the head of our System Department, had led the RCA team that designed the TIROS satellite, one of the workhorses of the space program. He had taught us a great deal about instrumentation and spacecraft design and integration, and many of his lessons were embodied in Uhuru. The instrument skin and the satellite support structure were one and the same, thus maximizing the detector area. The collimator was part of the rigid structure of the satellite and was not just stuck on to it.

We devised the simplest possible interface between the instrument and the spacecraft bus by reducing it to four bolts and a plug; this design allowed us to conduct separate thermal vacuum and shake tests of the instrumentation. (Because the spacecraft was intended to support other experiments in the future, the clean separation between spacecraft and experiment proposed by AS&E was very attractive to APL.) By using a simple spacecraft simulator, we could also carry out full electrical tests of our instruments.

The spacecraft consisted of an attitude-control system, a power system (including the solar panels), and a data transmission system. The experiment was designed with heavy emphasis on redundancy in many of its most important subsystems. The resulting instrument was capable of sustaining extensive simultaneous failures without seriously compromising the scientific objectives of the mission. The scientific payload was built with two almost identical sections mounted back to back (Figure 6.1). Each included a bank of proportional counters, an x-ray collimator, a sun and star tracker, and a shutter-activation detector. The electronic components were mounted in a separate box below these assemblies.

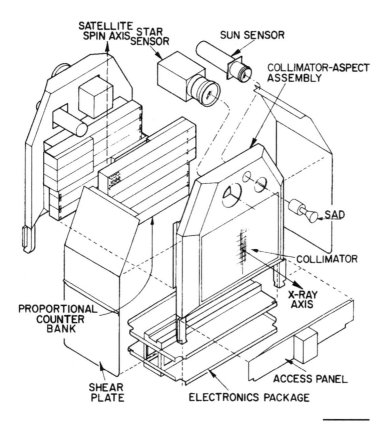

FIG. 6.1. *An exploded view of the Uhuru scientific instrumentation.*

Detector System

The x-ray detectors were assemblies of fairly conventional proportional counters with thin (2.5 mil) beryllium windows. The effective collecting area for x-rays was 840 cm^2 for each of the two sides. The pulses were 2 μsec long, and there was negligible dead-time loss, which allowed up to 50,000 counts/ sec. The x-ray signals were processed by a pulse-shape discriminator, which could distinguish x-ray pulses from those of particles and gamma rays. This technique, which had been invented by E. Mathieson and P. W. Sanford in 1963,[3] had been adapted for nighttime astronomy by Paul Gorenstein of our group and tested on a rocket flight in 1967.[4] The system could reject non-x-ray events with an efficiency of 90 percent. The two back-to-back counters could also be placed in anti-coincidence, which reduced the background by 75 percent when operated alone. The combination of the two techniques resulted in a non-x-ray background of 5 counts/sec in each of the counter assemblies. This rate was lower than that experienced with similar detector designs in rocket experiments; the improvement was due to the lower incidence of cosmic rays in the equatorial orbit of Uhuru.

Baffle collimators made up of rectangular cells (similar to an old car radiator) were placed in front of each detector bank, one with a field of view of 0.5×5 degrees and the other with a field of 5×5 degrees. The collimator with the narrow field of view permitted greater angular resolution and a small solid angle; it was best used in the study of individual point sources. The collimator with a wider field of view offered greater sensitivity for extended sources and the x-ray background.

The x-ray pulses generated by each counter were accumulated in different scalers for the broad-band counts and for the counts in eight spectral channels provided by a pulse-height analyzer. For the broad band, the accumulation times were 0.1 and 0.4 sec for narrow and wide fields, respectively. The integration times for the pulse height analyzers were 0.2 and 0.4 sec.

While the instrument was being built, Hewish and Bell discovered radio pulsars with pulse frequencies of 30 times/sec, which could not be observed with our integration times. We asked NASA for permission to add a simple mode to the data transmission. By stopping the engineering data stream during particular observations (for example, of the Crab Nebula), we could devote the entire telemetry stream of 1 kilobit /sec to direct transmission of the x-ray pulses without data binning, which would have increased the time resolution of Uhuru by a factor of almost one hundred. Funds for this modification existed in a special reserve, but Townsend decided that they should be used to further tighten up the specifications for the electronic circuitry beyond those originally agreed upon, and also beyond the standards adopted by other NASA or military programs. She appeared to be concerned that temperature variations could produce failures in our electrical components because of flexing of the terminals, and we had to put crimps at every soldered joint. Appeals to NASA arguing that this work was unnecessary and that the electronic components of Uhuru were in any case extremely well thermally isolated did not change the decision. We were told that the NASA project manager required a certain degree of discretionary power to be effective. The temperature inside Uhuru never changed, to my knowledge, by more than 1 degree. It is of course useless now to contemplate the substantial loss in scientific returns the decision entailed, but I can't help thinking of what could have been.

Aspect Measurement System

The star sensor system was by far more accurate than the sun sensor, which was intended primarily as a backup. There were two identical star sensors mounted on opposite sides of the spacecraft and independent of each other. Each sensor had an N-shaped slit, which produced three pulses for every 4th magnitude or brighter star that traversed the field of view. The slit was 10 degrees high, resulting in the detection of a star for each 10 degrees of satellite rotation. Sighting errors were less than 1 arcmin. The shutter actuation detector protected the star sensor from bright light (from the sun, earth, or moon) by closing a shutter in the star sensor. We used the time of occurrence of the star pulses to determine the azimuth and elevation of the star and therefore the positions in the sky of the axis of the collimators.

Electronics

As previously mentioned, we designed a great deal of flexibility into our electronics. On command, we could send the x-ray signals from one side to the electronics of the other side; if the electronics on one side failed, we could still process data from both counters. We could shut off the pulse-shape discrimination or anti-coincidence circuits in case of failure of their components. We could increase the gain of the star-tracker sensor by a factor of four if the efficiency of detection of visible stars diminished with time. We could carry out calibrations of the system electronically and with radioactive sources, thus allowing us to diagnose malfunctions. This built-in redundancy, which permitted soft failure modes, rather than complete loss of signal, was the only insurance we had against failure.

Toward the end of the tests, I became concerned with the silicon rubber process we had used for high-voltage insulation. This process, if not done properly, can leave little pockets of air trapped in the insulator; on reaching the vacuum of space, they might expand or burst and cause a breakdown. I went to the board of directors of AS&E and requested $100,000 of corporate funds to fix the problem. It was not a negligible amount for the company, since it took $2 million of work to earn that much profit. Thankfully they all agreed.

The Spacecraft Bus

The most important features of the spacecraft control system designed and built by APL included a rotor, magnetic torquing coils, a degausser, a nutation damper, and a coarse (3-degree) aspect system.

The rotor was spun parallel to the satellite spin axis and increased the satellite's angular momentum by a factor of ten over that from its rigid-body rotation. Control of the rate of spin of the rotor could be used to slow down or increase the spin rate of the spacecraft from the normal rate of 0.5 degrees/sec to as slow as 0.05 degrees/sec or as fast as 5 degrees/sec.

The torquing coils created dipole magnetic fields that reacted against the earth's field. They were used for large translations of the spin axis and to dump angular momentum when the vehicle needed to spin more slowly. Any part of the sky could be covered by maneuvering the spin axis by up to 90 degrees. This large translation required five hours to complete. The APL system was exactly what was required: by using magnetic torque rather than the conventional gas jets, the spacecraft used no consumables and could retain attitude control so long as there was power on board. The nutation damper was used to reduce the wobble of the spacecraft axis, the degausser to reduce residual magnetization, and the aspect system to provide a rough measurement of the local magnetic field direction.

APL's design of this system was crucial for the success of the mission. Once it was done, it looked easy; it certainly provided all the capabilities we needed to carry out the scientific program we had in mind from the beginning.

The power system of the spacecraft was rather conventional; it consisted of solar cell panels, batteries, and a charge regulator with a capacity of 26 watts, of which 10 were used by the experiment. The telemetry system used pulsed code modulation of the radiofrequency carrier at 1 kilobit/sec. The power level was 0.25 watt at 136.68 MHz. A tape recorder stored these data for 95.8 min (one orbit) and played it back when the spacecraft was over the ground station at Quito, Ecuador, with a playback time of 3.2 min.

Orbit and Operations

An equatorial orbit was chosen to avoid the regions with high densities of particles trapped in the Van Allen Belts, known as the South Atlantic Anomaly, thereby decreasing the background level in the counters. Primary cosmic-ray rates would also be reduced because of the high geomagnetic cutoff.

The launch was to take place from the *San Marco* platform of the Italian Space Agency, 3 miles off the coast of Kenya, which allowed insertion of the satellite into orbit within a few degrees of the equator. This trajectory had the further advantage of allowing the Scout rocket to carry a little extra weight into orbit. The remoteness of the site, however, increased the cost and complexity of the launch preparations compared to a launch in the United States.

In a sense, this difficulty inspired us to automate as much as possible the experiment test system. Ed Kellogg, Harvey Tananbaum, and John Waters designed a state-of-the-art test system based on the use of computer programs. This approach did away with the many hardwired panels of electronics that were normally used; it also permitted consistent, repetitive, and documented tests to be performed rapidly on command by only a few people. In fact, the entire AS&E staff in the field for two months in preparation for the launch consisted of Harvey Tananbaum, the project scientist; Stan Michiewicz, the electrical engineer; and Gerry Austin, the mechanical engineer. I could not keep away from the launch and joined them in Kenya a week before liftoff. After all, I was responsible (both as the principal investigator and as vice-president for the Space Research and Systems Division of AS&E) for Uhuru's conception, design, and execution. During the last few months, I was so concerned about its success that I would dream at night of wandering through the satellite's interior, testing every circuit, experiencing the cold and vacuum of space, and identifying as much as a man could with the machine we had constructed. I know some of my colleagues lived Uhuru as intensely as I did.

An important consequence of the computerization of testing was that it forced us to think early in the program about data retrieval, reduction, and analysis. We struck a deal with NASA that the data received at Quito would be recorded on tape and mailed to Goddard. There it would be reduced, recorded on IBM-compatible tape, and mailed to AS&E four to six weeks after reception. This schedule was fast compared to other programs of the times, but to us it seemed hopelessly too slow to permit informed decisions to be made about the scientific program. We asked that 20 percent of the day's data be relayed from Quito to Goddard on the day it was received and that a rapid telephone link should carry this portion of the data to AS&E. There we would process the data within a few hours and make them available for scientific analysis. The quick-look data would consist of

a rapid aspect solution to ±5 arcmin,
x-ray counts versus scan angle,
x-ray counts versus time,
energy spectra for selected sources or locations,
information on x-ray source variability, and
monitoring and housekeeping.

In a 1968 document in which we described our plan, we concluded:

The X-ray Explorer will provide a great increase in the total amount of x-ray astronomy data available. Therefore, the quick look data will enable us to take advantage without delay of these data in formulating subsequent sky scanning plans. It is hoped that data in usable form can be presented to experimenters within 24 hours after it is received. On special occasions, this time delay may be reduced to a few hours.[5]

The Moment of Truth: Launch, December 12, 1970

I spent the night before launch catching a few hours' sleep on the steel deck of the *San Marco*. The Italian crew could not have been nicer to me; they were proud that the scientist in charge was of Italian origin. Needless to say, we were concerned about last-minute accidents. To give an example, the beryllium windows of our counters could easily be corroded by the humidity, which at 3 degrees latitude and at sea is extreme. We insisted that the experiment be kept in a plastic shroud and continuously flushed with dry nitrogen gas. This continued on the actual launch platform (the *Santa Rita*) almost right up to launch time (Figure 6.2).

A problem with the APL power system batteries delayed the launch, resulting in a confrontation between APL and NASA on whether to change the batteries. Finally the APL people walked away, leaving Townsend to make

FIG. 6.2. *Harvey Tananbaum monitoring the final tests of Uhuru on the Santa Rita platform.*

the foreordained decision that the batteries must be changed. Not to do so would have been judged inexcusable negligence (if the batteries had failed after launch), even though the decision cost us a day's delay. (It is not clear in retrospect that the second set was better than the first one.) All I wanted was to get the satellite up into orbit and free from all the well-intentioned but risky handling.

The launch took place on a perfectly clear morning, so that the ascent of the Scout rocket could be readily observed. NASA's designation of SAS-A for the satellite was soon replaced by "Uhuru." December 12 was the seventh anniversary of Kenyan independence, and Luigi Broglio, head of Italy's Centro Ricerche Aerospaziali, was a good friend of Jomo Kenyatta, the first president of the independent nation; he suggested that, since Kenya was celebrating their *uhuru* ("freedom" in Swahili), we should name the satellite "Uhuru." I could not have been more enthusiastic, because for me, it also meant liberation from NASA's regimentation (Figure 6.3).

The launch was perfect, and Uhuru achieved a 540 × 520-km orbit with 3 degrees of inclination and a 96-min period, well within the range we had planned (Figure 6.4). Soon after launch, I received permission to go ashore to the base camp, and there, on the first pass of the satellite over the horizon,

FIG. 6.3. *The author (black shirt) and Luigi Broglio (white shirt) on the* San Marco *platform.*

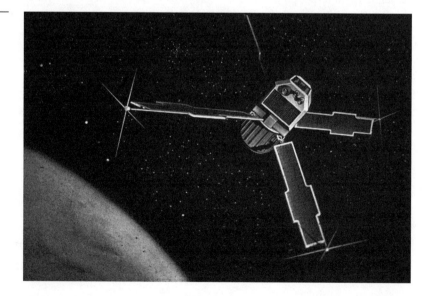

FIG. 6.4. *Artist's conception of the Uhuru satellite in orbit.*

we were able to turn on the power to the electronic circuitry and the high voltage and to observe the first x-ray scan of the sky in real time. We knew that it was against the rules—we should have waited for Goddard Control to do it in the prescribed manner. Soon after our glimpse, we turned the system off, hoping that the incident would not be noticed. What bothered me about the slow, deliberate approach taken by NASA was that one Uhuru orbit was equivalent to all the observing time obtained with x-ray rockets until then. One day of observing would have been ten times greater; even one week would have changed x-ray astronomy. I dreaded spacecraft failures that would prevent a clear scientific success for Uhuru before we had even gotten started. In any event, I obtained the demonstration I wanted—and spacecraft failures soon began to occur.

On December 27, 1970, we had the first telemetry failure. The cause was never fully explained, but it was speculated that it might have been related to the craft's orientation with respect to the sun, which produced the highest temperature in the spacecraft bus. Since the experiment section was thermally insulated, we did not observe any effects in the data. But the orientation of the spacecraft was changed to ensure a lower temperature, and we reacquired telemetry. From then on, the spacecraft was operated in this safer mode, which led to some restrictions on the original observation program.

On January 23, 1971, the tape recorder failed and never recovered. We could only receive data by direct transmission to the ground, and to obtain most of our data, many NASA ground stations along the equator had to cooperate. Cooperate they did, and I must say that NASA headquarters and the operations crews of all of these stations performed over and above the call of duty. We had additional failures later on. The famous batteries gave up, and thereafter we could only observe during the sunlit portion of the orbit.

On the experimental side, the only real failure we experienced was in the star trackers, which failed in November 1971. However, we were able to continue to navigate using as references strong x-ray sources, such as Sco X-1 and Cyg X-1, whose positions were accurately known. The mission continued to produce results through March 1973, when the craft ultimately failed.

Before I start discussing some of the early results, I pay tribute to the many scientists and engineers who worked on that mission. Most of the work was carried out by young people, and the project had some nonconformist aspects, perhaps influenced by the political atmosphere of the times. Yet we were still imbued with the can-do attitude of the Kennedy years. A year before the Uhuru launch, the United States had put a man on the moon.

First Flush: Initial Uhuru Results

FIG. 6.5. *Counting rate of the detector with the narrow field of view (0.5 degree) during the first scan of the galactic plane. The top and middle plots show weak sources; the bottom plot avoids saturation of the strongest sources.*

We were anxious to communicate our findings to the general astronomical community, and by agreement with Subrahmanyan Chandrasekhar, the great scientist and editor of *Astrophysical Journal,* we published four letters to the editor in the April 15, 1971, issue. We of course had to submit the letters in advance of this date, as early as February 13, two months after launch.

In the first letter, we described the satellite and the results of a scan of the galactic equator obtained in three hours of observation, which was the most complete survey to date (Figure 6.5).[6] The high angular resolution provided by the slow spin of the spacecraft had clearly paid off handsomely, allowing

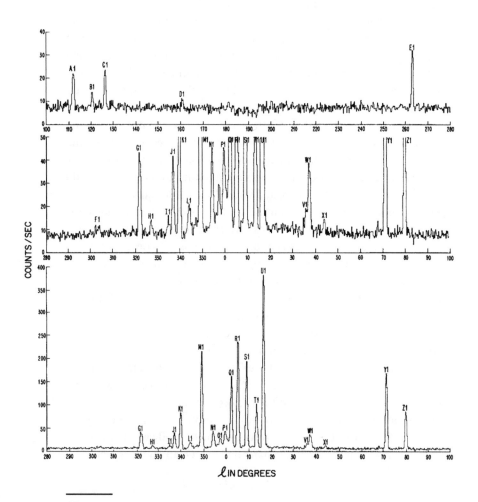

clear discrimination of sources in a relatively dense region. Twenty-nine sources were found, of which seven were previously unknown. Some of the sources detected even in this first survey had intensities (in the energy range of 2–20 keV) some three thousand times smaller than that of Sco X-1.

The second letter reported improved measurements for the locations of Sco X-1, Cyg X-1, Cyg X-2, and Cas A.[7] The third letter reported the discovery of a new type of extragalactic x-ray emitters: Seyfert galaxies.[8] In this paper, NGC 1275 and NGC 4151 were lumped together, although they had intrinsic x-ray fluxes differing by more than two orders of magnitude: 2.4×10^{44} and 1.1×10^{42}, respectively. In both cases, the x-ray emission equaled the optical emission. The fourth letter reported improved observations of the galaxies M87 and NGC 5128 and the quasar 3C273, which was the most intrinsically luminous of all x-ray sources.[9] The publication of these results created a great expectation of things to come. As the story unfolds, we shall see that the early promises were amply fulfilled.

The launch of Uhuru was a triumph for our group, for me personally, and, I dare say, for all of astronomy. We had truly started to probe "the hoary deep" in a new light, and Nature, always kind, rewarded us with dazzling sights. Seeing things that no one had seen before was an intoxicating feeling, and I can only say once again that I am grateful it was given to us to be first in space with an x-ray satellite. From that moment, work ceased to be a competitive struggle and became a privilege that I shared with my friends and coworkers. Had I but known it, those were the happiest years of my life.

SEVEN

Shedding Light on Binary X-Ray Sources—Extragalactic Sources—
Unresolved Questions: The X-Ray Background

Breakthrough: The Uhuru Results

To receive, record, reduce, and analyze the data from Uhuru, we had set up a data room on the sixth floor of 85 Broadway, one of the several buildings occupied by AS&E in Cambridge. They were mainly old milk truck garages owned by MIT and held by the university to await eventual demolition and expansion of the campus. Meanwhile, MIT rented them to AS&E as the need arose, and with a few coats of paint, they provided a wonderful, low-cost environment in which we could quickly set up new laboratories, offices, and facilities. The building at 85 Broadway was not a garage, but an adjacent, six-story office building. The top floor, which became our data room, was a world of its own, indistinguishable from any dedicated academic setting.

Herbert Gursky, Harvey Tananbaum, Ed Kellogg, and I were joined in the data analysis by other members of the AS&E staff, including Ethan Schreier, Paul Gorenstein, and Wallace Tucker. Minoru Oda, who had taken a leave from the Institute for Space Science in Tokyo to visit AS&E for a few months, became part of our team.

The first thing we did every morning, as soon as we arrived, was to examine the night's catch of quick-look data that had been transmitted by Goddard. Typically Harvey, Herb, or I would visually inspect all the data and then decide what we would do next, based on immediate evidence or previous plans. At the beginning of the mission and all through December and January, we based our science planning on a simple assumption: the possibility that the spacecraft would survive only as long as it had already lived—one more day, one more week, or one more month. We therefore considered it extremely urgent to make the best possible scientific assessments of what we were seeing, so that we could decide on priorities and follow up on any discovery.

At the start of the observations, we felt that we should make a detailed scan along the galactic equator to obtain the intensities and positions of the galactic sources. These results could be immediately compared with previous work done with rocket-borne instruments. We started the scan on December 27, 1970, but we had to interrupt it (because of a temporary telemetry failure) after having completed only three orbits. We did not resume the scan, because the subsequent failure of the tape recorder on January 23, 1971, made even more apparent the urgency of setting priorities. There followed a type of interaction among the scientists of our group that, little by little, developed into what we came to call the Uhuru spirit.

In addition to scheduled weekly meetings, every day or so we would hold impromptu discussions in which the most disparate opinions were aired and passionately criticized or supported without regard to their originator. Each of the scientists had prejudices and feelings about what was most important to do next and defended his point of view as vigorously as he could. I am happy to say that what prevailed in the end (in the words of Lawrence Summers, former president of Harvard) was "the authority of ideas rather than the idea of authority." These tough, no-holds-barred debates were most often resolved by consensus, and each of us would then carry out his part of the work to ensure the success of what we had decided to do. No matter how bitter the discussion, there were no personal attacks but only scientific conflicts. These ultimately did not separate us, but rather united us in a shared learning experience.

This type of behavior was characteristic of the Space Research and Systems Division, which was directly under my control, but was not typical of all of AS&E. It is fair to say that it was tolerated because we were successful, but we were not particularly liked. (What I later found quite disconcerting was that this sort of vigorous debate was not much liked at Harvard, either.) It did, however, produce results.

Looking back on that time, I recognize that those intense interactions were a rather unorthodox way of doing science and certainly not to everybody's taste. But we were young, enthusiastic, intoxicated by our daily glimpses into a mysterious new universe, and more than a little giddy with success. It was a unique period, brought about by a singular combination of circumstances, people, events, and instruments that occurs rarely in science. I have explained those days to myself by an analogy to thermodynamics: normal circumstances are those of thermal equilibrium, but occasionally there can be a spike in temperature when thermal equilibrium is violated, with a return to normal after a time. It is my opinion that such conditions occur during periods of breakthrough in all disciplines. Unorthodox it may have been, but I remember it fondly, particularly because, for the rest of my multifaceted career, I was never quite able to recapture that feeling, even though some of the Uhuru spirit did percolate down to my group at Harvard and later on to the Space Telescope Science Institute (STScI).

AS&E had grown rapidly during the period 1959–72, and a great deal of the growth had occurred in the Space Research and Systems Division. We had obtained, from NASA alone, forty-three new contracts for a total of $46.7 million. The staff had increased to more than four hundred people, most of them involved in hardware design and construction. Such an enterprise required strong management, and we had put in place all the structures and controls typical of an industrial organization. Engineering and program management had acquired a high level of professionalism, which was well recognized by NASA. I had been promoted to executive vice-president and in addition to my old division, I was responsible for the Education and Medical Products divisions. This is to say that I had a full-time job in addition to the science; all of the other scientists (save perhaps Wally Tucker) also had a functional "job," in the sense that they had to work at the tasks imposed by contractual obligations. In performing these tasks, the staff was extremely efficient, conscientious, and disciplined. The intellectual freedom I described above concerned personal research voluntarily undertaken and in large measure self-directed, but not what later in life we came to call the functional aspects of our work. This philosophy— everybody on the scientific staff working for a living and everybody doing research—was also transferred to the High Energy Astronomy Division at Harvard and to STScI.

When the mission was over in March 1973, we had increased the number of known x-ray sources from 30 to 340. They spanned a range of intensities of 20,000, from Sco X-1 at 3.4×10^{-7} erg/cm$^2 \cdot$ sec to the galaxy 3U 1237-07 at 1.8×10^{-11} erg/cm$^2 \cdot$ sec (Figure 7.1). We had surveyed most of the sky to a sensitivity of 2×10^{-10} erg/cm$^2 \cdot$ sec in the range of 2–6 keV, and we had

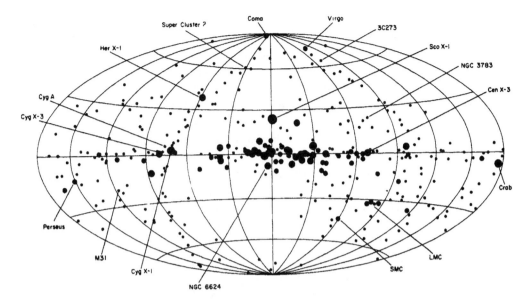

observed both galactic and extragalactic sources. The existence of the isotropic x-ray background was confirmed.

The galactic sources turned out to be either supernova remnants or binary stellar systems containing a compact object. The extragalactic sources included ordinary galaxies, clusters of galaxies, Seyferts, and quasars. The ratio of x-ray to visible-light luminosity for these objects spanned an enormous range, from one part in ten thousand for M31 (a galaxy like our own) to two hundred to one in Her A, a giant radio galaxy. Such differences implied the existence of many different production mechanisms at the source.

The detection of extended emission from clusters of galaxies was perhaps the most significant discovery, because it implied the existence of a component of the universe never previously seen (plasmas with temperatures exceeding 10^7 K), whose total mass exceeded that contained in the visible stars and galaxies. I am not able to give an adequate description of this disparate set of important discoveries in this book; I instead concentrate on a few aspects of the work.

FIG. 7.1. *The galactic latitude and longitude of the x-ray sources found with Uhuru mapped on the celestial sphere. The size of each dot represents the source intensity on a logarithmic scale.*

Shedding Light on Binary X-Ray Sources

The question of the time variability of the x-ray sources had always been at the top of our agenda. As discussed in the previous chapter, we did not have the time resolution we desired, but still, a resolution of 0.096 sec was adequate to investigate at least some of the phenomena of interest. Herb Gursky and I discussed how we could study sources that varied over short time scales, and it occurred to us that we could select such sources by visual inspection of individual passes over a particular source. During a pass, the source counts should increase gradually as the source comes into the center of the field of view and then should decrease gradually to zero, presenting an overall triangular response determined by the mechanical collimator.

Of course we did not expect to observe an ideal triangular shape because of the intrinsic statistical fluctuations in the number of photons impinging on the detector, but such fluctuations would be limited to the square root of the expected photon count. Only if there were real fluctuations on top of the statistical ones would we observe a clear effect. We thought such an effect could be recognized by visual examination by a trained observer; such a task would not be unusual for physicists trained in cosmic-ray research. We asked Minoru Oda to look into this aspect of the analysis.

X-ray variability in some x-ray sources had been reported by Walter Lewin and his associates on the basis of measurements obtained with balloon-borne instruments.[1] Moreover Byram, Chubb, and Friedman and Overbeck and Tananbaum had reported variations in the flux of Cyg X-1.[2] We had always considered Cyg X-1 particularly interesting, and one of our rocket flights had been devoted to measuring the spectrum and position of the sources in the Cygnus region. Cyg X-1 had an x-ray spectrum like that of the Crab Nebula, but there was no evidence of a supernova explosion or a strong radio source. While our rocket measurements had enabled us to identify an optical counterpart for Cyg X-2, we had not been able to do so for Cyg X-1. But the source was quite strong, only twenty times weaker than Sco X-1, and therefore bright enough for us to apply our qualitative analysis.

For this purpose, Oda used the counters with a wide field of view (5 × 5 degrees); at a rotation speed of one-half rotation every 720 sec, the source remained in the field of view for 20 sec. To his delight, it was immediately clear that the fluctuations in the x-ray flux of Cyg X-1, observed on the night of December 21, 1970, were much larger than would be expected on a statistical basis for a nonvariable source.[3] In our first report, we stated "this was an unexpected and surprising discovery particularly in view of the amplitude of the variations which appeared to be as large as 25% of the average source intensity in times as short as one second." None of the variable x-ray sources previously observed had exhibited such rapid and large variations in intensity except the x-ray pulsar in the Crab Nebula. Furthermore, the average intensity of the pulsar did not change over short times, whereas Cyg X-1 changed by a factor of two in 800 sec.

We then used one of Uhuru's design features—the ability to read each detector with either high- or low-resolution electronics—to sample the data more frequently (sample intervals ranged from 0.384 to 0.096 sec), while still observing for 20 sec at each pass with the detector with a wide field of view. The results are shown in Figure 7.2.

We were struck by the apparent periodicity of the data and tried hard to find a single period of pulsation that would fit all the data. I was so intrigued by the problem that I used to bring home the data and plot them by hand with Mirella. I remember many discussions in which I tried to convince her (and myself) that I was seeing something real. I must say that she remained unconvinced. She was, of course, right, because the search for a single period was in vain. The period, if any, seemed to be just below what was resolvable with our sampling rate, and while at times the evidence appeared convincing, at other times it was not—a frustrating will-o-the-wisp.

Finally we concluded that we could not, with our severely undersampled data, come to a definite conclusion; we offered a tentative period of 73 msec, which could explain our data but was not fully established. Soon thereafter a rocket flight by Steve Holt and his colleagues at Goddard showed that the pulsations were random in nature.

The mystery of Cyg X-1 had to wait. I return to this fascinating source later in this chapter.

Cen X-3

Three weeks after starting the work on Cyg X-1, we were applying the lessons we had learned to the study of other sources. This fast feedback loop from observation to working hypothesis to new observations was made possible by the use of the quick-look data. In retrospect, we did well in following the Galilean motto of "provando e riprovando," which does not mean try and try again, but rather measure, make a hypothesis, and measure to verify its predictions.

The first source we studied was Cen X-3, which had been shown by Ken Pounds and his colleagues to vary over periods of months.[4] We started with our observations of Cen X-3 on January 11 and 12, consisting of a 24-hour run using the 5×5-degree detector. Ethan Schreier noted the variability on the overnight strip charts, and in the morning during our check of the previous night's data, Harvey and I agreed that it showed clear variability. In fur-

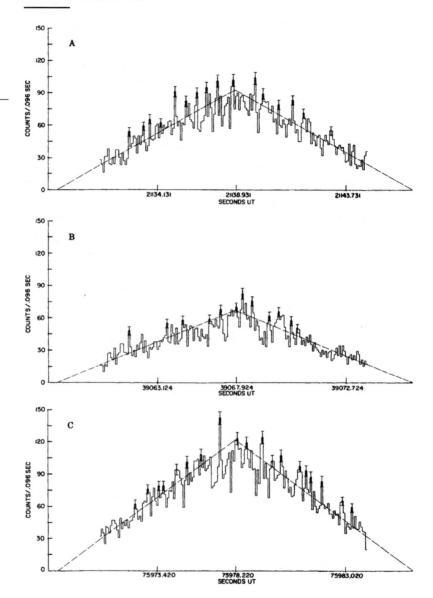

FIG. 7.2. *The variability of Cyg X-1 as observed in different Uhuru scans.*

ther observations, we found variability on the time scale of 5 sec, suggesting periodic pulsations;[5] there was also evidence of variability on scales of minutes and days. The phenomenology appeared so interesting that we planned a second run of observations on April 9, 10, 11, and 12. Ethan soon took the lead in the study of Cen X-3, while Harvey concentrated on the new pulsating source we had discovered, Her X-1.[6] I will not go through all the steps that gradually brought us from complete ignorance of the nature of the source to a clear understanding of Cen X-3, except to elucidate the process.

We first studied the pulsations occurring within a period of roughly 5 sec. We used the flexibility of Uhuru to obtain long-duration passes on the source lasting 150 sec (Figure 7.3). We could fit a sinusoidal function with harmonics to the data with a period of 4.876 ± 0.015 sec. This value was the average of the individual periods measured in each of twenty separate sightings obtained in January. The sightings were separated by the 720-sec rotation period of the spacraft and spanned four hours.

It occurred to us that we would much improve the precision of our measurement if we could follow the phase of the pulsation from sighting to sighting, even though the sightings were separated by an unknown number of pulses that we were unable to observe. (Although we reinvented this technique on the spot, it was already well known, as we found out later, to radio astronomers.) Once again, data came home with me to be plotted by hand for a first attempt at a solution. This time, however, even Mirella had to agree that we were on to something.

It turned out that the accuracy we had achieved in single passes was sufficient to compute the number of intervening pulses precisely, and consequently we were able to determine the period with a precision of a few tenths of a millisecond. The results left us perplexed. The period would remain constant for many hours and then change rapidly to another value that would itself remain constant for hours. It seemed at first that we were again confronted with a situation as frustrating as Cyg X-1 had been.

FIG. 7.3. *The regular nature of the Cen X-3 pulsations allows a sinusoidal fit to the data and the precise measurement of their period.*

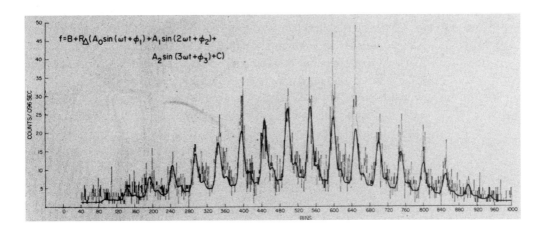

$$f = B + R_\Delta (A_0 \sin(\omega t + \phi_1) + A_1 \sin(2\omega t + \phi_2) + A_2 \sin(3\omega t + \phi_3) + C)$$

I was pretty sure that, in this case, there was some underlying regularity in what we were seeing. Herb, on the other hand, was convinced that I had a Ptolemaic turn of mind—that I was thinking in cycles and wanted to see periodicity at all costs. In one of our typical debates, he suggested that I wanted Cen X-3 to work with wooden gears like a cuckoo clock.

The solution of these problems came from the study of data obtained over a much longer period, from January to December 1971. We found that the diurnal variations in intensity of Cen X-3 were themselves occurring periodically with a period of about two days. The appearance and disappearance of the x-rays could best be explained by adopting the hypothesis that the star emitting x-rays was orbiting a companion in a binary system and was periodically occulted by that companion. The occultation's period could be measured with great precision, from the year-long observations, to be 2.08710 ± 0.00015 days.

Armed with this new knowledge, we went back to the study of the short pulsations. We improved the measurements by using long (100-sec) passes and the highest time resolution (0.096 sec), and at last we realized that the reason why the data refused to sit on a straight line was simply that they defined a sinusoidal function. This pattern is what we would expect if in fact the source was in a circular orbit about a companion; the frequency would be Doppler shifted because of the star's orbital velocity. The reanalysis of the pulsation data resulted in a perfect fit to a sinusoidal function (Figure 7.4); note that the errors of the measurements are smaller than the size of the dots.

The sinusoidal function had a period of 2.08707 ± 0.00025 days. The average period of pulsations was 4.84220 ± 0.00001 sec. Furthermore, the null points of the sinusoidal variability coincided with the centers of the occulted states and of the high-intensity ones with an accuracy of 0.003 ± 0.006 days.

Now we could explain everything we were seeing with a single working hypothesis. As we wrote in our paper, "we are dealing with an eclipsing binary system consisting of a compact object and a large massive companion. The short period pulsations (4.84 sec) originate from a rotating or pulsating compact star. The 2.087-day cyclical variations in the average intensity are due to occultations and the cyclical variations in the pulsation period are due to Doppler effects" (Figure 7.5).[7]

The ability to draw such strong conclusions based directly on high-precision measurements was new in x-ray astronomy. First, the opportunity to carry out observations over periods of months allowed us to come back repeatedly to a given object and study its transitions in their characteristic

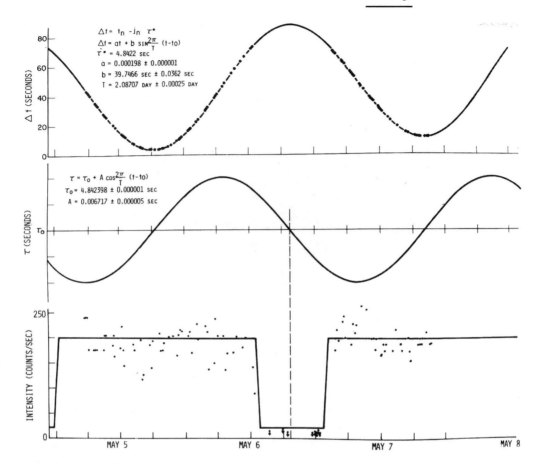

$$\Delta t = t_n - \bar{t}_n \quad \tau^*$$
$$\Delta t = at + b \sin \frac{2\pi}{T} (t - t_o)$$
$$\tau^* = 4.8422 \text{ sec}$$
$$a = 0.000198 \pm 0.000001$$
$$b = 39.7466 \text{ sec} \pm 0.0362 \text{ sec}$$
$$T = 2.08707 \text{ day} \pm 0.00025 \text{ day}$$

$$\tau = \tau_o + A \cos \frac{2\pi}{T} (t - t_o)$$
$$\tau_o = 4.842398 \pm 0.000001 \text{ sec}$$
$$A = 0.006717 \pm 0.000005 \text{ sec}$$

FIG. 7.4. *The period of the pulsations is also shown to follow a sinusoidal curve for Cen X-3. The zero points of this curve coincide with the midpoints of the maxima and minima of the total intensity of the source.*

times. Second, the abillity to slow down the spacecraft and observe high-frequency pulsations for long intervals (100 sec) made possible an ultimate precision of 1 μsec in the measurement of their 4.842398-sec period. Such precision, as we shall see, was critical in determining the source of the energy that produced the x-rays.

It was also amazing to me how precisely we could measure detailed parameters of the system. The phase delay of tens of seconds in the time of arrival of the pulses directly measured the size of the orbit; it turned out to be about 5 percent of the size of the earth's orbit around the sun. I felt as if I could almost touch this orbiting system.

The story of Cen X-3 was, however, not finished. We had determined that the x-rays were produced by a compact star (a white dwarf, a neutron star, or

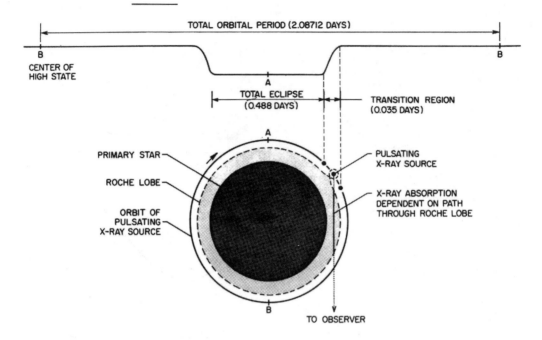

TOTAL ORBITAL PERIOD (2.08712 DAYS)

B
CENTER OF
HIGH STATE

B

A
TOTAL ECLIPSE
(0.488 DAYS)

TRANSITION REGION
(0.035 DAYS)

A

PRIMARY STAR

ROCHE LOBE

ORBIT OF
PULSATING
X-RAY SOURCE

PULSATING
X-RAY SOURCE

X-RAY ABSORPTION
DEPENDENT ON PATH
THROUGH ROCHE LOBE

B

TO OBSERVER

FIG. 7.5. *A schematic representation of the Cen X-3 system.*

a black hole; Figure 7.6) in orbit about a main sequence (normal) star. But what kind of object was our star? What was the source of the pulsations? What was the energy source that powered the x-ray emissions? The nature of the compact star could not be determined yet. The pulsations could be due to rotation or pulsations of any of the three candidates; given the long period of the pulsations, even the least compact (and thus largest in diameter) of the three—a white dwarf—was still plausible.

The energy source for the emission was also not known. Nuclear burning or accretion processes had been proposed by various authors, including George R. Blumenthal and Wallace H. Tucker; Kevin H. Prendergast and Geoffrey Burbidge; Josef S. Shklowsky, Yakov B. Zeldovich, and N. I. Shakura; and Alistair G. W. Cameron and M. Mock.

Apart from the detailed model calculations, there were even more fundamental issues regarding these binary systems. Neutron stars were believed to form in supernova explosions, as was the case for the pulsar in the Crab Nebula. Could a binary system survive a supernova explosion? If so, could the energy for x-ray emission be neither nuclear burning nor accretion but rather, as in the case of the Crab pulsar, the kinetic energy of rotation of the neutron star?

Her X-1

Some of these questions were solved by analogy after the discovery of Her X-1, which we first observed extensively in November 1971. It, too, was a regularly pulsating source but with a period of only 1.24 seconds, much shorter than Cen X-3.[8] As was the case for Cen X-3, we found that the period was varying sinusoidally with a period of 1.7 days (1.70017 ± 0.00004 days) and with drops in intensity having the same period. The data were interpreted, as in the case of Cen X-3, as occultations of a pulsating x-ray source in orbit around a normal companion.

However, the extremely short pulsation period of Her X-1 made it highly unlikely that the compact source could be either a rotating or a pulsating white dwarf. We were left with the conclusion that, for both Her X-1 and Cen X-3, the source emitting x-rays was a neutron star. This deduction was clear in the case of Her X-1 and probable in the case of Cen X-3.

What we had discovered was that neutron stars, no matter how they formed, could be found in binary systems. And furthermore, if they were formed in supernova explosions, as suggested by Baade and Zwicky, then the binary system remained gravitationally bound even after the explosion.

But we still did not know the source of the energy for the pulsations. This puzzle was solved by the long-term study of the period of pulsations in Her X-1 and Cen X-3. In a rotating pulsar, such as NPO 0531 in the Crab Nebula, the energy required to accelerate the relativistic electrons that produce

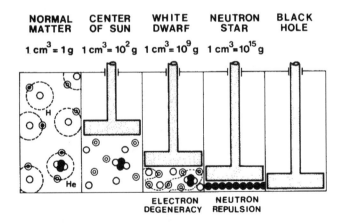

FIG. 7.6. *The density of different types of matter and the forces that prevent their collapse.*

the observed radio waves and x-rays is provided by the energy of rotation stored in the neutron star. Thus, NPO 0531's rotation slows down to compensate for the energy it radiates. Could the same thing be happening in Her X-1 and Cen X-3?

The answer, shown in Figure 7.7, is clearly no. We found that the rotation period of Cen X-3 had decreased by 3 msec in 18 months and that of Her X-1 by 6 μsec in 15 months. The x-ray sources were speeding up their rotations and therefore acquiring rather than losing energy!

The only remaining possibility was that the energy was furnished by the gravitational infall of gas accreting from the normal companion onto the neutron star, thus transferring angular momentum to the neutron star. Although this mechanism had often been discussed by theoreticians, it had never actually been observed. Because of its implications for all of astronomy, it is worth describing what this mechanism consists of in greater detail.

In a binary stellar system, the gas in the system is subjected to a gravitational field, which is the sum of the fields from the two stars. The field is generally illustrated by drawing its equipotentials, as shown in Figure 7.8. The

FIG. 7.7. *The pulsation period of Cen X-3 measured over four years. Note that the pulse period is getting shorter, not longer, with time.*

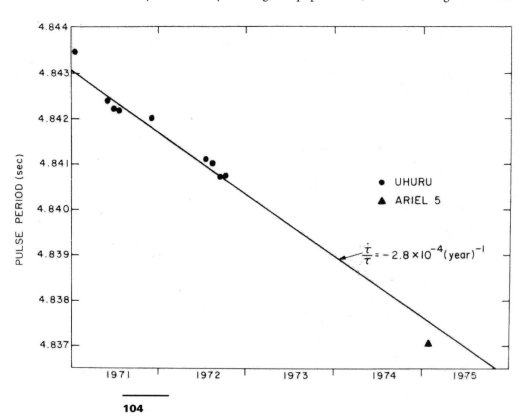

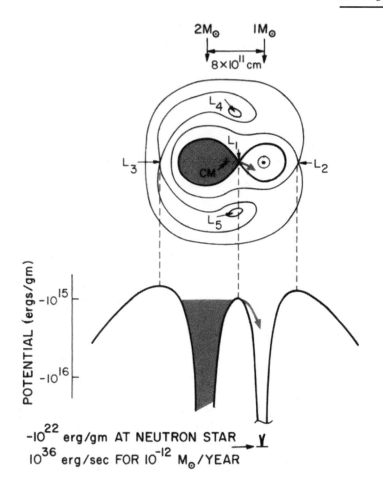

FIG. 7.8. *The gravitational equipotentials of a binary system containing a neutron star. In the top panel, a cross-section in the plane of the orbit shows equipotential surfaces as lines of contant gravitational force. The Roche lobe of the ordinary star is shown as filled with gas (gray shading). The bottom panel shows a cross section of the gravitational potentials of the ordinary star and the neutron star. The gas from the ordinary star escapes that star's potential and flows into the potential of the compact neutron star.*

equipotential surfaces that cross at L1 are called Roche lobes. As the gas in the normal star expands, it fills its Roche lobe (the gray lobe in the upper panel in the figure); its pressure makes it overflow onto the companion star. As it falls onto a collapsed object, such as a neutron star or a black hole, a nucleon in the gas acquires a maximum energy equivalent to one-tenth of the rest mass mc^2 per nucleon. This energy is much greater than that which can be extracted from matter through nuclear burning; the infalling gas is heated to ultrahigh temperatures (hundreds of millions of Kelvins), resulting in a plasma of electrons and protons. As this plasma approaches the neutron star, it is contained by the magnetic field of the star until it falls onto the star's magnetic poles, where it dissipates the acquired infall energy through x-ray emission. Rotation of the neutron star results in the regular periodic pulsations we observe, as one or the other pole comes into view from the earth. Because of the high energy acquired by each nucleon, a relatively low mass accretion rate of 10^{-6} to 10^{-8} solar masses per year can produce the observed luminosity of 10^{36} to 10^{38} erg/sec. Such a leisurely accretion rate means that the x-ray emissions can last for millions of years before the ordinary star stops supplying gas to its compact companion.

Thus, the energy for the x-ray emission comes through the heating of the gas accreting onto a collapsed star (Plate 1). The mechanism that we found to occur in stellar systems is now also generally accepted as the explanation for the emission from the most luminous objects in the universe—the nuclei of active galaxies and the quasars, which can radiate as much as 10^{45} erg/sec.

Cyg X-1

The initial announcement of the pulsations in Cyg X-1 discovered with Uhuru was followed within months by reports of variability on a 50-msec scale detected with rocket-borne instruments by the groups at MIT, Goddard, and NRL. While trains of pulses occurred in Cyg X-1, the pulsations were not periodic. In this the Cygnus source differed radically from Cen X-3 and Her X-1.[9]

The MIT group also measured the location of Cyg X-1 with greater precision than had been achieved with Uhuru.[10] Using both sets of data, Braes and Miley and independently Hjellming and Wade discovered a radio source, which they identified with Cyg X-1.[11]

The precise radio location (within 1 arcsec) led to the optical identification by Webster and Murdin and by Bolton of Cyg X-1 with the 5.6-day binary system HDE 226868.[12] The primary was a B0 supergiant, and conservative estimates of its mass (>20 solar masses) led to an estimated mass for the collapsed star of six solar masses, far in excess of the three-solar-mass limit for neutron stars computed by Rhoades and Ruffini.[13] The only possible conclusion was that Cyg X-1 was a black hole!

Given the importance of the discovery, many attempts were made to substantiate this finding. The pulsations were studied with increasing time resolution on rocket flights, particularly by the Goddard group. In 1974 they reached a resolution of 1 msec, and Cyg X-1 was shown to exibit pulsations at the limit of resolution (Figure 7.9).

Given its limited time resolution, Uhuru could not really help in these measurements; however, its unique capability to extend measurements over

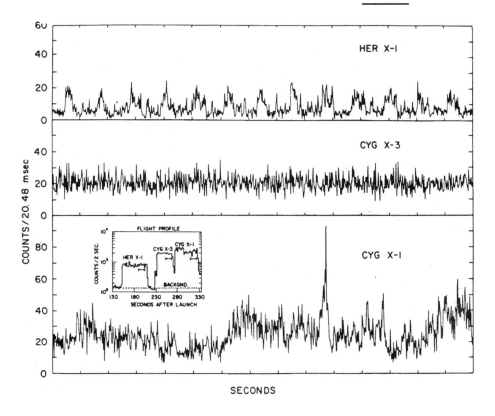

FIG. 7.9. *Time variability of three x-ray sources. Her X-1 exhibits periodic pulses, Cyg X-3 appears to be constant, and Cyg X-1 appears to show random noise variability.*

two years, during which radio data were also obtained, revealed a correlated transition in the source in the two wavelength bands. As the radio intensity increased by large factors, the x-ray intensity and spectrum also underwent a large transition. The spectral slope changed by a factor of 3.5 and the intensity by a factor of 10. This correlation clinched the identification of Cyg X-1 with the radio source and therefore with HDE 226868.[14] We also searched for a modulation of the x-ray emission with the same 5.6-day period as the orbital period of HDE 226868. We observed Cyg X-1 for 35 days in December 1971 and January 1972, but we found no evidence of such periodic behavior.

Still, the identification of Cyg X-1 with a binary system from optical observations strengthened the case for the identification of Cen X-3 and Her X-1 with binary systems. In addition, the discovery of an accretion process occurring in the two pulsating sources made it plausible to invoke the same energy source (accretion) for Cyg X-1. The difference was that accretion on a rotating neutron star produces pulsations due to the channeling of plasma onto the poles, while no such phenomena can occur at the critical surface of the black hole. The gas expends its energy just before its orbit crosses the event horizon—in the last orbit of the gas from which radiation can still escape the black hole.

To summarize, the conclusion that Cyg X-1 is a black hole is based on three main points: (1) HDE 226868 is the optical counterpart of Cyg X-1, (2) the mass of HDE 226868 is greater than twenty solar masses, and (3) the object emitting x-rays is a collapsed object with a radius of less than 10^8 cm.

The first point has been established on the basis of positional coincidence and the evidence of the correlated x-ray-radio transition. The second depends on a number of measurements by optical astronomers. A consensus seemed to emerge in 1974 that the distance to Cyg X-1 is greater than 2 kpc, and therefore its luminosity is that of a B0 supergiant; the mass of the compact companion must be greater than six solar masses. As mentioned above, this mass is much greater than the limit for the mass of a neutron star of three solar masses, established by Rhoades and Ruffini.

The third point, the compactness of the source, is based on the fact that the rapid pulsations (10^{-3} sec) are of very large amplitude, corresponding to changes on the order of 10^{36} erg/sec. Such large intensity changes, which are equal to the total output of the source, must involve the entire surface of the star. So the size of the star must be less than the distance light can travel in 10^{-3} sec, or 3×10^7 cm. The black hole's lack of a surface and of magnetic poles (as found in neutron stars) explains the absence of periodic pulsations (Plate 2).

Extragalactic Sources

In 1966 Byram, Chubb, and Friedman had reported the discovery of the first discrete extragalactic x-ray source, the radio galaxy M87 in the Virgo Cluster of galaxies. Among the more than sixty sources at high galactic latitude (>20 degrees), we found we could immediately identify fourteen with normal galaxies, radio galaxies, Seyfert galaxies, quasars, and clusters of galaxies.

For individual extragalactic objects, we measured definite low-energy cutoffs, corresponding to a full galactic path length of absorbing matter. Except for normal galaxies, the x-ray emission probably originated in small regions at the center, caused by the presence of an active nucleus. From some of these objects, the x-ray luminosity was a thousand times greater than in normal galaxies, and the ratio between x-ray and optical luminosities was close to unity. I mention some of the properties of these sources when discussing the results on the x-ray background obtained with Uhuru.

The study of x-ray emissions from clusters of galaxies produced one of the most unexpected and important discoveries of the mission. Clusters of galaxies are groups of hundreds of galaxies that are bound gravitationally in coherent systems over distances of millions of light-years. They are the largest aggregates of matter in the universe. The measured velocities of the galaxies in Coma, for example, are of the order of 1,800 km/sec; at that speed, it takes only 7×10^8 years for the galaxy to leave the cluster. But we believe that the clusters have remained bound for 9 or 10 billion years. The only explanation for the galaxies staying together is that there is enough mass in the cluster to keep it gravitationally bound. We know from optical data that the universe contains many clusters of galaxies distributed over large enough distances that they do not interact with one another.

Our study was carried out by Ed Kellogg, Herb Gursky, Harvey Tananbaum, and myself but with important contributions by two new staff members, Steve Murray and A. Solinger, and by a visiting scientist from the University of Rome, Alfonso Cavaliere.[15] The Uhuru sky survey resulted in the discovery of clusters of galaxies as a distinct class of x-ray emitters, but the most important aspect of the discovery was that these x-ray sources were all extended (instead of being pointlike), with angular dimensions greater than 35 arcmin. As we examined the four brightest x-ray clusters (Coma, Perseus, Centaurus, and Virgo), we found that the emission came from an extended region surrounding the cluster center. We interpreted the emission as coming from a low-density, very hot gas (temperatures of 10^8 K) pervading the space between galaxies (Figure 7.10).

We considered and rejected many alternative explanations for the x-ray emission, such as synchrotron emission from high-energy electrons or inverse Compton scattering off the 3-K microwave background, on the basis of lifetime considerations for the electrons. The theoretical predictions of Tucker and Blumenthal, which linked the x-ray emission to the measured velocity of dispersion of the cluster, appeared to favor a thermal bremsstrahlung process. Therefore, the best explanation for our observations appeared to be that the radiation was produced by thermal bremsstrahlung from a gas at a temperature of 10^8 K.

On this basis, we could determine a central density for the gas, a core radius for its distribution, and the total mass in the cluster. The masses of gas we found for Coma and Perseus were 5×10^{14} and 4×10^{14} solar masses, respectively. The total masses of galaxies and stars had been measured in the

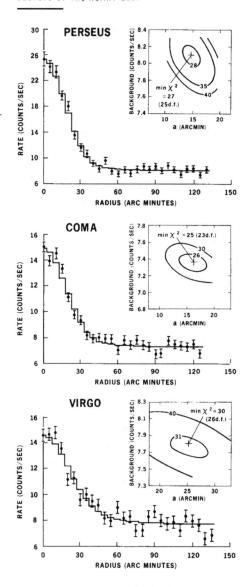

FIG. 7.10. *The angular sizes of galaxy clusters emitting x-rays, as measured by Uhuru.*

visible to be 6×10^{14} and 1×10^{14} solar masses, respectively, for these two clusters. The total mass in the gas was therefore comparable to or greater than the total mass in the galaxies. With this one measurement, if our interpretation held true, we had doubled the amount of mass contained in the clusters, by revealing the presence of the hot, low-density gas pervading the spaces between galaxies. This gas could only be detected by x-ray observations.

Soon after the Uhuru discovery, R. J. Mitchell and Len J. Culhane used the Ariel satellite of the United Kingdom to obtained evidence for an iron emission line with a collimated proportional counter. The existence of the line provided strong and direct evidence for the existence of hot plasma in the cluster. Furthermore, the overall spectrum was well described by a bremsstrahlung process. Peter J. Serlemitsos and his colleagues at Goddard found similar results for Virgo and Coma using OSO-8. The Ariel V group later found highly ionized iron emissions in Centaurus. Thus, our proposed explanation was completely confirmed.

Even having doubled the mass contained in each cluster of galaxies, however, there remains a large discrepancy between the mass detectable from its radiation and that required by the virial theorem to bind the cluster together gravitationally. This dark mass, which is generally believed to exist, is ten times greater in mass than the mass we measured. Its nature is not yet understood.

Unresolved Questions: The X-Ray Background

Since the very first discovery flight in June 1962, we had observed and reported an isotropic x-ray background that we believed to be of extragalactic origin. The low cosmic-ray background we had achieved with Uhuru meant that, when we observed the sky with the 5×5-degree collimator, 80 percent of the flux was due to the isotropic background radiation (Figure 7.11). We verified this fact by pointing the detector at the earth: 80 percent of the flux did indeed disappear. The flux per steradian was the same as that measured by rocket in 1962.

A further result that could be immediately obtained with Uhuru observations was to determine whether the x-ray background was smooth on smaller angular scales than had been previously attainable. The final analysis of the data constrained the fluctuations to less than 2 percent over a 3-degree field. This rate corresponds to a density of 1 source/arcmin^2, corresponding to more than one hundred million sources over the entire sky.

The compilation of the spectral measurements then available from a number of groups in the energy range of 1–100 keV seemed to suggest that the spectrum of the background had relatively more flux at high than at low energies than the flux of known extragalactic sources. These two facts taken together were considered evidence against a background made up of many discrete individual sources. The conclusion was that a diffused mechanism

FIG. 7.11. *A scan with the wide-field (5 × 5-degree) Uhuru detector. The dashed line represents the residual particle background. The remainder of the background is due to extragalactic x-ray sources.*

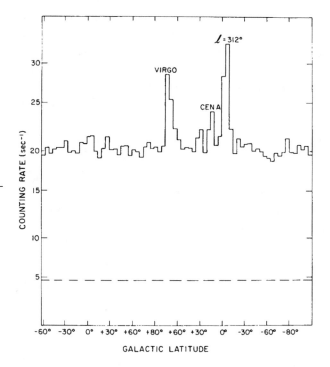

had to be operating (such as thermal bremsstrahlung from a very hot diffuse gas pervading the universe), and that to study the origin of the emissions, one had to measure the features of the diffuse x-ray spectrum with greater accuracy.

Not everybody agreed with those conclusions. I, for one, remained of the opinion that we could resolve this problem only by imaging the background. A density of 1 source/arcmin^2 was precisely what the 1.2-m telescope of my 1963 proposal to NASA (discussed in Chapter 5) was designed to resolve into individual sources.

I was encouraged in my thinking by the work of a young colleague, Terry Matilsky, who had recently joined the AS&E staff. He made a simple but powerful contribution while plotting the number N of sources of intensity greater than S against the intensity. Such plots (log N versus log S) were used at the beginning of radio astronomy to understand the distribution of sources. Terry's contribution was to separate the plot for sources at low galactic latitudes ($b < 20$ degrees) from those at high galactic latitudes ($b > 20$ degrees).[16] The result is shown in Figure 7.12.

If one is considering an isotropic distribution of sources in Euclidean space, the slope of the function defined by these plots should be 1.5. The low-latitude sources, presumably galactic stars, increase in numbers with that slope for a while, but then they increase with a much smaller slope. We interpret this change as being caused by the fact that we are running out of galaxy. As we look at fainter and fainter stars, we are looking right through the galaxy, and there are no more stars to be seen—thus the change in slope.

The situation is quite different at high galactic latitudes ($b > 20$ degrees); for these sources, the number continues to increase with decreasing intensity at the limit of our survey. The units of the plot are Uhuru counts equivalent to 1.7×10^{-11} erg/cm$^2 \cdot$ sec in the energy range of 2–8 keV. Assuming we could continue to extrapolate to lower fluxes using the -1.5 slope, we would reach the required number of sources and obtain the entire flux of the background by increasing our sensitivity by four orders of magnitude, that is, to a minimum observable intensity S_{min} of 1.7×10^{-15} erg/cm$^2 \cdot$ sec (the sensitivity we subsequently, in fact, reached with Chandra in the deep survey; see Chapter 20).

FIG. 7.12. *The intensity distribution of high- and low-latitude sources measured by Uhuru. N is the number of sources having intensity greater than S. The galactic latitude is indicated by b.*

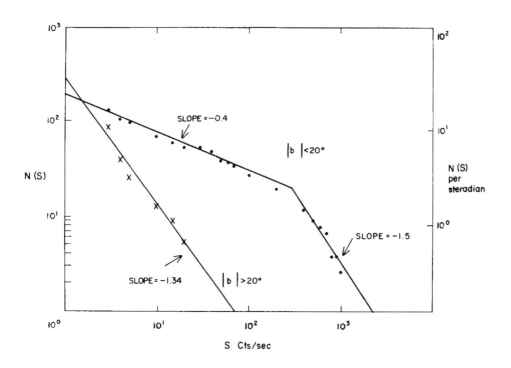

But why should the plot continue to rise with the same slope, given that relativity indicates that the universe is not Euclidean? This distribution of sources could occur if the number of sources existing in the past is greater than the number in the present because of the effects of quasar evolution, which would compensate for the effects of cosmological expansion. This behavior is in fact observed for quasar counts and was perhaps part of the motivation that led Lodewijk Woltjer and Giancarlo Setti to suggest that if all quasars were x-ray emitters like 3C273, the x-ray background could be due to the sum of their discrete contributions.

In any event my determination to image the x-ray universe with high-angular-resolution instruments, although not shared by other experimenters and opposed on theoretical grounds by some theoreticians, drove me to push hard for the development and use of an x-ray telescope.

EIGHT

Science and Technology of X-Ray Optics—Designing X-Ray Mirrors—
On the Steep Part of the Learning Curve: Experimentation, Fabrication, and Testing—
The Second Generation of Telescopes—The New Solar Telescopes—
Skylab and Solar X-Ray Observations—
Beyond Imaging the Sun: Plans for Stellar X-Ray Telescopes—
An Aside: Collectors of Light, Ancient and Modern

Constructing X-Ray Telescopes: Overcoming Technical and Institutional Hurdles

Science and Technology of X-Ray Optics

X-ray lenses or telescopes must use reflection rather than refraction because of the absorption of the radiation by matter. The optics using reflection must be designed, however, so that the reflection occurs at grazing incidence to obtain reasonable reflectivity. X-rays are efficiently reflected from surfaces at grazing angles of incidence by total external reflection, because the index of refraction of matter for x-rays is less than unity. In general, the refractive index can be written as $\mu = 1 - \delta - i\beta$, where δ and β are related to the atomic number and absorption index of the material at a given x-ray wavelength. Neglecting the imaginary part (which is normally quite small), Snell's law tells us that total reflection will occur when the grazing angle of incidence is less than $\theta_{critical}$, where $\cos \theta_{critical} = 1 - \delta$, or $\theta_{critical} = (2\delta)^{1/2}$ for small values of δ. Because there is an imaginary part, reflection will not be total, and there will be absorption of the incident power by the reflecting surface. In 1954 Parrat and Hempstead computed the theoretical values for different ratios of β to δ.[1] Compton and Allison developed a simplified formula to compute the reflection efficiency with sufficient accuracy for practical applications in 1963.[2] Our group used this formula to compute the efficiencies shown in Figure 8.1.[3] This figure clearly shows that a high efficiency of reflection requires small angles of grazing incidence for x-rays at wavelengths of a few angstroms—sufficiently short, that is, to reach the earth unabsorbed by interstellar gas. A gold mirror, for instance, will reflect 2-Å x-rays impinging on the mirror at 1 degree grazing incidence with 10 percent efficiency. The first challenge in the fabrication of x-ray optics is that one must polish a mirror surface fifty times larger than the desired collecting area. This factor comes about because the mirror surface is at a small angle θ with respect to the incoming rays parallel to the telescope axis, and the effective collecting area is given by sin θ times the area of the mirror. For θ of about 1 degree, sin θ is about 1/50.

All these calculations depend on the assumption that the classical laws of optics apply to x-ray optics, an assumption that seemed well justified by experiment. However, in 1963 there were still reports in the literature of the presence of two reflected x-ray beams from a reflecting surface rather than one; these reports were not put to rest until 1965. (As late as 1970, the group of x-ray experimenters at Goddard designed their optics to accommodate a nonexistent deviation from classical laws, which resulted in out-of-focus images from rocket flights.) I mention these facts to illustrate the rather immature theoretical foundation on which we were building in this field.

There was also the issue of how far the x-rays would penetrate into the mirror material. Again, theoretical work by Parrat in 1954 showed that the

FIG. 8.1. *Computed reflection efficiency for different reflection materials, angles of grazing incidence, and x-ray energies.*

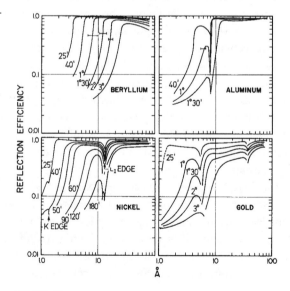

maximum depth reached by the radiation was on the order of the x-ray wavelength divided by $4\pi\theta_{critical}$.[4] For a nickel telescope, such as the ones we first flew, $\theta_{critical}$ was between 1 and 2 degrees for x-ray wavelengths of 4–8 Å, and so the maximum penetration depth was about 20 Å. To some extent, this penetration of the surface served to make the polishing of the telescope reflecting surfaces a little easier, because it averaged out some of the polishing errors. Glass flats for conventional light telescopes had been polished to this accuracy, and one could hope that with time the technology for polishing metal would catch up to our requirements. (Glass was too heavy and fragile to launch into space at this time; metal mirrors could be made lighter and stronger than glass ones.)

Designing X-Ray Mirrors

The paraboloid mirror that Rossi and I proposed in 1960 has the property that paraxial rays will be imaged to a point at the focus of the paraboloid, but the condition of grazing incidence requires that only the far zone of the paraboloid be used.[5] This system or any other single-reflection system cannot satisfy Abbe's sine condition and therefore will not produce a useful image of objects over an extended field of view. Consequently, we had titled our paper "A 'Telescope' for Soft X-Ray Astronomy," with the quotation marks around "telescope" to indicate that this device was not a telescope in the usual sense of the word. The difference between the paraboloid mirror system and a true image-forming telescope is important and is worth emphasizing because of the loose usage of the term *telescope* in x-ray and gamma ray astronomy. In this book, a telescope is a system of mirrors that can produce focused images of distant sources that lie within a finite field of view (that is, the field of focus is many times the size of the image of a pointlike source).

What Rossi and I had proposed in 1960, the system of confocal paraboloids, produced an image of a distant point source within a very small field of view. To obtain the image of an extended source, this device had to raster scan across the source. To my knowledge, this scanning approach was used in only a few experiments (such as our own instrument on the pointed section of OSO IV). The main advantage of the nested paraboloids is their ability to image faint point sources: if the instrument is aimed at a distant point source, it collects all the photons parallel to the axis of the mirror and reflects them to a single spot in the focal plane. The ratio between the mirror's collecting area and the area of the spot can be very large, up to one hundred thousand in the example we had discussed in our proposal. This number is the gain in sensitivity that can be obtained with the instrument. With such a device—one that is able to image only single pointlike sources unless used in a scanned mode—a reasonable astronomical application might be to aim it at a known optical or radio object and measure the object's x-ray flux. Had we not discovered the x-ray binaries (including Sco X-1), x-ray astronomy might have proceeded in that manner. We might have started looking at specific known objects one after another, because the device did not lend itself to surveying extended patches of sky looking for novel x-ray sources.

The great advantage of imaging is that, provided imaging detectors exist for the particular waveband of interest, one can focus and detect an entire region in the sky with a single exposure. Imaging detectors for x-rays were photographic emulsions until the development of imaging proportional counters in the 1960s and 1970s, followed by channeltron devices, and finally CCDs similar to the ones used in modern cameras. The advantage of a single exposure to raster scanning of an extended field is huge. As an example, the sun is 1,800 arcsec in diameter; an image with a resolution of 5 arcsec would be composed of 129,600 elements. The time required for a raster scan to obtain the same exposure for each of these elements would therefore be 129,600 times longer than for a single image. In retrospect, the importance of the nested paraboloid concept was mostly in showing the feasibility of very large gains in sensitivity using grazing-incidence optics.

Hans Wolter had shown that a two-reflection system could satisfy Abbe's sine condition, while no single-reflection system could.[6] He had carried out precise algebraic calculations on several possible conics configurations that would be able to produce a focused image of an extended field. The three configurations shown in Figure 8.2 were particularly promising: the first configuration consisted of a segment of a paraboloid followed by one of a confocal hyperboloid; the second used the same conics, but the hyperboloid and paraboloid were not confocal; and the third consisted of a paraboloid and an ellipsoid. Wolter failed to construct an x-ray microscope based on his designs, because the small physical dimensions of the system (on the order of millimeters) required impossibly high precision. We believed we could succeed in using his configurations for telescopes.

Of these three designs, the first had obvious advantages for use in space-borne instrumentation. Grazing-incidence mirrors tend to have long focal lengths. In visible light, one can fold the optical path onto itself using secondary mirrors within the telescope, but this compaction is not possible with x-rays. Therefore, an x-ray telescope must have a large ratio of length to

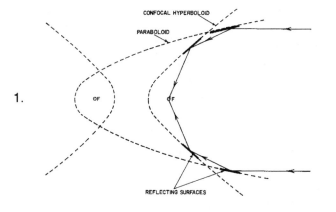

1.

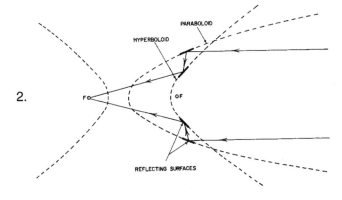

2.

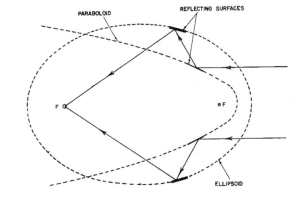

3.

FIG. 8.2. *Three possible configurations of image-forming x-ray optics.*

diameter, 10 to 1 or more. Wolter's type 1 configuration (configuration 1 in Figure 8.2) yields the minimum focal length. The mechanical structure of such a telescope can be made quite strong by physically attaching the two mirrors together and surrounding them with stiff, lightweight cylindrical structures. Mechanical strength is important both to ensure rigidity during the polishing process and to survive the stresses induced by the launch into space. Our group at AS&E was scientifically interested in obtaining the highest possible angular resolution over a wide field and therefore adopted the Wolter type 1 telescope optics throughout most of our x-ray astronomy program. Other specialized designs were used to reduce costs or satisfy specific scientific requirements, but I mention them only as they come up in the descriptions of particular missions.

On the Steep Part of the Learning Curve: Experimentation, Fabrication, and Testing

When we started our own laboratory program in 1961, we had to first obtain the necessary equipment and then learn the difficulties of the field by reproducing results already in the literature for other areas of research. We wanted to perform these tests with configurations that were germain to astronomy. As our first devices, we chose cylinders of glass and steel. Each had a conical hole drilled in its center that was polished. Industry could produce such devices quickly and cheaply, because they were used for chemical applications, and the polished cone was a sort of poor man's collector of photons. But as soon as possible, we switched to paraboloid-hyperboloid mirrors.

In the fashion typical of physicists, we wanted to go into the problem and evaluate the difficulties by doing experiments in our own laboratory. Norman Harmon, Richard Lacey, Zolt Szilagy, a vacuum technician, and a chemical technician joined me in this work.[7] We started the design and construction of a vacuum chamber with a long pipe attached to it. I designed and built a very simple, high-purveyance x-ray source, straight from an electron optics textbook.[8] This pointlike source of x-rays was placed at the end of the 400-inch-long pipe and produced a quasi-parallel beam at the chamber location (Figure 8.3). We placed a goniometer in the center of the vacuum chamber, so that we could measure the reflected beam for different grazing angles. Our detectors were Geiger counters with thin side windows, which we made in our laboratory. We also started the design and construction of two cone collimators, one made of glass and one of steel, and of several types

FIG. 8.3. *Initial laboratory setup at AS&E for the vacuum testing of designs and materials for prototype x-ray telescopes.*

of image-forming mirrors with a variety of approaches, which allowed us to evaluate different materials and fabrication techniques.

We learned immediately that highly polished glass would provide by far the best reflection efficiency (46%), in agreement with previous work. However, to be polished on the inner surface, glass shells must be thick enough to support the pressure of the polishing tool. They would also have to be thick enough to withstand the launch stresses, and would therefore be quite heavy to be launched into space.

An attempt to use a shell of polished aluminum gave varying and not fully understood or reproducible results. A polished aluminum telescope yielded a surprising 17 percent efficiency for the first few tests, but either the results were erroneous or the aluminum quickly degraded. The best we were able to obtain were efficiencies of 3.0 and 0.2 percent at 45 and 10 Å, respectively. Attempts at electroplating the surfaces with nickel and chromium were more successful, with efficiencies of 7.0 percent at 45 Å and of 1.5 percent at 10 Å. Epoxy mirrors cast from an aluminum mandrel performed quite well (4.5% efficiency at 45 Å).

Vacuum vapor deposition of metals, such as gold, on polished surfaces yielded some improvement of the efficiency at short wavelengths, but the results were critically dependent on surface preparation and on specific details of the vacuum vapor deposition process. We built and tested all these prototypes and measured an angular resolution of the image-forming telescopes of the order of a few arcminutes over a field of 30 arcmin. Even at this early stage of development, the telescopes provided a gain in sensitivity of at least a factor of ten over a pinhole camera of equal resolution.

The Second Generation of Telescopes

The results of these tests were still not satisfactory for astronomical use, so by 1963 we had developed a new technology, which yielded a much improved efficiency. The new approach consisted of electroforming nickel mirrors on optically polished stainless steel mandrels (Figure 8.4), a technology that was adopted and improved much later by Oberto Citterio at the University of Milan.

We built three telescopes with a focal length of 64 cm, 1-degree angle of incidence, angular resolution of a few arcminutes over a field of 40 arcmin, and 0.2 percent reflection efficiency in the band of 8–12 Å and 5 percent at 44 Å. In 1963 we flew these units on a sounding rocket and obtained x-ray pictures of the sun, but they were just indistinct images of the principal plage region that was active at the time. The work was done in collaboration with John Lindsay, and he mentioned our results at a symposium in Belgium in 1964. However, the quality of the images was rather poor, and we waited to publish any solar pictures until the flight of March 17, 1965, which used greatly improved telescopes.[9]

Meanwhile, the team at AS&E had changed: Norman Harmon, Richard Lacey, and Zolt Szilagy had been reassigned to other work. William Reidy, Giuseppe (Pippo) Vaiana, and Ted Zehnpfennig had just joined the group. Pippo Vaiana had come to the United States on a fellowship to Harvard that apparently had not worked out. He joined AS&E in 1964, and from 1965 on he increasingly assumed the scientific leadership of the solar x-ray astronomy group. Prior to his arrival, Reidy, Zehnpfennig, and I had just succeeded in fabricating the new electroformed telescopes, and Pippo jumped into the work of integrating the telescopes and cameras for the solar rocket payload of March 17, 1965.

The new electroformed telescopes were of the paraboloid-hyperboloid type and had a diameter of 7.6 cm, a focal length of 83.6 cm, and a collecting area of 1.6 cm^2. The slope of the first surface was 40 arcmin and that of the second surface 120 arcmin. The efficiency at 8.3 Å was about 15 percent (in contrast to the theoretically expected 69%). The quoted efficiency is

FIG. 8.4. (Top) *Early x-ray telescopes in cast epoxy and machined aluminum.* (Bottom) *The nickel electroformed telescope used to obtain the first x-ray image of the sun with focusing optics.*

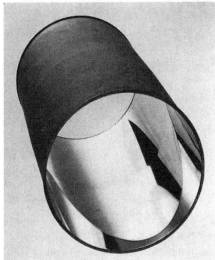

somewhat overstated, because it was measured by including all rays striking the focal plane within several arcminutes of the center of the image and not from the high-intensity central core alone. The angular resolution at the central core of the field was measured to be of 20 arcsec in visible light and 1 arcmin at 8.3 Å. This deterioration of the resolution in x-rays was the result of scattering by imperfections in the mirror surface. The rocket was equipped with an attitude control system and a biaxial pointing assembly. The combination of the two systems yielded a pointing accuracy of 1 arcmin for exposures of less than 10 sec. The recording medium was film, which was recovered

FIG. 8.5. *The first x-ray image of the sun, obtained during the 1965 AS&E-Goddard rocket flight.*

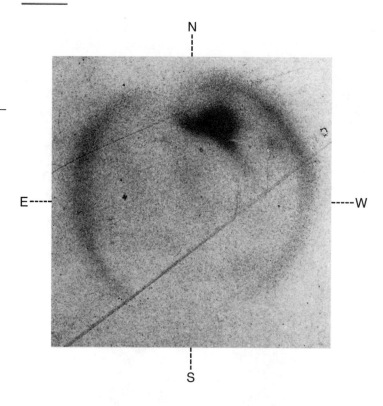

from the rocket after a parachute-aided landing. The first x-ray picture of the sun obtained with grazing-incidence optics is shown in Figure 8.5. To quote from our 1965 paper: "the photographs represent an improvement in sensitivity, in angular resolution, and spectral information when compared with photographs obtained with the pinhole camera technique. Of particular interest is the correlation of plage and x-ray emitting regions, the coronal latitude dependence, the observation of large regions of low activity and the spectral variation between these features."[10]

For the first time, we had obtained useful results in solar physics. This success after four and a half years of hard work was quite encouraging. Furthermore, we had established a laboratory capable of developing and testing a new technology in x-ray astronomy that could be immediately applied to solar astronomy.

We used the same type of telescope in the pointed section of the OSO IV, a project we had started in late 1963. The pointed section of OSO scanned the sun in a raster mode, which yielded a picture of 48 × 40 elements every

5.12 min. The spacecraft was successfully launched on October 27, 1967, and operated until May 12, 1968. The AS&E instrument was designed to obtain images of the sun at the raster frequency with a resolution of 1 arcmin at 12 Å and 4 arcmin at 2.5 Å. The instrument contained three filters that could be rotated in front of the detector to define a spectral band. The detector itself was a novel design, consisting of a cathode of cesium iodide on which the x-rays would fall and produce photoelectrons. These electrons were accelerated by a high potential onto a scintillator counter covered by a thin aluminum film, and they produced light pulses whose rate was telemetered to the ground. A system of baffles selected only those x-rays focused on the center of the telescope's field of view. The instrument had sufficient spatial and temporal resolution to measure the spatial characteristics of the x-ray emitting regions and to monitor the temporal behavior of individual active regions.[11] Frank Paolini led this program of construction and preliminary analysis. He subsequently left AS&E in the late 1960s to become director of research at Philips Electronic Instruments. Allen Krieger, Vaiana, and David Webb completed the analysis of the results in collaboration with Frank in 1971.

Notwithstanding these successes in solar astronomy, we were very conscious that these newly developed telescopes were far from reaching the level of performance required to study the fine structure of the x-ray emitting regions on the sun, which would have given us an insight into the physical processes responsible for heating the coronal plasma to millions of Kelvins. That would have required an improvement in the angular resolution of the telescopes from an arcminute to arcseconds.

The improvements needed for stellar x-ray astronomy were even more difficult to achieve; in solar observations, we could tolerate the relatively broad scattering wings due to surface or figure imperfections, which decrease the effective efficiency of the telescopes, because the large incident flux would still allow us to obtain the required angular resolution and sensitivity. In stellar observations, where the fluxes are a million times smaller, such problems would make imaging impossible. We again proceeded by first solving the problems of solar x-ray astronomy and later those of stellar astronomy.

The New Solar Telescopes

The institutional arrangement for the development of x-ray telescopes was considerably altered by the tragic and unexpected death of John Lindsay on September 26, 1965. Lindsay had been a member of the solar group of

Herbert Friedman at NRL until the formation of NASA. He had then moved to Goddard, where he became a strong and effective proponent of solar x-ray astronomy and was largely responsible for the start of the OSO satellite program. He supported our early start in telescope development; was the technical monitor on the first contract we had obtained; and was interested in collaborating with our group in the first use of the telescopes in solar research, a collaboration that continued over several years. The help of a NASA insider was quite important in those years to properly guide developments (for instance, the development of effective pointing controls) or to convince NASA headquarters of the utility of the development of x-ray telescopes. Possibly the great influence of NASA insiders on the scientific program was an unforeseen consequence of NASA winning the battle with the National Academy of Sciences for the control of space science.

Lindsay and I had become good friends as well as colleagues in the five years before his death. We had planned to propose jointly an advanced x-ray telescope for solar studies during the NASA Apollo Application Program. After he died, I visited Goddard and proposed to the members of his group, Bill Muney and John Underwood, that we continue to collaborate on the program. Goddard's management at some level, however, felt they had no need to collaborate with a mere contractor. As a result, the AS&E and Goddard solar physics groups became competitors, and telescope development in the two groups took somewhat different courses.

The Goddard group was the first to resume flights, on May 20, 1966, and on October 3, 1967, with a telescope of corrosion-resistant 440 steel manufactured by the Speedring Corporation of Warren, Michigan, and a new, higher-precision pointing control for solar rockets developed by the Ball Brothers Research Corporation of Boulder, Colorado. The optical surfaces were approximated by two cones, and the resulting resolution was 20 arcsec. The results were quite similar to those obtained in the AS&E 1965 flight.

At AS&E in 1968, Vaiana, Reidy, Zehnpfennig, Leon Van Speybroeck, and I developed instead a new set of x-ray telescopes of the Wolter type 1, specifically designed to achieve high angular resolution. We had found that angular resolution was limited more by surface tolerances than by surface finish. After Leon had made a detailed analysis of the fabrication tolerances to be achieved, we realized that we needed to build thicker mirrors to yield greater dimensional stability, and that we needed to achieve surface precisions comparable to those required for optical telescopes.

The resulting mirror specifications were 10^{-5} cm (a quarter wavelength in the visible) for the surface and roundness (conformity to a perfect circle) to approximately 2×10^{-4} cm over a 23-cm-diameter mirror. The reflecting surface of the telescope was a thin layer of Kanigen, a nickel alloy, which was deposited on a thick aluminum or beryllium support structure. The telescope was built in two separate pieces, the paraboloid section and the hyperboloid section, and then bolted together. The mirrors were ground, figured, and polished by the expert staff of Diffraction Limited, Inc., of Bedford, Massachusetts, who greatly contributed to advances in the fabrication of x-ray telescopes (Figure 8.6).

This new type of mirror was flown for the first time on June 8, 1968, by the AS&E group led by Pippo. The development of grazing-incidence optics with arcsecond resolution and the fine solar pointing control system developed by NASA for rocket observations, which had only a 2-arcsec jitter, permitted detailed studies of the structure of the corona with angular resolutions of 5 arcsec. This telescope had a collecting area of 34 cm^2 and a focal length of 132 cm, the largest ever flown; useful exposures as short as 2 sec could be made. Our group was lucky enough to catch a solar flare in progress.[12] The photographs revealed a strong correlation between hydrogen alpha emission lines and x-ray structures, the existence of loops interconnecting active regions, and the development of the x-ray flare along a neutral magnetic field line (Figure 8.7). These observations clearly showed for the first time the dominant role played by magnetic fields in the storage and release of energy in the

FIG. 8.6. *X-ray telescope with 5-arcsec resolution used in the June 8, 1968, AS&E rocket flight.*

FIG. 8.7. *The June 8, 1968, x-ray image of the sun during a flare. We observed a detailed correspondence between x-ray and hydrogen alpha line features.*

Reprinted from
9 August 1968, Volume 161, pp. 564-567

SCIENCE

X-ray Structures of the Sun during the Importance 1N Flare of 8 June 1968

G. S. Vaiana, W. P. Reidy, T. Zehnpfennig, L. VanSpeybroeck and R. Giacconi

FLARING X-RAY SUN

solar atmosphere and demonstrated the great value of high-resolution solar x-ray observations in the study of coronal physics.

The same type of telescope was used in five more rocket experiments, two in 1969, two in 1970, and one in 1973, which were spectacularly successful. This type of telescope was also used in the Skylab mission, with the improvement of nesting two confocal telescopes one inside the other.

Before discussing the Skylab mission itself, I mention some of the circumstances surrounding this mission as seen from our point of view at AS&E. John Lindsay and I were convinced that to make much more progress in solar astronomy, we needed to obtain high-resolution x-ray images over long periods, encompassing several solar rotations. We were both quite interested, therefore, when the possibility arose of placing a solar x-ray telescope in orbit as part of NASA's Apollo Application Program.

The sun is an extended x-ray source of about 40 arcmin; a full image at 1-arcsec resolution therefore consists of 5,760,000 pixels. In 1964, when we first started to work seriously on this project (AS&E's NASA contract NAS-5-9041), there were no x-ray electronic detectors capable of obtaining such images for transmission to the ground. Photographic film was the only medium available at the time both for intelligence surveillance and for scientific missions. Thus, the use of astronauts to retrieve exposed film seemed quite attractive.

However, as has often occurred in NASA's planning, priorities in the Apollo program changed with time, and that idea was abandoned. There was some interest in developing a robotic successor to the OSO spacecraft, because OSO's capabilities, in both pointing and data transmission, were no longer adequate for the evolving science requirements. This successor was to be called the Advanced Orbiting Solar Observatory (AOSO), and in size and complexity it was supposed to be similar to the Astronomy Orbiting Observatory, which had been devoted to measurements of stellar ultraviolet and visible light. This change of plans did not substantially affect our work on x-ray telescopes, but it prompted us to begin the study and development of imaging photon-counting detectors. The AOSO project was later abandoned.

During 1965–66, the proposal by the Marshall Space Flight Center to establish a small space station placed in orbit by a Saturn V rocket received approval from NASA headquarters. The station's main scientific instrument was a large solar-pointing platform designed to carry several instruments to study the sun in ultraviolet and x-rays. NASA requested proposals from the scientific community to undertake specific research programs, including the design and construction of the appropriate instrumentation.

Our group had been working on the basic elements of the program (namely, telescope and detectors) since 1964, but without any real certainty of a flight opportunity, and therefore we received the news with great joy. It was quite obvious that we had to win an opportunity to take part in this program if we wanted to continue working in solar x-ray astronomy. Given the rather aloof attitude of the Goddard group, we were particularly concerned that two competing proposals, ours and theirs, would result in our being excluded. To be chosen, we had to demonstrate that we were the best in the world in solar x-ray astronomy, and while we were clearly advanced in telescope development, we were worried it might not be enough. I thought that if we could add a spectroscopic capability to the imaging, we would have an unbeatable experiment, and I discussed this point with various people on the staff. It is still a wonder to me that within a week of my suggestion, Herb Gursky and Ted Zehnpfennig came up independently with the idea of transmission gratings that could be placed in front of the mirror. With the grating inserted, each source in the field results in a real image bracketed on either side by additional images corresponding to spectral lines of various orders. (This technology was later applied to great effect in stellar observations on Einstein and Chandra.) They were able to construct the gratings and test them in the laboratory in a few weeks. Whether or not this capability was the deciding point, we won a place on the Apollo Telescope Mount, the solar observatory flown on Skylab.

The solar group, led by Pippo Vaiana at AS&E, was deeply interested in the study of x-ray emission from the solar corona. For me, however, the main motivation was the opportunity to develop the instrumentation to be used ultimately in nonsolar x-ray astronomy. This concern actually dictated some of the design choices we made in executing the solar program. For example, the high solar x-ray flux did not strictly require the nested-mirror configuration we used on Skylab, but testing this technology was essential for stellar astronomy (Figure 8.8). The particular attention given to scattering problems and reflection efficiency at short wavelengths was also of greater significance for stellar than for solar astronomy. Much of the learning and technology development that made Einstein possible came from our solar work. Leon Van Speybroeck, who uniquely contributed to the development of our solar telescopes, was the scientist responsible for the Einstein and Chandra mirrors. He became recognized as the world's leading authority in this field.

Skylab and Solar X-Ray Observations

Skylab was launched on May 14, 1973, by a huge Saturn V rocket, the same type of vehicle that took Apollo to the moon. The first deployment was followed by three additional Saturn V flights for crew rotations on May 25, July 28, and November 16, 1963 (Plate 3).

FIG. 8.8. *The Skylab 30-cm diameter nested mirror system.*

A major component of the Skylab orbiting observatory was the Apollo Telescope Mount; our instrument, the x-ray spectrographic telescope (S-054 in NASA nomenclature), was part of this component. The mount contained several telescopes to study the solar atmosphere simultaneously at wavelengths from the visible to x-rays, with the spatial, spectral, and temporal resolution required to understand its structure and dynamics. The mount's axis was kept pointed at the sun with an accuracy of better than 1 arcsec over long time periods.

S-054 was in essence a large telescope assembly with a photographic detection system. The optics consisted of two coaxial and confocal grazing-incidence mirrors of the paraboloid-hyperboloid design, of 30 and 23 cm diameter, respectively. The focal length was 213 cm and the collecting area 42 cm^2 (Figure 8.8). The solar image was about 2 cm in diameter, and the angular resolution was about 2 arcsec over about 20 arcmin, degrading to 6 arcsec at the edge of a 40-arcmin field. Exposure times could be varied from 1/64 to 256 sec in steps differing by factors of four. A combination of filters and exposure times yielded a dynamic range in sensitivity of five million. This is sufficient to span the range of x-ray fluxes from the quiet sun, which differ by factors of one thousand, from faintly brightening spots to active regions, as well as the much greater fluxes during flaring conditions.

The essential characteristic of the instrument was that it combined the highest sensitivity and angular resolution ever achieved in x-ray solar observations with the long duration and operational flexibility of the manned Skylab station. Not only did the astronauts replace and recover our film for us, but they were able to repair major malfunctions in the station itself. For example, one of the wings of the solar-cell panels designed to provide power to the station was accidently blown away during ascent, and the astronauts performed miracles of improvisation to keep the station from overheating while an actual fix was figured out on the ground. The astronauts intervened in our experiment at a crucial time when the shutter got stuck, and by removing it, allowed us to continue the observations during the entire mission. This mission showed NASA's ability not only to launch but also to repair and resupply space vehicles, as well as the high degree of competence of the astronauts in maintaining complex observatories in space. This capability was to play a crucial role in the repair of the Hubble Space Telescope.

The Skylab mission provided a detailed view of the inception and dynamic development of structures on the sun, some of which, such as the bright points, were previously unknown. The advances in the quality of solar x-ray observations from the NRL pinhole camera picture of 1961 to the Skylab picture (Plate 4) were so great as to bring about a revolution in the theories of solar corona heating and of the role of magnetic fields in this process. Vaiana and Rosner summarized their conclusions in part as follows: "The observational and theoretical work suggests that the topological structure of the ambient coronal magnetic fields largely determines the physical state of the coronal plasma, and therefore implies that the introduction of structure is not to be regarded as a refinement of theories based upon spatial homogeneity, but rather a fundamental change in our understanding of the physics of the corona."[13]

Beyond Imaging the Sun: Plans for Stellar X-Ray Telescopes

Institutional Resistance

Despite the success achieved by x-ray telescopes in the study of the sun, the adoption of x-ray telescopes for stellar observations was slow. NASA rejected our proposals for the development of telescopes of higher efficiency, which was necessary because of the much lower stellar fluxes, as well as other proposals for exploratory flights on rockets or satellites. The only two exceptions were the inclusion of the Lockheed piggybacked x-ray instrument on OAO I, which experienced a complete power failure after launch on April 8, 1966, and the opportunity offered to Mullard Laboratory of the University of London to fly a small payload on OAO III. The U.K. experiment included small grazing-incidence collectors and proportional counters, which produced useful results.

One cannot help wondering why NASA was so slow in supporting the use of x-ray telescopes for stellar x-ray astronomy. NASA was not alone in its negative attitude toward the development of large x-ray telescopes. Some of the scientific advisory committees suggested that we should continue to carry out all-sky surveys with conventional detectors to determine the need for such telescopes. (This recommendation was made before the launch of Uhuru.) Others felt that x-ray astronomy might not turn out to be so interesting after all, and that funds could be better used to push optical and ultraviolet astronomy. Some theoreticians, Philip Morrison of MIT among them, were convinced on the basis of theory that x-ray sources would be mainly diffused, extended regions and that high resolution was unnecessary to study them. For the first time, I fully understood Galileo's remark that "our studies should be about the real world, not a world on paper." I realized that some theoreticians love to compute particular models not because they might be true but because they can be computed.

Without dedicated support, the technical development for stellar x-ray telescopes had to be piggybacked on the solar program and consequently proceeded at a slower pace than I had hoped. The lack of a specific flight opportunity also made it difficult to start even preliminary designs for flight hardware.

Marshaling Support for X-Ray Imaging in the Research Community

Notwithstanding these difficulties, we had already solved many of the fundamental technical issues, including optics design, nesting, polishing techniques, vacuum coating with high-Z metals, and electronic imaging detectors. We were confident that, given the opportunity, we could build a real x-ray telescope to study stars and galaxies. Scientifically I was as convinced as ever that we needed large x-ray telescopes in orbit to really advance the field.

In the summer of 1965 the x-ray and gamma ray panel of the Space Science Board of the National Academy of Sciences met under the chairmanship of Herbert Friedman at Woods Hole, Massachusetts, to plan future programs in high-energy astronomy. Rossi and Clark of MIT, Fisher of Lockheed, Kraushaar of the University of Wisconsin, Robert Novick of Columbia University, Frank McDonald of Goddard, and I participated in the discussions.

I took the opportunity to explain the limitations of conventional detectors with regard to both sensitivity and angular resolution. While the number of photons counted by a conventional detector increases with the collection area, the sensitivity increases only with the square root of this area because of the background. Thus, a detector a hundred times the size of Uhuru would gain only a factor of ten in sensitivity. (This gain is in fact what was achieved by the large NRL detector on the High-Energy Astronomical Observatory A [HEAO-1].) Furthermore, as the number of sources increased at higher sensitivity, their identification and study would be difficut and confusing because of poor angular resolution. These limitations did not exist in practice when one used an x-ray telescope. The sensitivity increased with the area because of the low levels of the x-ray background, and the confusion limit would not appear until one reached levels of sensitivity ten thousand times better than predicted for Uhuru.

On the other hand, gamma ray astronomers could not use focusing optics, and many x-ray astronomers had not become familiar with x-ray optics and felt more comfortable using traditional techniques. Eventually we did find a common denominator, namely, that high-energy astronomy missions needed heavy payloads, which could be simple in design and execution and therefore much less expensive than those used in optical and ultraviolet astronomy. In particular, x-ray astronomy with arcsecond resolution did not require high-precision pointing of the spacecraft. It had occurred to me that because of the small number of x-ray photons expected from celestial sources, it would be possible to tag each photon with a time of arrival and a location in the focal plane. Using aspect information from an optical sensor on board, a post-facto, precise reconstruction of the x-ray image on the celestial sphere could be achieved to the limiting resolution of the telescope (about 1 arcsec).

In the conclusions of its report, the x-ray and gamma ray panel stated that "x-ray telescopes, based on grazing incidence reflection, appear to be the only tools capable of many of the refined observations that will be needed beyond the early exploratory stage." The panel recognized the superior angular resolution of imaging optics, the increased sensitivity such optics provided, and the possibility of using telescopes with polarimeters and spectrometers. The panel also recommended that "a program of x-ray astronomy using total reflection telescopes should be started at the earliest possible time and pushed vigorously to exploit its ultimate capability." Finally, the panel recommended the use of an OAO spacecraft dedicated to x-ray astronomy.[14]

A consensus emerged at the Woods Hole meeting among cosmic-ray physicists who needed heavy payloads, x-ray astronomers interested in the utilization of large-area (100 ft²) detectors, and those hoping to use big x-ray telescopes that a series of large but simple and inexpensive spacecraft (so-called Super Explorers) could accommodate all their requirements. The spacecraft could be developed with a great deal of commonality and would be able to carry either survey instruments or telescopes. This simpler and cheaper approach was the result of the experience gained by high-energy astronomers in experimental space physics and of their knowledge of how best to utilize the capabilities of NASA and its contractors. Frank McDonald of Goddard, a cosmic-ray physicist, was among the earliest and strongest proponents of this approach. Apart from its technical merits, the strategy had the advantage of fitting in with NASA's love for series of spacecraft. I realize that, at least in part, this packaging of disparate missions under a single label was forced on NASA by the reluctance of Congress to approve programs consisting of many disparate missions selected by peer review, such as the Explorer program. This "Super Explorer" idea ultimately led to the HEAO series of spacecrafts.

Approaching NASA with an Integrated Plan

Encouraged by the scientific support from my colleagues, in June 1967 I wrote a letter to Homer E. Newell, NASA's associate administrator for space science and applications, which was widely circulated within NASA. In this letter I reiterated the need for an OAO mission to carry an x-ray telescope of modest size (30-cm diameter, 300-cm focal length, similar to that being designed for Skylab), to be followed by a more ambitious mission (1-m diameter, 10-m focal length) in the 1970s. I proposed that the smaller telescope be implemented as an experiment with a single principal investigator and that the larger telescope be implemented as a multi-experimenter facility. I offered our services as the lead group to provide the mirror optics and the x-ray cameras and to integrate other instruments into the payload.

Newell responded in August 1967, expressing interest in proceeding with the smaller telescope mission but added that it was NASA's intention to consider even the smaller version a "facility." At his direction, a meeting was held on October 9–10, 1967, under the chairmanship of Nancy Roman, with the participation of many active x-ray astronomers. The conclusion was that a large grazing-incidence parabolic focusing x-ray telescope would be necessary for future high-energy space astronomy programs.

As a result of these discussions and the explicit recommendations of the Astronomy Missions Board, an announcement of opportunity was circulated on February 16, 1968. The announcement read in part: "NASA is currently planning to use a parabolic focusing telescope for x-ray astronomy. This instrument is to be considered a facility rather than an experiment. At present we will welcome proposals for experiments to be performed at the focus of such an instrument, together with the necessary auxiliary instrumentation, but not proposals for basic telescopes."[15]

Of particular concern to me was the decoupling of the telescopes from the focal plane instruments. This separation was not just a casual remark but a clear indication of the direction in which NASA wanted to proceed: to rely on the aerospace industry for the design and development of the telescopes, with the experimenters allowed to compete only for the focal plane instrumentation. This approach would almost automatically exclude scientific direction vested in a principal investigator and would have allowed a NASA center to retain overall control of the mission. (The construction phase of Hubble was carried out in precisely this manner, and the lack of unified overall scientific control during the early years was painfully obvious as the program reached the commissioning and operation phase.)

This approach was not at all what I had hoped for. I was convinced that, to be successful, we needed an integrated design of the telescope, aspect system, instrumentation, and operational concepts. In early 1968 I convinced Bob Novick from Columbia and George Clark of MIT to join me in this view, and we sent a letter to Richard Halpern of NASA as a self-styled "Principal Investigator Group" proposing a Large Orbiting X-Ray Telescope (LOXT) as an integrated facility including telescope and instruments.

At this point, the possibility of a smaller precursor mission with OAO was no longer being discussed. The OAO program had problems of its own in carrying out the planned ultraviolet astronomy, with only two successful missions out of four launches. It was terminated in 1972, although OAO III (Copernicus) operated until late 1980. The only opportunity to proceed was to be part of the HEAO program and to express our interest in the mission with specific proposals.

A number of groups, including Clark's at MIT, Novick's at Columbia, Elihu Boldt's at Goddard, and mine at AS&E, were selected as potential investigators and asked to assist the Marshall Space Flight Center (Huntsville, Alabama) in defining the mission. The kickoff meeting of the interested scientists was held at Marshall on March 19, 1969, and chaired by Ernst Stuhlinger, associate director for science for the center. It became quite clear that it was not possible to set forth the design requirements for the observatory without knowing what science was going to be done.

The scientists involved decided to hold a series of meetings at their own institutions to define the scientific objectives of the mission, the experiments necessary to achieve these objectives, and consequently the design parameters for the telescopes. The results of these studies were presented to NASA on August 1, 1969, in a document titled "Preliminary Study: Telescopes and Scientific Subsystems for a High Energy Astronomy Observatory." In the report we stated: "The characteristics of the telescopes and experiments are so intimately related that we conclude that the specifications for the optics must ultimately be set by the users. In addition, the users must retain technical control of the implementation and testing of the devices."[16]

Following this initial study, the scientists at the four institutions decided to respond to the expected NASA solicitation as a single consortium with a single principal investigator who would be supported by a scientific steering committee. I was designated as principal investigator, and in May 1970, I submitted the joint proposal by the four institutions. The consortium had designed LOXT as a single integrated payload that would be built and operated by the consortium on behalf of NASA as a national facility available to all interested astronomers.

From the many proposals submitted by American and foreign groups, four missions had survived the brutal competition by September 1970. The first two consisted of a mixture of x-ray and gamma ray experiments, the third would carry LOXT, and the fourth consisted of cosmic-ray experiments. Only the first two missions were funded for flight; the other two were funded as development studies. Three years later, there occurred one of the disconcerting —but not unusual—changes in a NASA program: the third mission advanced to the second spot and was downsized to what was to become the Einstein Observatory.

An Aside: Collectors of Light, Ancient and Modern

In 1990, many years after starting work on the concept of an x-ray telescope, I came across the book *Lo specchio ustorio, ovvero trattato delle sezioni coniche* [The burning mirror, or treatise on conic sections], published by Francesco Bonaventura Cavalieri, Jesuato, in 1632. Cavalieri was born in 1598 in Milan, was a student of Galileo in Pisa, and was one of the most illustrious of the

FIG. 8.9. *Frontispiece and figures XXI and XXII of the book* Lo specchio ustorio.

latter's disciples. He was a leading mathematician of his time, whose ideas foreshadowed integral calculus, and was arguably the inventor of the reflection telescope prior to Newton. He was granted a professorship at the University of Bologna in 1629 at age 31, having previously been denied the post in 1619 on the grounds that he was too young.

I was struck by the drawing in figure XXII of his book (Figure 8.9), in which he shows the principle of the collector that I reinvented for x-rays some 330 years later. He refers frequently to the book of the same title as his written by Apollonius of Perga, the great mathematician who lived between 262 and 190 B.C. and made fundamental contributions to the study of conics. Cavalieri wonders about even earlier origins for the collector and suggests that the Vestals who kept the sacred fire in ancient Rome used similar devices to ignite the flame.

I felt a great kinship with Cavalieri for his love of projective geometry and his youthful enthusiasm for experimentation with focusing mirrors. His speculations on the seers of old Rome are reminders of the links that bind modern science to the ancient mysteries in the human search for truth in nature.

NINE

Plans for Space and Realities on the Ground: LOXT, Einstein, and NASA

LOXT, a Blueprint for a National X-Ray Observatory in Space

NASA's approval of the LOXT program in September 1970 marked a bold step forward in x-ray astronomy. Bypassing all intermediate steps of smaller telescopes and with only thirty x-ray sources known at the time (Uhuru had not yet flown), NASA initiated an ambitious development that was not completed until the launch of the Chandra spacecraft some 30 years later. The technical basis for the work was the solar x-ray program, although many technological improvements were still needed during the project. It was only with the launch of Chandra that telescopes and detectors of the required level of sophistication allowed us to fully realize the scientific advances that LOXT was designed to achieve.

However, the start of the project had a great impact on x-ray astronomy as a whole. It not only provided the technical concepts and developments that were utilized in subsequent missions, but also brought about a change in the sociology of x-ray astronomy research, moving the field from a series of individual experiments to the concept of national observatories open to all astronomers.

The LOXT project was also a training ground for the management of large cooperative projects, and many of the scientists who contributed to the realization of Chandra were alumni of that school. It certainly had a profound effect on my own views on the management and operation of observatory-class facilities, and much of what I was able to contribute later to the Hubble Space Telescope and Very Large Telescope projects was learned on LOXT. It is for these reasons that I describe this project, even though it never reached orbit.

The Management of LOXT

Although our proposal was considered technically and scientifically excellent, there were aspects of it that were troublesome to NASA. Entrusting to a single principal investigator an experimental payload of such substantial complexity and cost that required a large dedicated spacecraft (the size of a moving van) was contrary to NASA culture.

On the other hand, the consortium of the four institutions (AS&E, Columbia University, Goddard, and MIT) provided a strong basis for scientific leadership and had extensive experience in instrument development. Furthermore, the institution of the principal investigator, AS&E, had developed an experienced engineering staff that gave considerable assurance of success. The Space Research and Systems Division, which was directly under my management, had successfully carried out space programs for the NASA manned program and for the DOD, in addition to our own x-ray astronomy program. We had developed a fully integrated capability from hardware conception and design to fabrication and testing. Although Uhuru had not yet flown, at the time of the LOXT proposal it was completed and would be ready to launch in a few months, and the Skylab experiment was in advanced testing.

In connection with the work on the Skylab experiment and other experiments to be flown as part of Apollo and DOD programs, we had adopted and used all the necessary managerial tools for program planning, monitoring, and control—such as work breakdown structure, the Program Evaluation and Reporting Technique (PERT), and PERT/Cost—which were then becoming common in the aerospace industry. As a result, we were able to support the LOXT proposal effort with a great deal of technical work and management planning.

Although LOXT was proposed as an experiment with a designated principal investigator, the level of detail of the work breakdown structure included in our proposal was comparable to that used for spacecraft construction, thus helping to validate our cost and schedule estimates. The full cooperation between scientists and engineers, which had started in the AS&E rocket program and in the construction of Uhuru, was continued in LOXT. In my view, such cooperation is essential for the successful execution of missions as large as LOXT. At AS&E, we had the good fortune of combining scientific and engineering skills.

The perennial drawback, of course, was the concern of NASA and certain members of the scientific community (some sitting on the review committee) about entrusting the scientific and technical direction to a principal investigator in an industrial organization. The original suggestion that I assume the responsibility of principal investigator came from David Ellenbogen (the marketing director at AS&E), who urged me not to be shy about assuming the role. His argument was that we had set in place at AS&E the perfect tools to execute LOXT successfully, and that by assuming the lead role, I could ensure overall control of scientific specifications and guarantee performance. Apparently the argument was sufficiently compelling to convince the other members of the consortium. Their willingness to work together under an AS&E principal investigator and the creation of a high-level scientific steering committee to advise me during the execution of the program were instrumental in alleviating the concerns of NASA and its review committee. The idea of entrusting the design, development, construction, and operation of observatories to a scientific group representing a consortium of universities is a common practice for the National Science Foundation, even though it appeared to be novel to NASA.

I was highly honored by the confidence of my colleagues, but I also believed that our group at AS&E had all the necessary scientific, technical, and managerial skills to support me in the role of principal investigator for the LOXT mission. The work on rockets, Uhuru, and Skylab had prepared us to tackle even more challenging programs. Gorenstein, Gursky, Kellogg, Murray, Schreier, Daniel Schwartz, Tananbaum, Tucker, Van Speybroeck, and Vaiana supplied a formidable array of talents that any academic or research institution would have been proud to host, and I believe that this was generally recognized in the community.

Many individuals made significant contributions to the work on LOXT and Einstein. But the efforts of Harvey Tananbaum and Leon Van Speybroeck were essential and pervasive in these programs. I started working with Harvey in 1969, when he became project scientist for Uhuru. In that program, he demonstrated that he was not only an outstanding scientist but also an excellent manager of other scientists and engineers. As science program manager, he served as my alter ego on LOXT and Einstein from 1972 to 1981, and we were co–principal investigators in the 1976 proposal to NASA that ultimately led to Chandra. In 1981, when I left Harvard, he was more than ready to take on the task of head of the High Energy Division of Harvard's Center for Astrophysics. His high level of technical and scientific competence, managerial skills, ability to communicate, and quiet confidence provided steady guidance through the many years of work that were necessary to bring Chandra to successful conclusion. Leon was one of the most talented physicists I have ever met. He was the scientist who brought to the design of x-ray telescopes a deep understanding of surface physics, optics, mission requirements, and fabrication techniques. He led the effort at Speedring on the construction of the Skylab telescope and at Perkin-Elmer on the construction of the Einstein and Chandra telescopes, achieving x-ray optics of unparalled quality. It is clear that much of the scientific success of those missions would not have been possible without his contributions.

Apart from a feeling of confidence in the ability of our group, I was personally totally committed to the program, because it embodied the scientific approach that I had hoped to bring to realization from the beginning, even when very young: the construction of instruments powerful enough to solve physical problems by direct observation rather than by inference or speculation. I was in complete accord with George Ellery Hale's view that modern technology could and should be applied to the investigation of natural phenomena, and not only to military or commercial enterprises. In 1928 he wrote an article for *Harper's* stating his views: "From an engineering standpoint our telescopes are small affairs in comparison with modern battleships and bridges."[1] He noted that no advances had been made since the Lord Rosse 6-ft reflector had been completed in 1845, and that the time was ripe for better and bigger mirrors. These considerations led him to build the largest ground-based optical instruments in the world: first to construct the 60-inch and 100-inch telecopes at Mt. Wilson, and finally to initiate the construction of the 200-inch telescope at Palomar. In proceeding with LOXT, I was determined to construct the very best telescope for x-ray astronomy that technology and funding allowed.

My view of my own role in astronomical research was also changing. Although always troubled by a fundamental lack of self-confidence, after Uhuru I did not feel the need to continually prove myself. The competition with the

NRL group, which had spurred me on in the early 1960s, was over, and while analyzing Uhuru data, I came to love discovery for its own sake. I felt my greatest contribution to the field could be to build great instruments available to the entire astronomical community and to operate them in such a way as to maximize the scientific returns.

Scientific Direction

LOXT was created by scientific groups that had specific scientific interests and instrumentation capabilities. The responsibility for building the instruments was divided among the groups accordingly, and the work of each group was carried out under the direction of a principal scientist. The principal scientists were Herbert Gursky for AS&E, Bob Novick for Columbia, Elihu Boldt for Goddard, and George Clark for MIT. AS&E was responsible for the high-resolution imaging detector, the imaging proportional counter, and the monitor proportional counter. Columbia took charge of the objective crystal spectrometer and the polarimeter. Goddard handled the solid state detector and the flare alarm. MIT was responsible for the focal plane spectrometer and the focal crystal spectrometers. In addition, AS&E had the programmatic responsibility for the telescopes, aspect system, and integration of the instruments, ground support equipment, and data handling. I was named technical director by NASA and given the task of coordinating and integrating the activities of the participating organizations within the scope of their individual grants and agreements.

In addition to retaining the responsibility for specific instruments, the scientific groups participated in the overall scientific direction of the program. We established a scientific steering committee composed of the principal investigators and the principal scientists and augmented by two distinguished senior scientists, Bruno Rossi of MIT and Lodewijk Woltjer of Columbia University. The chairperson of the committee rotated yearly; the first chair was Bruno Rossi. The purpose of the committee was to ensure broad and objective scientific guidance and to be the forum for resolution of differences of scientific opinion.

As a group, we were quite conscious of the major impact that LOXT could have on astronomy as a whole: "The recent discoveries of the microwave background radiation, x-ray sources and pulsars have strengthened the conclusion that in our Universe high energy processes play a major and quite possibly a decisive role." X-ray observations were particularly suited to discover and study such processes. We further stated in our proposal for LOXT that "clearly the principal objective of x-ray astronomy is to develop physical models for the observed phenomena in order to relate x-ray observations to the main body of astronomical and physical theory."[2]

We felt we had a responsibility toward the astronomical community as a whole to ensure the participation of astronomers from every discipline. We therefore limited the exclusive access of each experimental group to data obtained with their own instruments to a maximum of one year. After this time, all instruments could be used by any of the consortium groups. Furthermore, we proposed to allot to guest observers (scientists not participating directly in LOXT) one-third of the observing time. In the proposal we also volunteered to assist guest observers in making observations and processing the data. In effect, we were proposing to act (on behalf of NASA) as a national x-ray observatory in the same sense as the National Radio Observatory or the National Optical Astronomy Observatory are U.S. facilities of the National Science Foundation.

As the steering committee discussed these far-ranging issues, I learned the precious lesson that while execution of a program is best carried out by a single responsible person, the overall scientific goals and objectives are best reached by open debate and consensus among the community of active users. I also became persuaded of the benefit to astronomy if the astronomers themselves took greater responsibility in the planning, construction, and operation of their space facilities, just as they had done for years for ground-based observatories.

Development of the Payload

The HEAO series of spacecraft was designed to utilize the maximum available volume and weight that were compatible with a Titan III-C rocket. The LOXT was a large payload about 10 ft in diameter and 40 ft in length. (An artist's conception of the payload is shown in Figure 9.1.) It consisted of two mirrors: the first was of the Wolter type 1 design (see Figure 8.2) to achieve the highest possible angular resolution of 1 arcsec but with a relatively modest collecting area of 1,000 cm² (still a factor of twenty greater than that of the solar Skylab mirror); the second was a modified Kirkpatrick-Baez design with a coarser angular resolution of 10 arcsec but almost twice the collecting area. (The Kirkpatrick-Baez configuration consists of a set of thin, nested one-dimensional paraboloids that provide a line focus, followed by another oriented at 90 degrees to the first.)

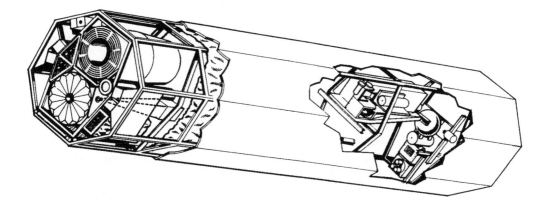

FIG. 9.1. *A view of LOXT showing the 1.2-m-diameter Wolter type 1 high-resolution mirror and the 1-m Kirkpatrick-Baez high-throughput mirror.*

The high-resolution mirror (telescope A) was quite expensive to manufacture because of the requirement for high accuracy and surface polish on shapes that are not usual for the optics industry. The thickness of the shells had to be kept small because of weight limitations. The glass chosen had to have a thermal expansion coefficient near zero to maintain its shape. The five paraboloid and hyperboloid shells had to be aligned to arcsecond precision.

The high-throughput mirror (telescope B) could be built of thin glass plates mechanically bent and held in place. Such surfaces are inexpensive to produce and easy to polish to very high finishes, using techniques similar to those used in making window panes. A larger collecting area and a better response at high energy can be achieved when it is not crucial to obtain the best possible angular resolution.

Telescope A had three focal plane instruments associated with it: a high-spatial-resolution x-ray detector, a crystal polarization analyzer, and a Johann-mount Bragg crystal spectrometer. In addition, a diffraction grating could be placed immediately behind the mirror and used with the imaging detector as an objective grating spectrometer. Telescope B utilized three focal plane instruments: an imaging proportional counter, a cooled solid state detector, and a polarimeter. A mosaic crystal mounted in front of the mirror could also be used with the imaging proportional counter to form an objective crystal spectrometer. Three other instruments not using the telescopes were also planned as part of the payload: an all-sky flare detector, a flat crystal spectrometer, and a monitor proportional counter. I do not describe these instruments in detail here, because ultimately some never flew and those that did are discussed later as part of the Einstein mission. A conservative summary of the specified LOXT observing capabilities is:

Energy range	0.2–4.0 keV (telescope A)
	0.2–7.0 keV (telescope B)
Angular resolution	<5 arcsec
Faintest detectable source	10^{-7} of the flux of the Crab Nebula in one day
Energy resolution	1 eV
Polarization	1% of the Crab Nebula in one day
Timing precision	0.1 msec

These capabilities represented a jump of many orders of magnitude not only with respect to what had been achieved until then but also over what would be achieved with Uhuru and all observatories not utilizing x-ray optics, including HEAO-1. The sensitivity quoted above would not be reached until three decades later with the flight of Chandra and X-Ray Multi-Mirror (XMM), with Chandra embodying the capabilities of telescope A and XMM approximating the capabilities of telescope B.

It is worth mentioning one particular aspect of LOXT's design that was later used in the Einstein and Chandra missions, having to do with how we reconstructed the position of x-ray images in the sky. The idea was to use an optical imaging system to take pictures of the sky every second. Through a reflection on the x-ray mirror and a corner-cube reflector, fiducial lights placed on the x-ray instruments were focused onto the optical camera (Figure 9.2). The system relates the position of arrival of each x-ray photon in the focal plane of the telescope directly to the star field with an accuracy of 0.5 arcsec, independent of misalignments or motions of the telescope, provided only that the rate of drift is less than 1 arcsec/sec. Thus, provided LOXT was pointed with sufficient accuracy (on the order of arcminutes) to keep the source under study in the field of view, small, slow drifts could be tolerated. This system, which is possible when the photon flux is low enough that photons can be individually counted, decreases by a factor of a hundred the accuracy requirements on the pointing system of the spacecraft. Although the idea may appear obvious in retrospect, it was the key in obtaining images with high angular resolution in x-ray astronomy with systems much less sophisticated and expensive than those used in optical space missions.

In our proposal, we were asked to consider two launch dates: July 1975 or mid-1976. We expected a start date of 1971, as in fact occurred, and the AS&E plan was to achieve the earlier launch date, but to accommodate potential delays up to 1976. The total estimate for the AS&E effort up to launch was $43.2 million. The funding for the experiments went directly from NASA to

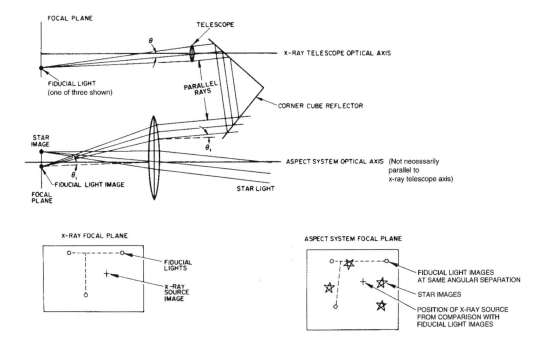

FIG. 9.2. *A self-calibrating system to align the center of the x-ray detector and the optical star field.*

the various institutions. Thus I can only estimate the payload cost to be $80 million. I do not have data on the cost of each of the three TRW spacecraft, but in 1973 a total cost for all three HEAOs was quoted by NASA at $450 million. On average, each HEAO (payload plus spacecraft) cost about $150 million, or one-tenth of Hubble or Chandra. The HEAO program was not expensive, did not have technical problems, and had experienced only modest overruns. There was no reason to drastically reduce HEAO, except to pay for Viking overruns and to start Hubble. These reductions, however, delayed progress in U.S. x-ray astronomy by 20 years.

I believe that the startup of the LOXT program symbolizes NASA's attitude in the late 1960s and early 1970s. The United States had just won the race to the moon and in doing so had built a can-do organization that was not afraid to take risks and was intent on maintaining its momentum in the scientific exploration of space. Nobody could then have imagined the disastrous choices that would be made in the U.S. program of manned exploration —the shuttle and space station programs. These programs were technically unsound and motivated more by political considerations than by any real benefit to science or exploration. The lack of a clear statement of realistic goals has turned NASA, in the past quarter century, into a poorly managed, sclerotic organization, more bureaucratic than visionary and totally averse to risk. In fact just recently a belated but very welcome decision was made by President George W. Bush to terminate the shuttle program and to reduce space station commitments to enable a real program of exploration of the moon and Mars. One can only hope that current lapses in NASA's funding for science will be overcome by a commitment to a rational program in the future.

Fallback: Einstein

The Einstein observatory was born of a retrenchment of LOXT in the face of intense budget pressure on NASA.

Work had started on LOXT in 1971, with particular emphasis on the development of the 1.2-m telescope A and the high-resolution imaging cameras. We made good progress. However, in January 1973, the HEAO program underwent major changes. The cost of the total program had risen to about $450 million. The increase was modest enough (I believe 15%) that a reduction in the scope of the program, which would have brought us back within target, appeared feasible within contingencies.

The real problem was that the large overrun of the Viking program forced NASA to cut the HEAO program in half. This reduction implied scaled-down spacecraft and three rather than four missions. In a series of meetings at NASA headquarters called by Jesse Mitchell (John Naugle's successor as associate director for space science and applications), we were told that the program had temporarily been canceled; to save the sinking ship, some passengers would be thrown overboard, and he urged those in the water to swim along and push.

The most grievous cut was the cancellation of HEAO-4, which was to carry the cosmic-ray experiments, including that of the Stanford University group led by Nobel Laureate Robert Hofstadter. For x-ray and gamma ray experimenters, the situation was a little less grim. The case of Novick's group at Columbia provides an example. Although the reduction in cost and capabilities forced the cancellation of polarimetry, his group remained part of the consortium and fully shared in the use of the telescope. The science that was lost, however, still remains to be done.

LOXT suffered the loss of the high-throughput telescope B and the reduction of the high-resolution mirror (telescope A) to half its original size. Several other instruments or modes were canceled, including the polarimeters, the objective crystal spectrometer, the flare detector, and half of the monitor counter. The only bright spot was that in the reshuffling of the missions, we were advanced to mission two. Given NASA's tendency to cut the last mission of a series, we naturally welcomed the change.

We were then asked to submit a proposal to demonstrate how we could do valid science with these new constraints. A special review of the new configuration was held on September 13–14, 1973, by a NASA committee. The consortium discussed the capabilities of the new observatory HEAO-2 (renamed "Einstein" after launch) in light of the recent advances in x-ray astronomy, particularly the Uhuru findings. Cutting the diameter of the high-resolution telescope in half (from 120 to 60 cm) entailed a quartering of the instrument's collecting area. However, through improvements in polishing techniques, we could achieve higher reflection efficiencies (we were hoping for a factor of four to five), which would compensate for the loss in area; we could also use longer exposure times by reducing the total number of targets (five thousand were ultimately observed in the Einstein mission).[3] We showed the committee the solar x-ray images obtained from Skylab to demonstrate the power of focusing optics. We emphasized our commitment to design and operate Einstein as a national facility that would be open to guest observers. As a result of this meeting, the HEAO-2 program was approved.

The most serious restriction in the revised HEAO program was the imposition by NASA of a fixed life for all three HEAO spacecraft. This was unusual, as most spacecraft were allowed to operate well beyond their minimum planned lives, provided they continued to produce science. Uhuru, although designed for a lifetime of six months, continued to operate for years, and these continued operations were crucial in obtaining some of the most important scientific results from the mission.

Apparently this new policy had been adopted by NASA as a result of considerable pressure by the Office of Management and Budget (OMB). Their intent was to limit the overall cost of programs by presetting the time allotted for post-launch operations and data analysis. From a scientific point of view, this policy was so absurd that most scientists could not believe it would actually be implemented. The cost of constructing an observatory is so high that the idea of not using it while it is still productive seems an extremely inefficient way to spend research funds. Nevertheless, the policy was in fact carried out on HEAO. Of course, OMB is not charged with maintaining the scientific productivity of the nation; what was lacking was a strong reaction by NASA or any scientific advisory body, including the National Academy of Sciences. This policy was later abandoned for all major programs.

There followed a long and largely futile struggle on our part to ensure that Einstein's life could be prolonged. NASA had decided to place the HEAOs in relatively low Earth orbit, so that the shuttle could reach them, although there was no possibility of servicing them. A hook was belatedly placed on Einstein so that the shuttle could grapple it and possibly raise its altitude to prevent reentry resulting from atmospheric drag. Because the shuttle was delayed and was still being tested when Einstein reentered, this plan could not be executed, but the choice of a low Earth orbit ensured an early reentry. In addition, choosing gas jets for momentum dump (rather than the magnetic torquing system used on Uhuru) meant that the spacecraft would be limited by the available gas reservoir to a one-year life.

Experimenters from HEAO-1 and HEAO-2 asked NASA headquarters to reconsider this policy of treating the missions as short-lived experiments without regard for results. In a letter to John Naugle (then acting associate administrator for space science and applications at NASA) on June 20, 1974, I pointed out that "the HEAO-1 and -2 missions should be considered as more than just individual experiments. They are the only opportunities now existing for significant x-ray observations. They represent a unique national resource in astronomy. It is clear that they would continue to give important results for at least several years." In the same letter I stressed that "X-ray astronomy deserves and needs the commitment of resources that will establish permanent X-ray observatories in space."

Because these requests were not granted, we tried every trick in the book to reduce the expenditure of gas by the gas jets, to reduce atmospheric drag, and to retain pointing capabilities notwithstanding gyro failures. But despite our efforts, Einstein ceased operations after 29 months. The premature demise of Einstein and HEAO-1 in the early 1980s resulted in the absence of major U.S. x-ray observatories for the next 20 years—until the launch of Chandra in 1999. Even allowing for the fact that NASA is a mission agency, space astronomy has become so important a tool in the study of the universe that the agency should be conscious of its responsibility to ensure continuity in its research programs.

Many other cutbacks occurred in the Einstein program during its execution. Hardware was cut, and even the calibration program at the 1,000-ft x-ray facility in Huntsville, Alabama, was reduced from the planned six months to one month. We compensated by working in shifts 23.5 hours per day during that month. In general, any financial problem encountered by the program was solved by reducing scientific capabilities. In spite of these constraints, the Einstein observatory was the most powerful x-ray observatory ever flown until the launch of the Roentgen Satellite (ROSAT; a German program with U.S. and U.K. participation) in 1990. Einstein operated with a sensitivity of 10^{-14} erg/cm^2 · sec, in the energy band of 1–3 keV, a flux that is a million times smaller than that detected from Sco X-1 in 1962.

The Construction of Einstein

In constructing Einstein, we followed a simple design philosophy based on three points: a unified scientific approach, the establishment of spacecraft requirements at the minimum acceptable level, and the idea that failures should be soft rather than hard, so that the science might be reduced but would not be shut off by malfunctions.

The most important aspect of the unified scientific approach was to make all scientists shareholders of a common observatory, meaning that all would share in the data from all instruments. This philosophy greatly facilitated any decisions on priorities or trade-offs that had to be made.

We minimized requirements on the spacecraft by a variety of means. I have already mentioned the relaxed pointing requirements. We eliminated on-board data processing by excluding all instruments that did not detect individual photons, thus eliminating the need for integration either in the detectors or in memory elements. All raw data could be transmitted directly to the ground for reduction and analysis. This design avoided on-board complexity but also proved an inspired choice when we identified (after launch) a problem with the star trackers caused by residual magnetic fields in their magnetic shields. We were able to correct for these effects by calibrating the distortions in orbit and by data analysis on the ground. This work-around would have been impossible had we done on-board data processing.

Soft failure meant that we had to place redundancy where it was clearly needed (for instance, using three rather than just two star trackers) but also required that we use designs and materials that would permit continued operations even if a system failed. Our x-ray optical train was built of thermally matched components, so that even if the active thermal control system failed, the images would be little affected. Soft failure also meant the use of redundant focal plane detectors to prevent catastrophic loss of science stemming from the failure of a single instrument.

While most of the above appears obvious now, it is important to note that many missions have failed because of lack of communication among experimenters, needless complication of hardware components (leading to greater chance of failure), and hard single-point failures caused by inadequate system engineering analysis.

Einstein's Mirror

The most demanding technical part of the Einstein program was certainly the construction of the mirror. It was twice the size of the Skylab mirror, and if it was to be useful for stellar astronomy we could not afford the large-angle scattering wings that were present in that solar instrument's mirror. These wings were due to imperfections of the surface and resulted in a significant loss of efficiency. In the solar telescope, only 10 percent of the incident beam fell within 5 arcsec of the narrow central peak of a point-source image. We needed to improve this performance by a factor of four or five. Such improvement seemed possible by using glass mirrors, which could be polished to a high degree of smoothness.

An experimental program was set up to explore the performance of fused silica with a reflecting surface of evaporated nickel or platinum. It was found that for flats we could reduce the scattered beam to negligible amounts outside of an arcsecond aperture. The choice of fused silica substrate, however, presented significant engineering problems, because this material has poor stiffness characteristics and required a carefully designed support structure.

A more general problem was the integration and testing of the mirror assembly in our gravity environment for use in orbit at zero g. Engineers and scientists at AS&E and at Perkin-Elmer, guided by Leon Van Speybroeck, designed (after many iterations) an invar and graphite epoxy support structure that was quite rigid and stable. The telescope consisted of four confocal concentric mirror pairs, so that each ray would strike first a paraboloid and then a hyperboloid. Focal length was 3.44 m, and the diameters of the mirrors varied from 0.34 to 0.58 m. A German firm, Schott AG in Mainz, provided the fused quartz for the mirror (a material called Zerodur, a special ceramic glass with a thermal expansion coefficient of almost zero). The same type of glass was used for the mirrors of Chandra and for the 8-m mirrors of the European Southern Observatory's Very Large Telescope at Cerro Paranal, Chile.

The fabrication took place at Perkin-Elmer in Danbury, Connecticut, under the leadership of Peter Young in collaboration with Van Speybroeck. The mirrors had to be polished in the "free" state, because any rigid constraint on the mirrors tended to distort their shape beyond tolerance. This free state is, however, difficult to achieve, and the mirrors had to be floated on mercury to permit measurement of their status, an idea that I believe Leon suggested. The most difficult tolerances, those for axial slope and surface roughness, were achieved. The mirrors were coated with chromium and nickel after polishing, assembled in a gravity-free support configuration, and then bonded to the support cell (Figure 9.3).

The Marshall Space Flight Center had designed and built a special facility for Einstein's testing and calibration at Huntsville, Alabama. The facility consisted of a microfocus x-ray source, a 1,000-ft-long vacuum pipe to yield an almost parallel x-ray beam, and a 20-ft-diameter by 40-ft-long vacuum chamber to house the experiment (Figure 9.4). The point response function as measured by the flight high-resolution imager is shown in Figure 9.5. The full-width half-maximum of the focused beam is 3.8 arcsec, and 50 percent of the power is contained within a circle of radius 5.6 arcsec. This efficiency is five times better than that of the Skylab mirrors. The performance of the mirror in flight was indistinguishable from that on the ground.

Instrumentation

In 1973, I accepted an offer to become a professor of astrophysics at Harvard and head of the High Energy Astronomy Division of the Harvard-Smithsonian Center for Astrophysics (CFA). Eight members of my AS&E group came with

FIG. 9.3. *The Einstein mirror being assembled in the clean room at Perkin-Elmer.*

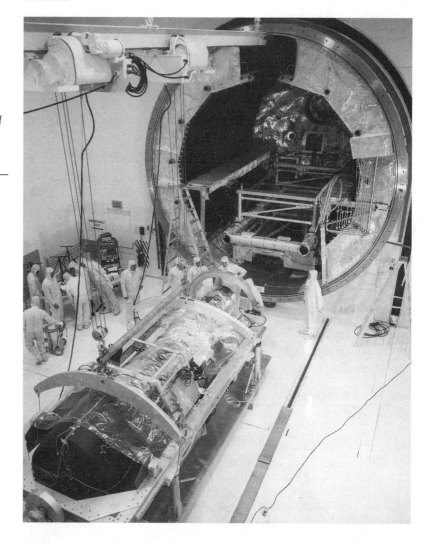

FIG. 9.4. *The integrated mirror and instrument system of Einstein being placed on the optical bench in the vacuum chamber at the end of the 1,000-ft Marshall x-ray facility.*

me: Paul Gorenstein, Herbert Gursky, Edwin Kellogg, Steve Murray, Ethan Schreier, Daniel Schwartz, Harvey Tananbaum, and Leon Van Speybroeck. We retained scientific and technical direction of the Einstein program, and we built and tested in the CFA laboratories the prototypes of the instruments for which we were responsible, while AS&E built the flight hardware. (I discuss my reasons for this transfer in a later chapter.)

Figure 9.6 shows an exploded view of the configuration of the Einstein observatory and the placement of the instruments. The high-resolution

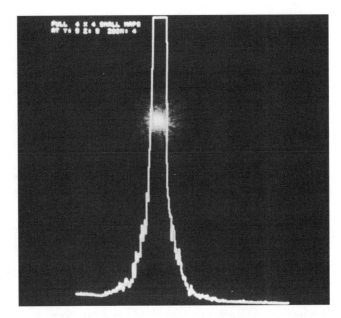

FIG. 9.5. *An image of the point source of x-rays in the facility obtained with the flight HRI.*

FIG. 9.6. *Exploded view of the Einstein observatory showing the placement of instruments.*

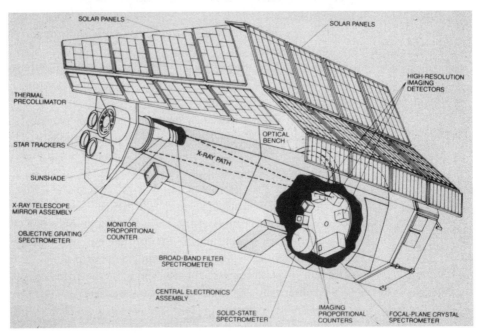

imager (HRI) was developed at CFA under the leadership of Steve Murray. The HRI provided 2-arcsec spatial resolution over the central 5 arcmin of the focal plane. X-ray events are detected individually by measuring their time of arrival (with a precision of 8 μsec) and their arrival position over a field with a diameter of 25 arcmin. Although the image of a point source can be reconstructed to a resolution limited by the telescope itself, the centroid of the distribution can be measured to a precision limited by statistics and knowledge of the aspect to about 0.5 arcsec. The HRI had the highest sensitivity to point sources of any instrument on board during long exposures. Three HRI instruments were included in the Einstein payload for redundancy.

The development of the HRI, which was crucial to the success of Einstein, was quite challenging. After many sophisticated designs were tried unsuccessfully through industrial contractors, we decided to develop the detector ourselves. Steve and a graduate student built a functioning prototype of the detector in the laboratory at CFA, and the design was later given to AS&E technicians, who would build the flight units. Steve started from an approach that had been invented by Ken Pounds's group at Leicester for use in one-dimensional detectors suitable for spectroscopy. The basic idea was to make a detector system in which the ultimate resolution was not limited by the scale of the spacing of the individual electrodes. By allowing the cloud of electrons produced by each x-ray photon to spread, one could find its centroid, which is the position of the incoming photon, with sufficient accuracy to obtain the required resolution in the focal plane of the telescope. This approach simplified construction enormously.

The final detector was extremely simple: it consisted of two micro-channel plate image intensifiers placed in cascade and separated by 38 μm. The input face of the micro-channel plate was coated with magnesium fluoride to yield a photoelectric efficiency to the incoming x-rays of 10 percent. Nickel electrodes were applied to the front and back of each micro-channel plate, and a difference of potential of 300 V was applied between them. As the electrons proceeded down the channels, they struck the walls and thereby created a cascade, so that after crossing the two channeltron multipliers there were 10^7 to 10^8 electrons for each photoelectron emitted in the front.

This beam of electrons was allowed to spread (as shown in Figure 9.7) and was then collected by a cross grid of wires that were kept at a positive potential of 250 V with respect to the back end of the micro-channel plate. The wires of the collector grids were at 90 degrees to one another. They were connected to a resistor chain; every eighth wire was connected to a charge-sensitive preamplifier. The reflecting plate at the back of the grid was used to

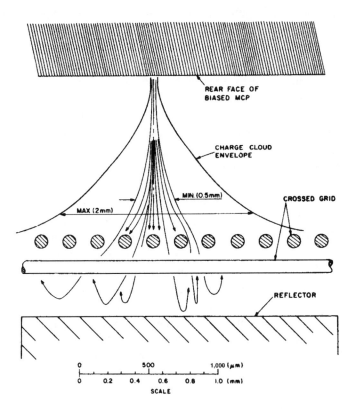

FIG. 9.7. *Schematic diagram of the Einstein HRI. The electron cloud exiting from the micro-channel plate spreads over the crossed grid assembly. The centroid position of the cloud is measured in two dimensions from charge division.*

improve the collection efficiency and to shape the electric field for optimum imaging. The cloud of electrons spread up to 2 mm in the vicinity of the cross grid, striking several wires. Electronic processing of the signals determined the centroid of the distribution in two dimensions by using the ratio of the charge collected in the different wires to a precision corresponding to a small fraction of the wire spacing.

The detector was enclosed in a vacuum housing so that it could be tested on the ground using ultraviolet sources. A door opened to space when the instrument was placed in orbit. A parylene ultraviolet/ion shield protected the detector from particle or ultraviolet contamination. The detector was extremely simple to construct and very rugged, but its development required intensive effort and some help from the Leicester group. In recognition for this help, Ken Pounds and some of his collaborators were invited to CFA to share in the use of the first Einstein data. The simplicity of the design, which still allowed us to achieve high angular resolution, is typical, in my prejudiced view, of the contribution that experimentalists can make in solving sophisticated instrumentation problems.

Measurement of the non-x-ray background in orbit yielded 5×10^{-3} counts/arcmin^2 · sec, a factor of ten above the diffuse x-ray background. This noise was due to imperfections in the channeltron devices, cosmic rays, and residual ultraviolet transmission.

The imaging proportional counter (IPC) was a position-sensitive counter that provided higher efficiency, by a factor of five, than did HRI, full coverage of the focal plane (about 60 arcmin), a moderate angular resolution (1.5 arcmin at 1.5 keV), a time resolution of 63 μsec, and an energy resolution ($E/\Delta E$) of 2 at 3 keV. The IPC had the highest sensitivity for short exposures because of its higher efficiency, but at long exposures its poorer angular resolution made it background limited and less sensitive than the HRI. The IPC was also developed at CFA, under the leadership of Paul Gorenstein.

The IPC consists of a gas-filled proportional counter containing a plane of anode wires that is centered between two orthogonal planes of cathode wires. An electron avalanche at the anode results in localized induced signals in each cathode that propagate at both ends of the plane. The difference in risetime of the signal measures the x-ray position, according to a method first developed by Borkowky and Koop in 1972. The spatial resolution of the detector is 1.0–1.5 mm, and the centroid can be determined to 20 arcsec. Particle background is rejected by a fourth wire plane in anti-coincidence and by the rejection inherent in the technique of time-delay measurement to determine position. The background counting rate in orbit is 5×10^{-4} counts/mm^2 · sec in the energy range of 0.25–3.00 keV, or approximately half that experienced by the HRI. Most of these counts are from the diffuse x-ray background.

The solid state spectrometer (SSS) was developed by Steve Holt's group at Goddard Space Flight Center. It used silicon crystals doped with lithium for the observation of x-rays in the range of 0.8–4.0 keV in a non-imaging mode. The energy resolution of the detector is 160 eV over the entire range, thus yielding an average $E/\Delta E$ of 10. At energies of 1 keV, the effective area of the SSS is similar to that of the IPC and much greater than that of the focal plane crystal spectrometer (FPCS) and the objective grating spectrometer; thus it can give a relatively good spectrum for very weak sources.

The detector consists of a silicon (lithium) chip that is 9 mm in diameter and 3 mm thick. The detector is kept at an extremely low temperature (103 K) by a solid ammonia/methane cryostat. Two detectors are flown for redundancy. By its nature, the SSS has a finite lifetime because of the expenditure of the cryogen, which evaporates after 11 months in orbit.

The FPCS was developed by the MIT group. It utilizes the beam produced in the focal plane by the telescope. An aperture–filter wheel assembly selects a region of interest in the sky. The beam strikes a curved diffractor, and x-rays satisfying the Bragg condition are reflected by the crystal and detected in a position-sensitive proportional counter. Six different crystal diffractors could be used; the most frequently used were those that covered the energy bands 0.7–1.2 and 0.5–0.8 KeV, which include OVII and OVIII (lines of ionized oxygen). The crystal was mounted in a Roland circle configuration. The effective detector × telescope area of 1 cm² was much smaller than the effective areas for the IPC or the SSS (about 100 cm²), but it provided the highest spectral resolution ($E/\Delta E$) on Einstein: 50–100 at energies below 1 keV and 100–1,000 at higher energies. The observations could be carried out with different slits or apertures (3×30, 2×30, and 1×20 mm, and a round aperture 6 inches in diameter), and scans were performed on extended objects.

Two auxiliary instruments were the objective grating spectrometer provided by the University of Utrecht, whose dispersed spectrum of point sources was read out by the HRI, and a monitor proportional counter pointed coaxially with the telescope. The counter had an area of 667 cm² and a field of view of 1.5×1.5 degrees.

Observatory Operations and Data Processing

The planning and operations of the Einstein observatory took place during a time of change in x-ray and space astronomy: the use of individual experiments was giving way to the concept of national facilities. Although the Einstein project was considered by NASA to be a program with a principal investigator (implying by custom that the data are proprietary), the scientists involved in the project recognized that it was potentially a powerful observatory; in fact, it became a widely used astronomical facility with hundreds of guest observers.

Consequently, we were faced with the need to define an operating philosophy, specify the detailed requirements necessary to implement it, and develop the necessary databases, software, hardware, and procedures. The solution was to create, under the direction of Ethan Schreier, a stand-alone, dedicated facility at CFA to carry out the following functions: simulations of experiments, analysis of calibration data, planning of missions and operations, generation of ongoing observing programs, accounting of observations, quick-look monitoring of experiments, development of software, reduction and analysis of data, and archiving and distribution of data.

The computer system we had available was quite modest and was procured entirely with Einstein project funds. I had some difficulty in persuading George Field, the director of CFA, that a space mission required high priority in the execution of all the above-mentioned functions and that we could not carry out these activities as part of central computing at CFA. In the end, we proceeded with a dedicated minicomputer purchased and operated by a scientific staff led by Ethan Schreier.

The first test of the software and hardware development came at the time of Einstein's x-ray calibration. Up to that point, no x-ray testing had been done on the mirrors, and the only test we could afford to do, given the time limitations, was an end-to-end test carried out on the fully integrated system ready for flight. The calibration data were therefore recorded exactly as we would measure them in flight when observing a star. As a result, the entire automated data handling and analysis system had to be ready at the time of calibration. We were able to achieve this goal by relying almost entirely on the scientists themselves for the development of the data analysis software, mainly because of the lack of funding by NASA for mission operation planning and implementation.

The calibration facility was built at the Marshall Space Flight Center by a team of scientists and engineers from the consortium institutions, Marshall, and industrial subcontractors. The challenge of carrying out at this facility in one month the testing that we had planned to do in six was successfully met only because we had sufficient computer support to permit Leon Van Speybroeck to develop an observing plan that took into account the requirements for different wavelengths, many angles of incidence, the three temperatures that had to be maintained for the nine on-board detectors, as well as the restrictions imposed by the test facility. Azimuth or elevation corrections could be done in minutes, temperature changes required hours to a day, and instrument reconfiguration required minutes to hours. Leon was able to combine all of these requirements in a computer program that optimized the testing. A schedule of 1,397 separate tests was developed, which had to be run 24 hours a day for 18 days, leaving 12 days in reserve. We actually achieved 23.5 hours a day in shifts for 30 days. Harvey Tananbaum and I split the time (12 hours each), so that one of us would always be present during testing.

The preparation for the science operations and data analysis systems (which took three and a half years) was essential for the analysis of the large volume of calibration data, the equivalent of more than a month in orbit. It also became the key to providing a mature data system capable of digesting a day's worth of data from Einstein in less than a day. The system was thus able to furnish consortium scientists and guest observers with reduced and calibrated data ready for further analysis on an ongoing basis. This capability was significantly in advance of any other system available at the time in any branch of astronomy. It contributed significantly to our ability to obtain important scientific results immediately after launch.

TEN

Einstein's Launch—Normal Stars and Stellar Systems—
Normal and Active Galaxies—Clusters of Galaxies—
A New Paradigm for Astronomical Data Management: The Einstein Databases

The Einstein Results:
Observation Collides with Theory

Einstein's Launch

The High Energy Astronomy Observatory (HEAO-2) was launched into orbit by an Atlas Centaur rocket on November 13, 1978. Shortly after launch, the consortium scientists (from AS&E, Columbia University, Goddard, and MIT) renamed it "Einstein" in honor of the centennial of the great physicist's birth.

The hours before the launch at Cape Kennedy were filled with anxiety for all of our team at CFA. We had worked on this project for many years, and we were conscious that the satellite carried (in the words of Harvey Tananbaum) "our lives' work under the nose cone." For the first time I had brought my son Marc with me to watch an actual launch; he was then 15 years old and he enjoyed it as much as our visit to Disney World after the launch.

The launch went perfectly. It is difficult to convey the emotion of seeing for the first time on a monitor screen actual x-ray pictures of the stars streaking across the field of the telescope, prior to its settling on Cyg X-1. For me it was the realization of the 1960 dream of celestial x-ray images finally embodied in reality. I knew then that we of the consortium had made an advance in observational astronomy such as had rarely occurred in its history, and that in the future even more powerful x-ray observatories based on the

162

same design would be built. I had no doubt that x-ray astronomy would make unique contributions to the study of the universe.

The Einstein mission fulfilled and exceeded our expectations. The observatory operated successfully until April 1981. At the most basic level, focusing optics enabled us to achieve sensitivities at least five hundred times greater than those obtained with any large-area counter with mechanical collimation, including those flown by NRL and Goddard on HEAO-1. The NRL experiment had an area of 1 m^2, some twelve times larger than Uhuru, but predictably had only three- or fourfold times better sensitivity. Thus their new catalog of sources contained only 842 x-ray sources over the entire sky, a modest increase with respect to the 360 sources found with Uhuru. Einstein, on the other hand, had a sensitivity that was 200 times greater, and we observed sources with a space density of 2 per degree^2, equivalent to 80,000 sources over the entire sky.

The sensitivity of x-ray observations had improved in the 20 years since the first rocket-borne detectors by such a degree that we could detect sources as faint as 10^{-12} of the solar x-ray flux or 10^{-6} of that of Sco X-1. Comparable gains had occurred in optical astronomy, going from the naked eye to Palomar, over a period of centuries.

From a scientific point of view, this development was significant in that it opened up to x-ray observations all known types of celestial objects, not just the brightest or the most exotic. In the course of more than five thousand observations during the mission, Einstein was used to study some of the objects nearest to us, such as comets and planets, as well as the most distant quasars then known in the universe. It changed x-ray astronomy from the study of exceedingly luminous and peculiar stellar systems (such as Sco X-1, Cyg X-1, and the Crab Nebula), of interest mostly to high-energy astrophysicists, to an observational branch of astronomy on a par with optical or radio astronomy, and therefore of general interest. X-ray observations would bring complementary and unique contributions to all disciplines of astronomy.

In the expectation of these results, I considered it vital to provide on-line data calibration and reduction, so that useful early findings would be available to all astronomers while follow-up observations could still be planned and executed during the mission lifetime. The data system was also designed to provide a homogeneous set of data that could be used for archival research after the end of the mission.

A total of 450 guest investigators from 150 different institutions observed 2,000 out of the 5,000 targets studied; the fraction of observing time devoted to them during the mission increased steadily with time—up to 40 percent during the last few month of Einstein's operation—and averaged 25 percent of the total available time.

Approximately a thousand papers based on the Einstein data were published in refereed professional scientific journals in the first 10 years after launch. It is impossible to summarize here the important contributions made to astronomy by this body of work; they were well presented in the symposium held in Cambridge, Massachusetts, on November 13–15, 1988.[1] In the next several sections, I sketch a summary of a few subjects in which I was personally most interested.

Normal Stars and Stellar Systems

The most fundamental discovery in the study of main sequence stars was that nearly all classes of stars were x-ray sources. Conventional theory had attempted to explain the heating of the solar corona as stemming from acoustic waves generated in the convection zones of the star. The sun was thought to be a particularly strong x-ray emitter, and the predictions for other stars depended on an understanding of their convection-zone formation. This theory had already been shown to fail when applied to the sun to explain the Skylab observations by Pippo Vaiana and his colleagues in 1973. They had shown a tight spatial correlation between closed coronal magnetic structures and the x-ray emissions above sunspots and active regions. The theory completely failed to explain the copious x-ray emissions from O- and early B-type stars (stars that are more massive and much hotter than the sun), which do not have significant convection zones and therefore could not produce acoustic waves.

In a more general way, the contemporary theories of stellar formation and evolution from the initial nebula had never taken into account the need to dissipate both angular momentum and magnetic field during the collapse of the protostar. These processes could not be accounted for by any spherically isotropic model and would presumably result in anisotropies in the inflow of condensing material that, in turn, could give rise to high magnetic fields and shocks. High-temperature plasmas and x-ray emissions could result from these conditions. Therefore, we could not predict from current theory what would be observed by Einstein.

Vaiana was well aware that prior to Einstein, stellar x-ray astronomy was limited to the detection of a few rather special types of stars: luminous compact binaries, dwarf novas, cataclysmic variables, and binary systems containing main sequence stars (RS CVns). He was confident, however, that the five-hundred-fold increase in sensitivity provided by the imaging telescope would make it possible to detect many more stars. Yet he was almost alone in our group in being deeply interested in this subject, and we did not have sufficient funds for data analysis to mount a serious effort. But Pippo was determined to find a way. He had left AS&E to join the solar physics group of CFA in 1974 and became director of the Palermo Observatory in Italy in 1976. There he found an institution with no research funds and only one aged astronomer. In a short time, he was able to recruit a number of young Italian astronomers who dedicated themselves to the in-depth study of the data generated by Einstein. The Palermo Astronomical Observatory G. S. Vaiana (as it was named after his death in 1991) is today a well-known institute that specializes in x-ray astronomy with emphasis on the study of the sun and stars.

From the Einstein group's point of view, Vaiana and his colleagues (the "Foreign Legion," as they were jokingly named) brought unexpected but very welcome help. They were given access to and support during observing time and had access to all the proprietary data of our group. They were soon joined in their work by observers and theoreticians from many other groups both within and outside the consortium. Many discoveries rewarded their efforts. Jeffrey Linsky and Vaiana gave useful summaries of the field in their contributions to the 1988 symposium.[2]

With Einstein, we detected x-ray emission from all classes of late-type stars (F, G, K, and M) with luminosities ranging from 10^{26} to 10^{31} erg/sec. The existence of the hot coronal plasmas arises from the heating of the corona by magnetic waves generated by convection within the star and from energy release by magnetic fields. These processes are amplified by the dynamo process. Magnetic fields also play a role in the confinement of the plasmas that are generated.

X-ray emission was also discovered from younger stars (pre–main sequence or very young stellar objects) with luminosities ranging from 5×10^{29} to 3×10^{31} erg/sec. Among the youngest stars we detected were T Tauri and "naked" T Tauri stars with x-ray luminosities of one thousand to ten thousand times that of the sun. The "naked" T Tauri stars were the first new class of stars identified on the basis of Einstein's observations. It is thought that these objects are highly luminous in x-rays because they are still in a state of collapse, they are rotating rapidly, and they may be highly convective or turbulent throughout.

One of the most unexpected and startling discoveries was x-ray emissions from early-type stars (B and O) at levels as high as 10^{33} erg/sec, a million times greater than that of our sun! These stars were believed to be too hot to have convective zones, and therefore, the mechanism invoked to explain coronal heating in late-type stars could not work. A complete understanding of the physical processes at work would have to await more powerful observatories, such as Chandra and XMM.

Even from this brief summary it is clear that imaging x-ray telescopes gave us a powerful new tool to study the formation and evolution of stars by imposing hard new constraints on any theory of stellar evolution, constraints that until Einstein were not even imagined. Einstein also provided an abundance of data on the later stages of stellar evolution. High-resolution images of the historical supernovas, including Cas A, the Crab Nebula, Kepler, Puppis A, and Tycho, enabled us to discriminate between the contribution of the pulsars and that of the remnant itself (Plates 5 and 6). F. Seward of CFA listed forty-two other supernova remnants observed with Einstein, some previously known from radio observations and some newly discovered.[3]

Most important for the understanding of the composition and temperature of the shell of material blown off by a supernova were the spectroscopic measurements obtained with the Goddard solid state spectrometer (SSS) and the MIT focal plane crystal spectrometer (FPCS). Figure 10.1 shows the SSS spectrum of Tycho's supernova remnant; somewhat surprising was the relatively low abundance of iron, which was expected to be produced abundantly in a type I supernova explosion. For instance, the Fe XVII ionization line at 895 eV is not observed at all, even though it should be observable within the SSS energy range; the resolution of this and other questions had to await the launch of Chandra.

Figure 10.2 shows the FPCS spectrum of Puppis A. This high-resolution spectrum has been used to identify iron, oxygen, and neon emission lines and to deduce the electron temperature of the plasma, the ionization state of the elements, and their relative abundances. In Puppis A, oxygen is several times more abundant than in the sun, a feature that appears to be present in several other supernovas.

Einstein's observations helped clarify the highly complex phenomenology of x-ray binaries. Starting from the Uhuru observations of the early 1970s, a more complex picture had emerged from the data of the ANS (Astronomische Nederlandse Satelliet; launched in August 1974), those of the British satellite Ariel V (launched in October 1974), and finally those of SAS-3 (launched in May 1975).

The simple view of the brightest x-ray binaries containing a neutron star or a black hole had evolved to encompass systems containing a white dwarf as

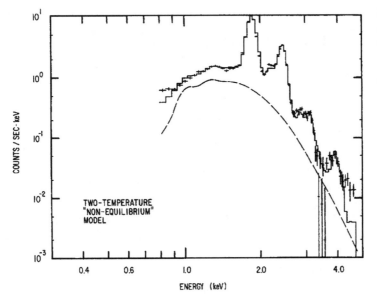

FIG. 10.1. *Pulse-height spectrum of the Tycho supernova obtained with the SSS.*

FIG. 10.2. *Dispersed spectrum of the Puppis A supernova obtained with the FPCS.*

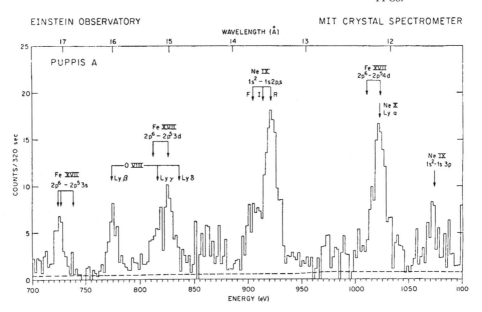

the accreting object, and also systems in which the "normal" star providing the accreting material could be a white dwarf. The different types of binary systems that could result led to a wide variety of behaviors, including the x-ray bursts discovered by Josh Grindlay and John Heise with ANS and the rapid burster discovered by Walter Lewin and his MIT colleagues with SAS-3.

Einstein contributed to the studies of these systems by discovering a new class of low-luminosity accreting x-ray sources some thousand times less luminous than the low-mass x-ray binaries (LMXRBs), which are believed to be cataclysmic variable stars. Einstein's high-resolution images led to the optical identification of a number of LMXRBs, the globular cluster source in M 15, and the x-ray burster 4U1415-05. Accurate x-ray positions were also used to measure statistically the masses of x-ray sources in globular clusters. Timing measurements of 4U1415-05 led to the discovery of occultation dips in that source (the dimming of its luminosity as the source is occulted by its companion), confirming its binary nature. Other x-ray bursters were also discovered by timing analysis. Einstein's results amply demonstrated that the in-depth study of binary x-ray sources required the high angular and spectral resolution provided by imaging optics to complement the timing measurements that were being carried out by other satellites.

Normal and Active Galaxies

Uhuru was able to detect x-rays from a few clusters, one quasar, and one Seyfert galaxy (that is, from only the brightest objects in each class), but the low-luminosity galaxies were beyond its reach. Many of the unidentified objects at high galactic latitude seen by Uhuru (which I had named "x-ray galaxies") would later turn out in fact to be galaxies. Einstein opened up extragalactic astronomy, just as it did stellar astronomy.

Einstein could observe stellar x-ray sources in external galaxies. The Large Magellanic Cloud, for instance, was studied by the Columbia University group with enough sensitivity to detect sources with luminosities of 3×10^{35} erg/sec over the entire galaxy. Classical x-ray binaries and known supernovas could be detected; new supernova remnants were discovered, as well as diffuse emission from the interstellar gas. In M 31 (the galaxy in Andromeda), Leon Van Speybroeck of our group discovered 120 individual sources (Figure 10.3).

Some two hundred normal galaxies were observed with Einstein, some of them in enough detail to study their morphology, individual source distribution, and spectra. These characteristics could then be compared to their

FIG. 10.3. *The center of the galaxy M 31 obtained with the HRI, showing the x-ray emission from individual high-luminosity stellar sources.*

optical, infrared, and radio properties. The major finding was that normal galaxies of all types are sources of x-ray emissions. They are spatially extended sources with luminosities between 10^{38} and 10^{42} erg/sec. The result gives important clues to the evolution of x-ray sources in galaxies and their relation to star formation, cosmic-ray production, and the formation and heating of the interstellar medium. With such data, researchers could begin to study the difference in evolutionary scenarios between our own and other normal galaxies.

Before the launch of Einstein, only three quasars had been detected, along with a few active radio galaxies. Within a year of the launch of Einstein, Tananbaum had detected thirty-five of the optically known quasars, and I had detected five previously unknown quasars through their x-ray emissions at cosmological redshifts as large as 2 (or 10 billion light-years away).[4] Schreier had carried out a detailed study of the x-ray structure of Cen A and found an x-ray jet joining the nucleus and the northeast inner radio lobe.[5]

Over the entire Einstein mission, the increased sensitivity of the observatory led to the detection in x-ray wavelengths of hundreds of previously known active galactic nuclei (AGNs) and quasars, and to the discovery of hundreds of hitherto unknown ones through their x-ray emissions. The deepest x-ray surveys by Murray, Primini, and me reached a sensitivity of 1.3×10^{-14} erg/cm^2 · sec and resolved approximately 25–35 percent of the x-ray background (XRB) in individual sources. The predominant contributors to the background were found to be quasars. The deep surveys extended the log N–log S relationship for extragalactic x-ray sources (the number density of sources as a function of the minimum detectable flux) by three orders of magnitude. A remarkable and unexpected result was that the Euclidean slope of the relationship, which we had discovered with Uhuru, continued at fluxes one thousand times smaller than those previously observed. Because the deep surveys probed distant regions of the universe, where Euclidean approximations were no longer valid, this result implied a positive evolution in the number of quasars, different from what had been observed at other wavelengths.

The intermediate sensitivity surveys carried out by Tommaso Maccacaro and Isabella Gioia filled in the gap between 10^{-11} and 5×10^{-14} erg/cm^2 · sec. They were able to separately plot log N–log S for AGNs and clusters (Figure 10.4), which plot on distinct paths because of the differences in their evolutions. They also examined the differences in the Hubble diagram of AGNs

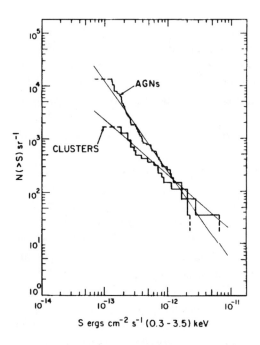

FIG. 10.4. *The number-luminosity relation for all extragalactic sources and for clusters of galaxies and AGNs separately, showing the different evolution of the two classes of objects.*

selected by ultraviolet and x-ray observations and studied the luminosity function and evolution of a particular type of quasar, namely, the BL Lac objects.

Apart from their intrinsic interest, these survey results became the basis for a vigorous debate on the nature of the XRB that continued for the next 10 years (until the launch of ROSAT and finally Chandra). The controversy was generated by an observation made by Goddard's A2 experiment on HEAO-1, which appeared to indicate that the spectrum of the XRB was indistinguishable from the thermal bremsstrahlung spectrum of a 40-keV gas.

In addition, the average spectrum of the AGNs, which had been measured with the same instrument, yielded a value of approximately 0.7 for the power law index. Other measurements at lower energies indicated an even steeper slope of 1.2. These values could not be reconciled with the observed spectrum of the XRB, which fit a power law of slope 0.4 up to 10–20 keV. For this reason, hundreds of papers were written to discuss the possibility of the origin of the background from diffuse, hot intergalactic gas. If diffuse, the gas would be the predominant component of barionic matter in the universe and would provide enough mass for closure of the universe, according to cosmological theory. Many arguments against this hypothesis were raised on the basis of the large amount of energy required to heat the gas. George Field tried to validate the hypothesis by constructing a model that posited heating by distant quasars (that is, those at large redshifts of $z = 3$) and clumped the diffuse gas (to decrease its heating requirements), but the model was unsuccessful.

Studies by David Helfand and others showed that the fluctuations in the XRB require a large number of AGNs (about 1 per arcmin^2) to explain its smoothness. This high density seemed to contradict the extrapolation of the number counts of known optically selected AGNs and their evolution. Nevertheless, this density of AGNs and quasars is precisely what was ultimately found in the Chandra deep field surveys.

In my opinion, too much was made of a fortuitous coincidence (valid only for the restricted band of energies from 3 to 40 keV) between the XRB spectrum and that of bremsstrahlung. Even more was made of the fact that Einstein and HEAO-1 measured different ranges of energies, implying that a different population of sources would be observed in the two regimes and forgetting that natura non facit saltum (nature makes no leap). Indeed, the two sets of observations could not be directly compared without taking into account the observational bias introduced by the much lower sensitivity of HEAO-1. Einstein could therefore detect sources undetectable from HEAO-1 and whose spectrum could not be measured by the Goddard group.

Although there was no direct evidence for a truly diffuse component of the XRB, there was clear evidence for a point-source contribution, which ranged between 50 and 70 percent of the total background. There is nothing more difficult for people to accept than what they do not wish to believe. Steven Holt of the Goddard group wrote: "I must confess to some personal satisfaction in the probable demise of quasars as the primary contribution to the XRB."[6]

At a certain point, I considered the many theoretical papers about the XRB as exercises in futility, of interest only to graduate students as possible subjects for PhD theses. The typical article would point out that none of the suggested theories could satisfy all the results unless new classes of objects were discovered or particular forms of evolution of known ones were to occur. Most authors then advocated the launch of observatories with the capabilities of Chandra or XMM to resolve the problem. Because I had firmly advocated Chandra from the start, I could not seriously disagree with this last statement, although I was not really surprised when the Chandra deep surveys resolved at least 90 percent of the entire XRB into point sources, completely confirming and extending the Einstein and ROSAT findings.

Clusters of Galaxies

As already discussed, Uhuru's all-sky survey discovered that rich clusters of galaxies were powerful, extended x-ray sources.[7] The luminosity was much greater than could be expected from the summed contribution of individual normal galaxies, and our team at AS&E concluded that the emission was due to bremsstrahlung in the hot (10^8 K) plasma pervading the cluster.[8] The total mass of this gas exceeded the mass contained in the visible galaxies, and although it was not enough to provide the virial mass necessary to keep the cluster gravitationally bound, it revealed a new and previously undetected component of the universe. Subsequent spectroscopic x-ray measurements detected iron emission lines from the hot gas, confirming the early interpretation of thermal bremsstrahlung. The detection of iron lines also showed that the gas had been processed and enriched through stellar nuclear reactions.[9]

Einstein's increased sensitivity and angular resolution extended the study of clusters to much greater distances (redshift $z = 0.5$) than previously possible and permitted the detailed study of the morphology of clusters.[10] Thus, the properties of x-ray clusters could be studied using a much larger sample of objects (hundreds rather than the few discovered with Uhuru), extending in luminosity from 10^{42} to more than 10^{44} erg/sec. Morphological studies revealed that x-ray emission from the interstellar gas is a powerful tool to study the substructure of the cluster gravitational potential. Figure 10.5 shows the isointensity x-ray contours detected from four clusters superimposed on an optical photograph of the field. In the optical domain, they had been identified as single clusters, because the optical galaxies appeared to be at the same redshift (distance) and the number statistics were too poor to reveal clumping. In x-rays, it was obvious that there are two different masses of gas that are collapsing to form a single gravitationally bound system. Because these

FIG. 10.5. *Contours of constant x-ray intensities for some binary clusters superimposed on their optical images. These clusters appear to be forming in the present epoch.*

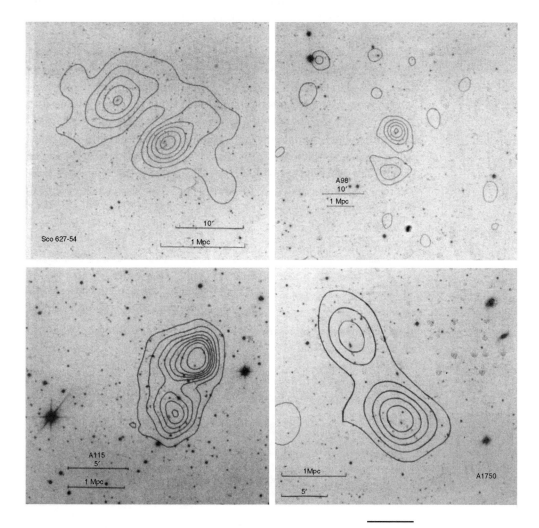

binary systems are relatively close to Earth, our team at CFA came to the conclusion that clusters are still in the process of forming.

Morphological studies of individual clusters can give information about their dynamical evolution and lead to two-dimensional classification schemes, as proposed by William Forman and Christine Jones of CFA.[11] They classified the four clusters of Table 10.1 according to whether they contain a massive central object and by their different stages of gravitational collapse (indicated by the smoothness of the x-ray contours). Interestingly, the presence or absence of a central massive object (such as M 87 in the Virgo Cluster) is independent of the state of dynamical evolution of the cluster. Thus, the formation of these objects, normally identified as supermassive black holes, occurs independently of cluster formation and evolution.

It took me a long time to grasp intuitively the nature of clusters. The concepts were clarified for me by the treatment of gravitational collapse first introduced by Linden Bell to explain the formation of globular clusters. Figure 10.6 shows a schematic cross section of the gravitational potential across a cluster. The small dimples are the gravitational wells of the individual galaxies. The deeper the dimple, the higher the escape velocity required for material to escape the gravitational potential. The gas produced by stellar evolution in each galaxy can be expelled through supernova explosions or retained in the potential well of the star's galaxy, provided the kinetic temperature of

TABLE 10.1. *Two-dimensional cluster classification*

Characteristics	No x-ray-dominant galaxy (large core radii)	X-ray-dominant galaxy (small core radii)
Low x-ray luminosity ($<10^{44}$ erg/sec) High spiral fraction (>40%) Low central density Cool gas (a few keV) Irregular gas and galaxy distribution	A 1367	A 262
High x-ray luminosity ($>10^{44}$ erg/sec) Low spiral fraction (<40%) High central density Hot gas (6 keV) Regular gas and galaxy distribution	A 2256	A 85

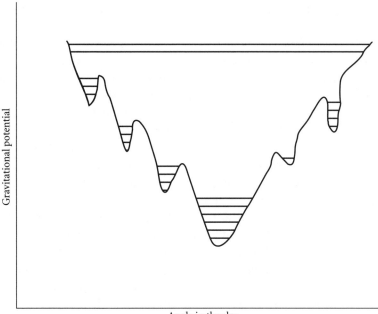

FIG. 10.6. *Schematic diagram of the gravitational potential of individual galaxies and of the cluster as a whole. Each dimple represents the potential well of an individual galaxy in the cluster. The horizontal lines in each dimple represent the energy levels of gas particles trapped in the gravitational potentials of the individual galaxies. The horizontal lines at the top of the entire complex show the energy levels of gas particles trapped in the gravity of the cluster as a whole. The cluster can trap higher-temperature plasmas than can the individual galaxies.*

the gas results in particle speeds that are less than the galaxy's escape velocity. If expelled from the source galaxy into the cluster, the gas will be bound to the cluster even at much higher gas temperatures (those corresponding to particle speeds less than the escape velocity from the cluster as a whole). As the cluster shrinks (because of the mutual attraction of its member galaxies), the change in potential raises the energy of each nucleon in the cluster and therefore raises the temperature of the plasma that is still contained in the potential well of the entire cluster.

For optical surveys, the detection of a cluster depends mainly on finding a higher density of galaxies in a particular spot in the sky. The total optical luminosity is equal to the sum of the luminosity of the galaxies, and nothing in the optical data provides dynamical evidence that the system is gravitationally bound. In contrast, the overall x-ray luminosity of a cluster is evidence of the gravitational interactions of its member galaxies.[12] A dynamically evolved cluster of galaxies results in extremely hot temperatures of its shared intergalactic gas, and therefore the x-ray luminosity for the cluster is much greater than the sum of the luminosities of the member galaxies. It is as if the cluster waved an x-ray flag to call attention to the fact that it is a bound system. Once I understood these properties, I became enthusiastic about the study in x-rays of the formation and evolution of clusters at very early times in the life of the universe.

The researcher's selection of clusters for study using x-ray images can lead to a different sample than selection using optical observations. The latter tend to result in the selection of centrally condensed, evolved clusters rather than irregular or fragmented clusters that are still in the early stages of formation. Alfonso Cavaliere of the University of Rome, with whom I had often discussed the subject, was enthusiastic about computer modeling of cluster formation, which would permit comparisons of predictions and observational results. For instance, one can use x-ray images to map the structure and mass distribution in clusters. Because simulations provide time-dependent statistical distributions of these quantities during the evolution of a cluster, the comparison of models with observational data yields estimates of an observed cluster's age.

By studying the density and temperature distribution of the gas, one can determine the mass of the central galaxy and the interaction of the hot gas and the radio-emitting plasmas. One can observe the Sunayev-Zeldovich effect (that is, the inverse Compton scattering of the 3-K radiation off the hot plasma electrons). By this scattering process, the photons of the microwave background acquire energy, leaving hotter patches on the otherwise rather uniform background. In addition, comparison of radio and x-ray data yields a value of Hubble's constant (an indicator of the recession velocity as a function of distance, from which one can derive the age of the universe) that is independent of its estimation using optical observations.

Clusters of galaxies are the largest known gravitationally bound aggregates of matter in the universe, and given their high luminosity, they can be used to trace the large-scale distribution of matter in the universe. The evolutionary properties of clusters are tied to the chemical and dynamic evolution of their stars and the mass-loss history of their galaxies, a property that enables researchers to study the history of star formation and compare it with that observed at other wavelengths.

Interest in these studies has remained with me, somewhat unfulfilled, to this day. In part the slow pace of discovery in this field came about because theorists (including Jeremiah Ostriker of Princeton University, who should have known better) had struck again and convinced themselves—and more importantly, NASA selection committees—that there were no clusters to be found at great distances because (according to theory) they had not yet formed. Therefore, NASA saw no point in looking for them. Once again the theorists were wrong.

A New Paradigm for Astronomical Data Management:
The Einstein Databases

In the decade after the Einstein launch, the computer resources of the CFA group improved by an order of magnitude. This leap in capability inaugurated a new approach in the presentation and distribution of the data: they were organized into a series of databases available to the entire community in hard copy and on line, and also via CD-ROM and magnetic tapes.

The largest of the databases is the Einstein IPC source catalog, which includes all IPC sources obtained from standard calibration and processing algorithms as well as x-ray contour maps of all the fields. An example of a map from the Harris et al. compilation is shown in Figure 10.7. The hard copy of the catalog contains four thousand pages of IPC observations, one per page. In addition, written descriptions are given of the method used for catalog construction, the parameters used, the procedures to convert from counts/sec to flux, and maps for 150 fields obtained by merging several observations. Auxiliary information is also given, such as exposure time, number of detections, and background rate.

An Einstein observatory stellar objects catalog was derived by Pippo Vaiana and his colleagues from the IPC catalog in the decade after launch (1979–89). The 4,000 observational fields of the IPC database contained 16,000 distinct x-ray sources; 5,000 of these have been identified with radio and optical quasars, galaxies, and stars. Of these identifications, 2,500 are with stars. These matches were obtained by running specialized software for selected Einstein fields to refine detection statistics and location errors, search for identification with objects in the IPC master catalog, provide upper limits if no coincidences were found, and archive the findings. These results were then ported to a distributed database system implemented on Digital Equipment Corporation's VAX minicomputers and Sun workstations. The final result of this huge effort by Vaiana and his collaborators is a real contribution to stellar astronomy.[13] It yields information on the positions, fluxes, and spectra of the x-ray sources detected by Einstein; it identifies optical counterparts in the Yale Bright Star Catalog, the Wolley Catalog, and the Garmany Catalog; and it supplies upper limits on the x-ray emissions of cataloged stellar objects.

The HRI source detection catalog is similar to the IPC catalog and contains several thousand sources. Images of the full HRI field are also available. Special databases for Einstein observations of supernova remnants were pre-

FIG. 10.7. *A sample page of the data provided for each field surveyed with the Einstein IPC database.*

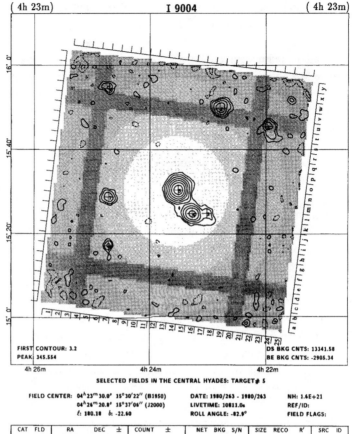

(4h 23m) **I 9004** (4h 23m)

FIRST CONTOUR: 3.2
PEAK: 345.554

DS BKG CNTS: 13341.58
BE BKG CNTS: -2905.34

4h 26m   4h 24m   4h 22m

SELECTED FIELDS IN THE CENTRAL HYADES: TARGET# 5

FIELD CENTER:	$04^h23^m30.0^s$ $15°30'22''$ (B1950)	DATE: 1980/263 - 1980/263	NH: 1.6E+21
	$04^h26^m20.8^s$ $15°37'06''$ (J2000)	LIVETIME: 10811.0s	REF/ID:
	ℓ: 180.18 b: -22.60	ROLL ANGLE: -82.9°	FIELD FLAGS:

CAT #	FLD #	RA (1950)	DEC (1950)	± ''	COUNT RATE	±	NET CTS	BKG CTS	S/N	SIZE COR	RECO	R'	SRC FLG	ID
1022	1	04 21 58.7	15 45 32	51	*0.0094	0.0017	39.6	11.4	5.6	1.8	1409	26.6	AH	*
1027	2	04 22 47.2	15 49 46	48	*0.0256	0.0023	129.5	11.5	10.9	1.3	906	21.9	AH	S
1030	3	04 23 00.5	15 24 57	41	0.0071	0.0012	51.9	22.1	6.0	51.5	0	8.9		
1032	4	04 23 15.7	15 24 51	38	0.0397	0.0024	300.6	25.4	16.6	9.0	0	6.5	AH	S
1034	5	04 23 30.1	15 30 24	31	0.2101	0.0051	1690.8	23.2	40.8	1.8	0	0.2	AH	S
1045	6	04 24 42.1	15 54 26	55	*0.0063	0.0016	24.2	13.8	3.9	1.0	1106	29.8		
1046	7	04 24 42.6	15 28 42	48	0.0138	0.0017	83.7	18.3	8.3	1.2	0	17.5	H	S
1047	8	04 24 42.8	15 17 11	51	0.0059	0.0013	31.4	17.6	4.5	0.9	0	22.1		

pared by Seward. Fabbiano prepared a database for normal galaxies (a sample page is shown in Figure 10.8).

An Einstein database for optically and radio-selected quasars was compiled by Tananbaum and collaborators. An all-sky slew survey archive was constructed by Martin Elvis. The Einstein extended-medium sensitivity database of Gioia et al. contains 835 serendipitous high-latitude objects. Francis A. Primini prepared a deep survey database with 178 IPC and 202 HRI sources.

M ap

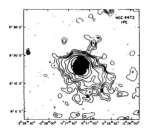

FIG. 10.8. *A sample page of the data provided on each object listed in the Einstein x-ray normal galaxy catalog.*

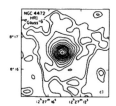

Galaxy position
$\alpha = 12^h\,27^m\,14^s$ $\delta = 8°\,16^m\,42^s$
Einstein Observation
IPC : I4308 (July 1979)
HRI : H7068 (Dec 1979 and Jan 1981)

Radial Profile

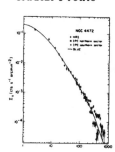

Isodensity x-ray contours for NGC 4472. The lowest contour level is a 2 σ level over the background. Values for 3, 4 ... σ are then plotted.

X-ray and optical surface brightness profile for NGC 4472. The scale for y-axis refers to the HRI data. For the IPC points, the value on the y-axis should be multiplied by a factor of 6.

Spectral Analysis

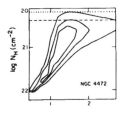

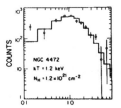

Parameters

IPC gain = 13.9
PHA channel = 2 – 11
Energy range = (0.2 – 5.4 keV)
KT = 1.2 keV (0.5 – 2 keV)
$N_H = 1.2 \times 10^{21}$ cm^{-2} (2.6 – 12 x 10^{21})

Flux

$F_x = 1.2 \times 10^{-11} \pm 2.6 \times 10^{-13}$ erg sec^{-1} cm^{-2}
$L_x = 5.6 \times 10^{41} \pm 1.2 \times 10^{40}$ erg sec^{-1}

Within radius of 780″ and energy range of 0.2 to 4 keV and at distance of 20 Mpc, and assuming thermal spectrum with kT=1 keV.

For detailed data analysis and discussion, see Trinchieri, Fabbiano and Canizares (*Ap. J.* 1986, **310**, 637)

SECRETS OF THE HOARY DEEP

All these astronomical resources were and are open to astronomers from the entire world. They were initially made available in 1978 through the Einstein CFA group, and later through the High Energy Astronomy Archive of NASA, which wished to exercise control over the distribution of x-ray data (1992).

The considerable effort involved clearly demonstrates the commitment of the AS&E and CFA teams as a group to make x-ray astronomy data available to all astronomers in a usable form. The large number of refereed papers published in professional journals in the first 10 years after Einstein's launch could not have been written without this support.

To my knowledge, this was the first organized effort of its kind; this approach to analysis and distribution of the results has since been adopted as the standard for national observatories in space and on the ground. Ethan Schreier, Rodger Doxsey, and I first transferred the methodology from Einstein to Hubble for use at STScI, the European Coordinating Facility (ECF) at the European Southern Observatory (ESO), and the Dominion Astrophysical Observatory in Canada. Harvey Tananbaum and his group at CFA continued the Einstein tradition in the execution of the Chandra data system. Piero Benvenuti (director of ECF), Peter Quinn from STScI, and I introduced the new methodology to the ESO Very Large Telescope Program. And the Atacama Large Millimeter and Submillimeter Array being built in Chile as a joint European, North American, and Japanese project has adopted the same principles since its start. In my opinion, this change in the handling of the massive datasets generated by modern observatories has had a profound influence on the way astronomy is done. It enables massive cooperative efforts among researchers and ensures widespread dissemination and organized archiving of the precious information gleaned from modern observational instruments.

Transitions: From American Science and Engineering to Harvard

Growth and Divergence at AS&E

When people are young, they are quite happy to be given the opportunity to work in an interesting field where they can learn new and important subjects. Provided the organization in which they happen to work can sustain their personal and professional growth, there are few reasons for discontent. In my own case, I found working at AS&E wonderful for more than 10 years. After feeling (in my fellowship years) underutilized, or unwanted, I finally had a job that was worthwhile, for which I was paid a reasonable salary—so that I could support my family—and in which I could learn a profession. With the help of the company, my visa status was changed so that I no longer had to leave the United States, which I had come to love.

The company's classified work was not a burden to me; on the contrary, the climate of the early 1960s made it seem a patriotic duty. In the revisionist atmosphere of the era that followed John F. Kennedy's tragic death, one often forgets his firm and courageous stand against the USSR's threat, particularly during the Cuban missile crisis in October 1962. His decision to counter the USSR's atmospheric nuclear tests, carried out the previous year, with high-altitude tests in the summer of 1962 was the prelude to the test ban treaty that followed.

The work at AS&E to develop better educational approaches to the teaching of science was also extremely attractive. There was hope in the early 1960s that one could make science education more attractive for our children by following the empirical hands-on approach introduced by Jerrold Zacharaias and others at MIT, and AS&E was involved in designing and producing inexpensive learning kits for elementary school children. In addition, the Boston Children's Museum was trying to apply some of the educational approaches advocated by Jean Piaget, such as role playing, to a variety of subjects, and it had started programs in which we at AS&E were collaborating.

The development of medical instrumentation appeared to be an exciting area, with good prospects for growth and clear societal benefit. Herbert Gursky and I had actually come up with the idea of using "good geometry" techniques in x-ray radiography. This approach is the basis for the scanning devices produced by AS&E for airport security and border protection, which even today constitute the main business of the company. Even more significant has been the further application of the technique to medical tomography (computerized axial tomographic, or CAT, scanning), which I wish we had developed.

The field of research I had been asked to concentrate on, space sciences, was opening the window to an unexplored universe, which promised an almost limitless wealth of discoveries. I have mentioned in Chapters 4 and 9 the rapid development of the Space Research and Systems Division at AS&E into a vertically integrated organization, capable of executing technically challenging and ambitious programs.

During these years of challenging and frenetic work, I too had changed. I was no longer the starry-eyed young man, grateful for the opportunity to do any work at all and content to follow what seemed an almost predetermined path. Martin Annis, the president of the company, had been increasingly relying on my managerial support with respect to the most general corporate issues. In 1968 he had insisted that I take management and sensitivity training, then much in vogue, and as early as 1969 I was given increasing responsibility for the management of AS&E. I negotiated all aspects of the contracts for my division, which had come to provide some 90 percent of the company's income. This responsibility included the development, together with the company's Accounting Office, of corporate targets for general and administrative expenses, divisional overheads, material handling, budgets, and fees. These budgets included marketing expenses plus expenses for independent research. I had to defend our estimates to the government auditors on a yearly basis. It was necessary for me to learn about such administrative issues as compensation, benefits, and insurance provisions for the staff, and to contribute to the formulation of policies regarding hiring, yearly evaluations, rewards, and terminations. I also became deeply involved in the planning and construction of a new office building in Cambridge, Massachusetts, including technical plans and specifications and financing. In 1969 I was named executive vice-president of AS&E and given responsibility for both the Education and Instrumentation divisions in addition to my own Space Research and Systems Division, which was split into a Research Division and a Systems Division. Gil Davidson became vice-president for the Instrumentation Division, Herbert Gursky for Research, and I retained the Systems Division. In the same year I became a member of the board of directors, and I was involved in our first public stock offering and in discussions about potential mergers and acquisitions.

Even though I had become quite good at delegating responsibility, I was so busy with all this managerial work and with the preparation of the LOXT proposal that I almost gave up going to the Uhuru launch, although that little satellite was my creature and the apple of my eye. I am very glad now that I did go. I only regret that I spent but a few days in Africa and that I did not take my family with me.

The heaviest responsibility for me was the need to ensure government funding on a more or less continuous basis if I did not want to lose the precious team I had put together. This task was particularly onerous because my division brought in such a large part of the AS&E work that there was little opportunity to reassign staff across the company. The need to provide research funds was not unique to AS&E, as most U.S. research institutions have had to fight for a research budget every year. Most academic and governmental institutions, however, have some built-in security, which allows them to ride the ups and downs of the congressional appropriation process. Research departments in large corporations, which normally account for only a small fraction of the corporate business, are also somewhat protected; a splendid example used to be Bell Labs.

In the late 1960s AS&E was still a reasonably fragile company that had not yet developed any substantial line of products, nor had it been particularly successful in competing for pure engineering contracts not connected to our own research. This uncompetitiveness was partly due to the relatively high overhead costs that resulted from the bulk of the computational support and data analysis being funded by the company rather than by existing contracts, and from independent research efforts. Because none of us at AS&E was willing to abandon this high level of research, continued funding of research was a survival requirement for the company as a whole.

Work prospects for the 1970s seemed to be fairly good, mainly because of the Skylab, HEAO-1, and Einstein projects, but what would happen after completion of these projects was not clear. In the long run, the only areas where there appeared to be substantial prospects for growth and stability had to do with the development of instruments based on x-ray technology for defense, security, and medical applications.

As for me, it seemed that I would become a full-time manager and drift away from science. But concurrently with the progress of my career as a manager, there came the incredible scientific rewards of the year 1970, with the launch of Uhuru and the approval of the LOXT project under my scientific leadership. In 1971, I was elected a member of the National Academy of Sciences, an honor that moved me deeply. The net effect was to give me some personal reassurance of my ability to contribute to science and to strengthen my resolve to strive for excellence in all that I would undertake in the future.

The importance and beauty of the Uhuru data captivated me. I found that I was able to provide leadership to a formidable group of young scientists and to spur them to achieve their highest potential in research. The atmosphere in which we worked, the so-called Uhuru spirit (described in Chapter 7), was much more a communion of free spirits than a traditional corporate environment.

I had not personally carried out classified research since the early 1960s, but AS&E was still involved in it under the leadership of Jack Carpenter, one of the earliest and most senior of the AS&E scientists. With the continuation of the Vietnam War and the advent of the Nixon administration, there was also a real clash of views between the young researchers in my group, many of whom were actively against the war, and the older staff, including the president of AS&E. That made the environment at AS&E less friendly than it had been in the early years.

By 1972 I had become convinced that x-ray astronomy was going to grow into an important branch of astronomy, particularly when the potential for imaging telescopes became a reality with the launch of LOXT. Observations in x-rays would then become relevant to all of astronomy, and I started wondering whether AS&E would be a proper setting to carry out this research. I had also found that I felt much closer to scientists than to business people: all my friends were astronomers. These circumstances led me to reflect on what other environment might be better suited to carry out my work and to formulate some ideas about institutional settings in which our group could operate.

First Thoughts about a National X-Ray Observatory

During the early 1970s I started reflecting on my past endeavors. It was clear to me that, in addition to ability and persistence, Dea Fortuna had played a considerable role in my achievements. The discovery of Sco X-1, the successful launch of Uhuru, the approval of Skylab and LOXT, all had come about through a fortunate combination of circumstances that presented great opportunities for success. Opportunity was the critical factor that enabled important discoveries and yet the most elusive. It followed that if one was given an opportunity to carry out an experiment, it would be criminal to waste it because of lack of commitment, incompetence, or—worse yet—laziness.

I had created my own recipe for how to execute a scientific program: I would always work at the very limit of what I could conceive and what could be done within the available resources. I would make sure that I would do useful science even in case of partial success. I would design and build instrumentation that was as simple as could be used to attain specific results, thus reducing both cost and the risk of complete failure. I would use the best available and most professional engineering and managerial talents to execute my programs; this would increase the chances of success and decrease cost; any savings thus achieved could be plowed back into the science. I would try to finish the development phase of a program as fast as it could be done; this policy reduced costs and allowed my group to be first in exploring a new field.

Because the programs I was involved in were typically quite large, many scientists had to work together as a group. It was my experience that talented people always do best when convinced of the value of what they are doing and when they share a common vision. The Uhuru spirit of ruthless intellectual honesty, truthfulness in the pursuit of excellence, respect of others' opinions, tolerance of foibles, and mutual support would be the glue that would ensure the cohesiveness of a premier research group at the cutting edge. In our group, we knew that the purpose of what we were doing was to solve scientific problems, not to build clever things. If the program was successful in obtaining data, their rapid and thorough scientific exploitation would be as essential as all that had been done before. Making our data available to other scientists for their analysis would increase, not decrease, the impact of our work.

When I compared this ideal of doing research with the conditions in which we were working under NASA sponsorship and management control, I could see a great chasm. Perhaps the most serious problem was NASA's propensity to overcommit itself to an excessive number of programs and to underestimate their costs. There followed huge overruns and delays, which would affect guilty and innocents alike, by forcing slowdowns and cost increases across the board, sometimes resulting in outright cancellations. Because these negative effects were spread uniformly across all programs, there was no incentive to work effectively, thus compounding the problem. Scientists exposed to these vagaries risked working for years, only to see their diligent efforts come to nothing. I desperately wanted to do science, and I did not want to waste my life in useless activities, no matter how well remunerated or secure. For these reasons, my thoughts turned toward alternative solutions to the pursuit of a life in research.

NASA's approach to the implementation of the research projects it sponsors is quite different from that adopted by other government agencies, such as the National Science Foundation, the Department of Energy, or the National Institutes of Health. Since World War II it has been the norm for these agencies to encourage the scientific community to create institutions responsible for the development and operation of national facilities required for research in various scientific fields. These institutions are funded by the agencies but are independent of them. Examples in physics are the Brookhaven National Laboratory, Upton, New York, and the Fermi National Accelerator Laboratory, Batavia, Illinois; in astronomy, the National Radio Astronomy Observatory, Charlottesville, Virginia, and the National Optical Astronomy Observatory, Tucson, Arizona. These national institutions were created in the conviction that scientists do better research when they take direct responsibility for the design, development, and operation of facilities that are typically state of the art.

The staff of such an institution is normally responsible to a consortium of universities through a board that sets the overall goals and policies of the organization and that hires and fires the director. This approach has proven very successful in the past 50 years and has resulted in outstanding discoveries, many of them recognized with Nobel Prizes.

Why NASA did not follow this approach is difficult for me to fathom. To answer that question would require a candid, in-depth examination of the circumstances that led President Dwight Eisenhower to establish NASA. It is natural to assume that foreign policy, as well as military, industrial, and institutional considerations, played a crucial role. I am not at all sure that scientific interests were properly taken into account.

As for today's NASA, I can only speak from my own personal experience, based on work over more than 40 years with three NASA centers (Marshall, Johnson, and Goddard) and with NASA headquarters.

I was very impressed in the 1960s by the high degree of engineering competence of Wernher von Braun's group at the Marshall Space Flight Center. Von Braun was actually interested in the scientific research he was making possible with the development of powerful space engines, and during the Skylab program he participated in some of the discussions, trying to understand our needs and requirements. But he clearly saw that his major task was to provide access to space for robotic and manned missions. Marshall has continued in that tradition of engineering competence over the past 30 years and has managed successfully the construction of some of the most important of NASA's astronomy missions, including Skylab, HEAO-1, Einstein, HEAO-3, Hubble, and Chandra.

Johnson Space Flight Center is an entirely different world. There, one is almost overwhelmed by the enormous systems engineering effort that is required to place man in space as safely as humanly possible. The incredibly tight quality control requirements placed on instruments to be flown on manned spacecraft were a real eye opener to me in the 1960s and 1970s. I got to know them well because my group at AS&E had detectors flying on the Apollo missions.

The tragedy of the manned program resides, in my opinion, not in any technical or even managerial incompetence, but in the wasteful overall misdirection of the program. Specifically, the decision by NASA headquarters to develop the Shuttle Transportation System as a reusable cargo and people carrier was technically unsound and unjustified based on any realistic projection of the expected number of flights (sixty per year were assumed). The shuttle's initial problems were compounded by the unrealistic expectations created in Congress and among the public about its operational readiness and safety. A similarly wrongheaded decision was made with regard to the International Space Station, which has not been and probably will never be a step toward the manned exploration of the planets. It is remarkable that in the midst of grave international problems, President George W. Bush found the time to finally call an end to these mistakes in 2004. When a technically sound program is set in place for manned exploration, the expertise of the Johnson Space Flight Center will still be needed.

There is therefore no doubt in my mind that both Marshall and Johnson provide engineering and management capabilities that are essential to carry out the most ambitious of the NASA programs. However, I have great reservations about the mission of the Goddard Space Flight Center, since they provide capabilities in science and engineering which are no greater than those already existing in academic institutions or industry. During the time of the AS&E series of rocket launches, Goddard rocket engineers provided great support to a large number of other research groups. However, the Aerobee rocket, the workhorse of the program, had been developed not by NASA but by a group at APL of the Johns Hopkins University under the leadership of the Nobel laureate Jim Van Allen. During the construction and launch of Uhuru, Goddard provided a program manager to supervise the work of two organizations, AS&E and APL, that would have done equally well or better by themselves, with either one being in charge. I have described my interaction with Goddard on the solar astronomy program, which resulted in an unnecessary duplication by Goddard of the AS&E experiment on Skylab. As for the Einstein satellite, Goddard appointed a project scientist whose greatest virtue, in my eyes as the principal investigator, was that he would agree with me most of the time.

This is not to say that Goddard did not have significant strengths in engineering and operations. In the second part of the 1990s, Goddard did a fantastic job under its director Joe Rothenberg in advocating and executing, together with Marshall, Johnson, and STScI, the repair mission to Hubble. Astronomers owe much to Frank Cepollina and his team at Goddard for their contribution to this work.

When it came time in 1971 to choose a NASA center to manage the LOXT program, I was asked my opinion by NASA headquarters. There followed discussions with the Jet Propulsion Laboratory (JPL), Goddard, and Marshall, which had the following results: JPL expressed an interest in LOXT only if they had complete responsibility for the entire program, both the science and the engineering (with some consulting with my group); Goddard wrote a two-page answer expressing mild interest; and Marshall sent a group to discuss with us how we could work together and how our scientific capability and their engineering expertise could be combined. We were very glad when Marshall got the job.

I had little problem with Goddard in their initial role in the early 1960s, which was to help the outside community to do their research, but I became increasingly concerned with the shift from this supportive role to a preoccupation with their own research. Beginning in the 1970s, I have increasingly come to perceive Goddard as a competitor to the community rather than an enabler of its work. Although this characteristic became more obvious in later years, my interactions with Goddard in the late 1960s and early 1970s convinced me that scientists at the center saw their mission as doing research of interest to themselves rather than serving the community. This philosophy would be tolerable, if not ideal, if those scientists did not use their government position to favor their own programs.

With these considerations in the back of my mind, I started asking myself whether a national x-ray astronomy observatory would not be the best home for this new field of research. My concept was of an institution working closely with academic research teams and responsible for the development of x-ray technology, the design and development of space facilities (such as LOXT) and their operations, and the selection of observations (through a competitive peer review process). The institution would also be responsible for the development of the software to operate the facility, calibrate and reduce the data, perform first-cut analysis, and finally archive and distribute the data. Such an institution would be funded primarily by NASA and would be governed by a board responsible to a consortium of universities, as is the case for Brookhaven or Fermilab. The institution I had in mind was essentially of the type that was later created to manage the science program of the Hubble Space Telescope, namely, the STScI. The existence of such an x-ray astronomy observatory would ensure continuity of commitment and advocacy for this new branch of astronomy.

I had long talks on these ideas with Pippo Vaiana between 1970 and 1973. He and I had become close personal friends as well as colleagues, and in our rambling discussions we covered a number of topics. We agreed on the desirability of a national x-ray observatory, but we also recognized that it would be difficult to launch such an initiative from a corporate setting. We understood that x-ray observations would soon become relevant to all astronomy, and because none of the scientists in my group at AS&E had been trained in traditional astronomical disciplines (with the notable exception of Wally Tucker), we would need a great deal of work to catch up, and above all contact with "real" astronomers. Although we had tried to increase the extent of our collaborations with optical, radio, infrared, and gamma ray astronomers through joint symposia, such as the one I helped organize in Madrid in 1972, and at summer schools, it seemed clear to us that this intellectual discourse could best take place in an academic setting.

Pippo and I also discussed the changes occurring at AS&E, which seemed to be steering the company away from research. In the Space Research and

Systems Division, my group and I had created an oasis of sanity and camaraderie, which had been an essential element in the success of the projects we had undertaken. As head of the division, I had been able to shield my staff from arbitrary and capricious actions by the president, who seemed to have lost contact with the staff. Given the growth of the division and its involvement in so many aspects of AS&E, this was becoming increasingly difficult. A particular episode, which for me was the last straw, was the firing in 1972, on flimsy grounds, of Tom Quinn, a longtime and valuable employee who was responsible for the facilities and who reported to me. All my pleas that the decision be reversed were for nothing. I felt I should resign, but I was held back by the commitments I had, to both my colleagues and my family. I decided, however, that I would soon leave AS&E in an orderly manner and that never again would I be forced to acquiesce to such an injustice.

I had also come to a parting of the ways with Bruno Rossi, a person I had admired and trusted since our first meeting in 1959. I have always recognized my debt of gratitude to him for the initial suggestion to investigate the possibility of x-ray astronomy. However, when Uhuru was successfully launched, I was a mature and independent scientist with a group that looked to me for direction and inspiration. Bruno had not been involved in the project and did not participate in the planning of the Uhuru scientific program or in the data analysis. Still, my group and I were extremely open in sharing the early results and prepublication information with him, even though he was not a coauthor on any of the papers. We were sure he would properly credit our work and, in fact, aid it by divulging the results at various meetings and summer schools. The fact that he was reporting our work was well understood by astronomers in the United States; as the senior member of our group, I received all the awards that the American Astronomical Society and the Astronomical Society of the Pacific could bestow.

The situation was less clearly understood in Italy, where astronomers did not appreciate the difference between the AS&E and the MIT groups and where, following the Italian gerontocratic tradition, all the credit would naturally be given to the eldest scientist in any way associated with the work. Old-school astronomers unfamiliar with our work were propagating this view at meetings and conferences and in the popular press. This caused great distress to my mother in Milan, who happened to listen to some of the talks and who felt that her son's contribution was being ignored.

I kept reassuring her by letter that everything was fine, until I came across the notes of lectures given by Bruno in February and March 1972 under the auspices of the Accademia Nazionale dei Lincei; they were based entirely on our published and unpublished Uhuru papers, though he was not a co-author, but he mentioned AS&E only as the builder of the Uhuru payload. Part of the blame may be due to the poor job of transcription of his talk, but still, Bruno was responsible for the version published by the Lincei in 1974.[1] I would never have expected this behavior from him and I was keenly disappointed and deeply hurt. My feelings made leaving AS&E much easier.

In later years, I was struck by the view expressed by Bruno that the discovery of x-ray stars was entirely due to his intuition. This view neglected the efforts of several groups that had set out to find x-ray stars even prior to the work at AS&E. It also ignored the role of my group's inspired ideas and hard work in the initial discovery and further advances in the field. But Bruno was so convinced of his point of view that in his acceptance speech for the Rumford Medal of the American Academy of Arts and Sciences in 1976, he referred only to his intuition and never recognized, named, or thanked any of the people who had actually done the work. This attitude was tantamount to saying that everybody else was only doing a menial job of execution. It was terribly unfair and hurtful.

I decided that I would never act that way myself. In my Nobel lecture, I took pains to recognize his contributions, and to name and thank my coworkers as I thought was just and proper. George Clark made every effort to bring Bruno and me together at the Einstein launch in 1978 by addressing Bruno as the grandfather, and me as the father, of x-ray astronomy, and later in writing the motivation for the Wolf Prize that was shared in 1987 by Herbert Friedman, Rossi, and me.

Culture Clash: Introduction to Harvard

Some time in 1972, I went to visit Leo Goldberg, director of the Harvard College Observatory, whom I had met and worked with on the Skylab program. I wanted his advice, as an elder statesman of astronomy, about what I should do and what opportunities might be open to me. Leo was very sympathetic, with regard both to the science and to my own situation. He told me that Fred Whipple, the director of the Smithsonian Observatory, had decided to resign after many years of service and that perhaps I could have his position. He himself was leaving Harvard to become director of the Kitt Peak National Optical Astronomy Observatory. But he also told me that a joint Harvard Observatory–Smithsonian Astrophysical Observatory committee had been formed to decide how the two Cambridge observatories should be run and whether they should be united as one organization. He thought that there would be a great deal of interest at Harvard in initiating a research program in x-ray astronomy and promised that he would discuss the matter with the faculty.

I was soon involved in discussions with Alex Dalgarno and George Field about the committee plans for a joint Center for Astrophysics (CFA), of which George would become director. After this discussion in the spring of 1973, my main point of contact became Dean John T. Dunlop. He was a distinguished economist, labor negotiator, and advisor to President Nixon on labor matters. At the time he was dean of the Faculty of Arts and Sciences at Harvard, and while I found him at the beginning rather awe inspiring, it turned out that he was extremely friendly and thoughtful. I was offered a full professorship at Harvard and the position of associate director of the High Energy Astronomy Division of the Harvard-Smithsonian CFA. Four tenured Smithsonian positions would be made available for my most senior associates, and additional positions could be supported by my own grant and contract funds. One of Dean Dunlop's concerns was whether I could carry out my work as part of CFA, or whether I would need to set up a new enterprise with its own budget, overhead structure, and administrative functions. Perhaps naïvely, I was convinced that it was important for our group to be fully integrated with the life of the observatory, and that any difficulty in adjustment could easily be resolved.

I discussed with NASA what I was planning to do, namely, to retain responsibility as the principal investigator for the scientific programs I was directing, in particular the Uhuru data analysis and HEAO-2. For the latter program, I asked that 5 percent of the funds be made available to me as principal investigator, and that AS&E continue to carry out the engineering efforts with the remaining 95 percent of the funds, under the leadership of Arthur Vallas, a great manager and a friend.

The Skylab solar astronomy program was to continue at AS&E under the direction of Pippo Vaiana (who left about a year later to join the Solar Physics Division of CFA). John Naugle of NASA was again very supportive and approved of the arrangements.

When things seemed to be in order, I discussed my plans with members of the scientific staff at AS&E and was able to assure them that they would continue to be supported whether they joined me at CFA or continued on at AS&E. Eight scientists moved to CFA with me in 1973; they were Paul Gorenstein, Herbert Gursky, Edwin Kellogg, Steve Murray, Ethan Schreier, Dan Schwartz, Harvey Tananbaum, and Leon Van Speybroeck. Wally Tucker had semi-retired to Strawberry Hill Farm in Bonsall, California, but he spent a part of the year (I guess outside the strawberry season) with us at CFA.

Given the experience and the renown of the team, the scientific interest of the research subject, and the grant and contract support we were receiving, the High Energy Astronomy Division immediately became the dominant group in the observatory. This recognition promptly sparked resentment from preexisting groups, which was to last for many years, a situation not helped perhaps by our frenzy of activity. We found that CFA had a caste system in which the Harvard faculty and staff occupied the top rung of the ladder; below them came the Smithsonian staff, and finally all the staff on soft money. Because our group consisted of excellent scientists hired under all of these categories, the cohesiveness of the group was extremely important to us, and on practical and moral grounds we found these distinctions unacceptable. In addition, we had a very different style of work compared to other researchers at the observatory—more aggressive, more critical, and more helpful to one another—but we quickly found out that the Uhuru spirit of joyous intellectual combat was definitely not encouraged.

I remember a journal club meeting for faculty and postdocs at which I asked the speaker (a young postdoc) whether the a posteriori model he had elaborated to explain some observational data had any predictive power. I was immediately reproached by senior astronomers for asking such a question, as it was well known that this was rarely a requirement for astrophysical theories.

I found at Harvard an atmosphere resembling that of a private club. There were perks for the senior faculty, including the use of tennis courts on the observatory grounds, private bathrooms attached to each of the full professors' offices, teas, and so forth. I created a little scandal when—oppressed by the dingy walls on the fourth floor of the CFA building, after requesting that they be repainted and having waited a while—I finally suggested that CFA provide the paint and the members of the High Energy Division, including me, do the painting. The facility administrator was quite upset and sternly informed me that "we at Harvard do not allow the scientific staff to do such things."

The faculty seemed more interested in elaboration of accumulated knowledge than in discovery of new phenomena. This attitude was very much at odds with my own style of work, which I described to myself as working with an axe rather than a scalpel. I preferred to trailblaze in virgin territory rather than elaborate and embellish previous work. The metaphor of the axe also conveyed to my mind the desire to attack physical problems with overwhelming experimental capabilities. At Harvard I felt that my drive to bigger and better experiments was considered somewhat unseemly by most of my colleagues.

In my opinion, the level of much of the faculty did not, in fact, justify their sense of uniqueness and their arrogance. By this time, I had met enough very good people—including John Bahcall, Neta Bahcall, Geoffrey Burbidge, Margaret Burbidge, William Fowler, Jeremiah Ostriker, Allan Sandage, and Steven Weinberg in the United States and Fred Hoyle, Hendrik van de Hulst, James Pringle, and Martin Rees in Europe—to be able to make my own judgment of competence. I was particularly struck by the vitality and depth of the Russian school of theoretical astrophysics: I had met Vitaly L. Ginzburg, Roald Z. Sagdeev, Josef S. Shklowsky, Rashid A. Sunyaev, and Yakov B. Zeldovich and was quite keen to learn from them. With some surprise I found that astronomers at the observatory were not particularly interested in what my group and I were doing—but always with notable exceptions, including George Field and Alan Lightman.

Some of the people working in high-energy astronomy whom I had inherited asked to be transferred to other divisions, while Josh Grindlay continued on with great success. It is fair to say that Harvard was a good address, but that our collaborations were mostly with people at other institutions all over the world, rather than with the people next door.

A more difficult problem that concerned me had to do with the management of research. I had expected that CFA would be run with some sort of participative management style, in which issues would be discussed among senior people. Instead I found that all administrative decisions had been delegated to an associate director of administration, and that neither he nor the director was interested in discussing general issues of the observatory with me. I learned a great deal under these circumstances about what I would do differently as a director.

Although we were charged a considerable overhead on our contracts, we received little in return, and the research support actually available was less than that we had had at AS&E. I must also have annoyed people by demanding service, accountability, and efficiency from such CFA facilities as the machine shop, computer center, and administration. I found that although the nominal overhead costs were lower than at AS&E, the actual cost of doing something was higher at CFA, because much of what I had considered overhead cost had to be provided directly from contract funds. The supporting groups at CFA seemed to feel that they were working in a privileged place, where cost and time had little significance. I had always been of the opinion at AS&E that whatever had to be done should be done at the lowest possible cost, so that the "profits" could be used to do more science. CFA's rather relaxed approach to management reached the point that one year, the radio astronomy group forgot to request the funding to run the George R. Agassiz Radio Telescope of Harvard Observatory, Fort Davis, Texas; the funding was of course provided by CFA at the expense of other existing or new programs.

I have read that George Field remembers those days as a struggle, but I believe that, being a theoretician, he underestimated the work necessary to complete Einstein in less than five years and to do so with limited resources. I hope that, with time, he came to appreciate our point of view. I also trust that much has changed in the 25 years since I was at CFA.

For the first time since my university days in Italy, I taught a course—a six-month course on x-ray astronomy together with Josh Grindlay—that was rather fun. I learned a lot from him and I hope he learned a little from me. We had six to eight people taking the class, mostly members of the staff.

In time, though, it became clear to me that what I really wanted to do was to carry out difficult and ambitious projects in observational astronomy, and I came to feel somewhat uncomfortable at Harvard. Years later I found a quotation by Freud that explained how I felt better than I ever could have: "I am actually not at all a man of science . . . but a conquistador . . . with the curiosity, daring, and tenacity characteristic of a man of this sort."[2] On second reading, I think that, while Freud's statement may have been appropriate for him, and it fully expressed my feelings in the 1970s, today it sounds a little too Wagnerian for me. In fact, I have always felt a little unsure of myself and never left a job before having another one lined up, so that the boldness was entirely confined to the realm of ideas and mountain climbing.

While our group was adapting to our new home, we were very conscious of the need to recover the ground lost with the cancellation of the LOXT program. Einstein was still being built, but we knew that when launched in late 1978, it might last only one or two years; after its demise, no comparable U.S. x-ray telescope would exist for some years. We also knew that larger telescopes would be required to follow up on Einstein's discoveries, and we had set our sights on a 1.2-m telescope as originally envisaged in LOXT. Harvey Tananbaum and I submitted a formal proposal to NASA in April 1976 for an x-ray telescope national space observatory with a proposed launch in 1983.[3] After many more studies and with increasing support from the astronomical community, NASA was persuaded to initiate the program that became the Chandra Observatory, to be launched in 1999, some 16 years later than we had hoped for.

At the same time, I started working toward the realization of a national x-ray institute on the basic premise that it would actually take over responsibility for the operations of Einstein while studying and ultimately implementing the next generation of x-ray telescopes. I started informal discussions with my colleagues on the Einstein project and with other interested scientists, including Herbert Friedman of NRL and William Kraushaar of the University of Wisconsin. Their response was generally positive, with some expressions of concern about NASA's reactions.

Several of us—Clark, Gursky, Kraushaar, Vaiana, other colleagues, and I—jointly drafted a proposal for a National X-Ray Astronomy Institute in October 1975, and we were invited to present our thoughts to NASA's High Energy Astrophysics Management Operation Working Group, chaired by Al Opp. A month later I wrote a letter to NASA's administrator, James Fletcher, outlining a plan for the management of x-ray observatories through a national institute. Fletcher responded on February 17, 1976, stating that in view of the HEAO-2 overruns (on the order of 15%), he was delaying assessment of the plan.

Notwithstanding this rather cool reception, we continued our work and asked the University Research Association (URA) that was responsible for Fermilab whether they would be interested in managing the x-ray institute. A new draft proposal of February 1976 identified URA as the organization planning to submit the proposal; discussions continued between URA and our group into the early summer of 1976.

Meanwhile a meeting had been called by Gerald Tape of Associated Universities, Inc. (AUI) on July 7 to discuss specific organizational arrangements and management methods that could best facilitate the interaction between NASA and the scientific community involved in the major space observatories. Tape circulated a letter summarizing the results of those discussions. This activity was, however, overtaken by the study carried out July 19–30 by the Space Science Board of the National Academy of Sciences, meeting at Woods Hole, Massachusetts, under the aegis of the Assembly of Mathematical and Physical Sciences, regarding institutional arrangements for the Space Telescope.

In their report, the committee clearly stated that "an institutional arrangement, which we call the Space Telescope Science Institute (STSI), is needed to provide the long term guidance and support for the scientific effort, to provide a mechanism for engaging the participation of astronomers throughout the world, and to provide a means for the dissemination and utilization of the data derived from the ST."[4]

I was delighted by this strong conclusion and by the rest of the report, which was closely akin to my own thinking. It is unfortunate that this case was one of the few in which the academy took on a substantial issue regarding the implementation of NASA programs. In part this passive stance may be because academy work is funded by the agencies it advises, and these agencies set the agenda of what can be discussed with strict guidelines, which are specified in each request for a study. In this particular case, Noel Hinners of NASA left sufficient freedom that the concept of an independent institute carried the day.

However much the academy and the scientists liked the concept, not all of NASA did, particularly the people at Goddard, who considered optical and ultraviolet astronomy to be their fief. During a conversation I had with Fletcher about the X-Ray Astronomy Institute, he told me that, as far as he was concerned, he already had one institute too many: STScI. And indeed it was only with some pain that STScI was given the green light.

Undeterred by these difficulties, I next tried to convince the Smithsonian Institution to take on the cause of the X-Ray Astronomy Institute. But the Smithsonian Astrophysical Observatory Visiting Committee was not at the time supportive of the idea of having an x-ray astronomy institute at the observatory because of its anticipated size. This opinion, communicated by David Challinor of the visiting committee to Smithsonian Director S. Dillon Ripley, killed any opportunity we had to proceed. George Clark and I tried one more time in a letter of September 15, 1979, to convince George Field (the director of CFA) and Thomas Jones (vice-president for research at MIT) to create a joint Harvard-MIT institution similar to the Sloane School of Management, but without success.

By this time it was quite clear that neither the 1.2-m x-ray telescope nor the X-Ray Astronomy Institute would materialize in any reasonable time. The very success of a space mission in a particular wavelength was seen by NASA not as a spur to continue studies in that wavelength, but as a signal to give other projects a chance at being funded. This approach was inspired not so much by scientific considerations as by political and social ones.

The same was true for successful groups, such as ours. When some of my colleagues submitted a proposal for a new independent project, they were usually discouraged by Nancy Roman of NASA headquarters on the grounds that my group had already received support. I do not know whether this policy was in fact formulated at a high level or invented on the spot, but it was quite discouraging, particularly as it seemed to be applied mostly to my group. Over time NASA's hostility convinced me that a new start for LOXT would have little chance of being approved if it was identified with me. I began to withdraw from the study committees that I had led.

The Ventures of Others into Celestial X-Ray Gazing

While my group and I at CFA were completing the analysis of the Uhuru data and constructing the Einstein observatory, other groups and other nations had started to fly x-ray satellites of their own. The first to produce new and significant discoveries was the ANS of the Netherlands, which carried ultraviolet spectrometers supplied by the University of Groningen, x-ray detectors from the University of Utrecht, and a high-energy x-ray detector from AS&E. ANS was launched in August 1974 and operated until 1976. The most important discovery of the mission (by Josh Grindlay and John Heise) was the detection in December 1975 of an x-ray burst from NGC 6624.

SAS-3, a satellite in NASA's Small Astronomy Satellite series, was launched from the *San Marco* platform, Kenya, in May 1975. The instrumentation, built by Clark's group at MIT, included both collimated counters and modulation collimator detectors, which could yield a source location measurement accurate to arcminutes. The spacecraft bus and control system were almost identical to those of Uhuru. The glory of this mission was the discovery of the rapid burster in the globular cluster Liller 1, which produced explosions at the rate of a thousand per day. SAS-3 contributed greatly to the detailed study of accretion processes in x-ray binaries, the determination of the mass of neutron stars through accreting torque measurements, and the discovery of a binary x-ray system containing a highly magnetized white dwarf (Am Her). SAS-3 operated until 1979.

The satellite closest in capabilities to Uhuru was the United Kingdom's Ariel-5, which was spin stabilized and had detectors both perpendicular and parallel to the spin axis. The instrumentation was provided by Ken Pounds's group in Leicester. The collimators included modulation collimators to improve measurements of source position to about 2 arcmin. The payload also included a Bragg crystal spectrometer. Ariel-5 was also launched from the *San Marco* platform in October 1974, and it operated until 1980. Its major contributions included the discovery of x-ray novas and of iron x-ray emission lines in various supernovas. In addition, the researchers involved with the project established Seyfert 1 galaxies as a class of x-ray emitters.

All of these efforts took place before the launches of HEAO-1 and Einstein, and they clearly showed the richness of the field that still remained to be investigated even at x-ray fluxes three hundred times greater than would be attained by Einstein. This type of work continued in the 1980s and 1990s with great success, leading to the identification of many types of x-ray binaries, x-ray bursters, and gamma ray bursters.

My own scientific interest was, however, focused on studying the "hoary deep" at cosmological distances with the greatest sensitivity and angular resolution I could achieve, in the hope of solving the mystery of the x-ray background and studying the formation and evolution of the earliest structures in the universe. While waiting for Einstein data, and trying to ensure the early start of its larger successor, in the early 1980s I became involved in the new German mission ROSAT, which was being planned by the Max Planck group in Garching led by Joachim Trümper. This observatory, a somewhat scaled-up version of Einstein, had a telescope that was 0.8 m in diameter (rather than Einstein's 0.6 m); a focal length of 2.4 m (rather than 3.4 m), implying a sensitivity to a lower energy range; a field of view of 2 degrees (rather than 1.25 degrees); and a collecting area of 2,500 cm^2 (rather than 500 cm^2).

The mirror resolution was only a little worse than that of Einstein (5 rather than 4 arcsec). However, the detectors that the group was planning to fly, which were imaging proportional counters, could not resolve images with this resolution. By then I was convinced that we needed high angular resolution, but even more importantly, high positional accuracy to identify the weakest x-ray sources detectable in deep surveys of selected fields in the sky. We knew in early 1979 that the optical counterparts of the x-ray sources of 1.3×10^{-14} erg/cm^2 · sec were of magnitude 23 or fainter, very difficult to identify even with deep plates. To change the design of the satellite that the Max Planck group had already started to build would have been difficult and expensive.

Trümper and I came up with the idea that NASA could fund the Smithsonian Astrophysical Observatory to furnish an identical copy of Einstein's HRI and provide the launch vehicle. Max Planck would modify the optical bench to achieve greater rigidity and the star trackers to achieve higher precision and would integrate the HRI into ROSAT. The mission would thus become a cooperative venture between Max Planck and NASA, and the observing time would be split into two separate programs. To make this possible, it was essential that the Smithsonian observatory start construction of a second HRI on a sole-source procurement basis at the earliest possible date.

To convince the people reluctant to slow down other Explorer programs, the Smithsonian Astrophysical Observatory offered to waive all data rights that normally would have accrued for their effort. We were successful in convincing our colleagues and NASA of the great opportunity that this program represented for U.S. astronomy, and the new ROSAT became a reality. There was only one sour note in this entire venture. Trümper and I had proposed that the existing Einstein facility at the Smithsonian observatory should han-

199

dle the distribution of the data and administer the program for guest investigators. A decision on these points was deferred, and it ended up that both responsibilities were awarded to Goddard by NASA fiat and without any discussion with us. Because we were discussing data from an instrument we had designed, built, and calibrated at the Smithsonian Astrophysical Observatory, and because NASA had no previous experience in handling stellar x-ray images, this decision is a clear example of injustice and of Goddard's self-serving attitude.

Nevertheless, I am happy that I was able to contribute in a significant way to the ROSAT program. After its launch in June 1990, I was privileged to participate in the use of the German proprietary data to pursue the study of the x-ray background. This collaboration continued through the 1980s and 1990s. I have fond memories of the times when Richard Burg, Günther Hasinger, Maarten Schmidt, Joachim Trümper, and I would meet at the Augustiner (one of the oldest beer halls in Munich) to conduct our most inspired and enthusiastic planning sessions with the help of liters of Augustiner Brau. As a memento of our collaboration, Joachim gave me a beautiful x-ray picture of the moon obtained with ROSAT (Plate 7).

The Hubble Space Telescope and the Space Telescope Science Institute

TWELVE

Politics, Science, and Priorities: The Start of the Space Telescope

One of the difficulties in convincing NASA to proceed with a 1.2-m x-ray telescope in the 1970s was the overwhelming desire of some people at NASA to start with the Space Telescope, regardless of its technical readiness or cost. The scientific potential of a telescope in space had been understood by one of the most colorful of the space pioneers, the Russian Konstantin E. Tsiolkovsky, who, as early as 1911, foresaw that space observatories would revolutionize astronomy. In the United States, Lyman Spitzer was the first to point out the great advantage of placing an optical telescope in space in a 1948 Rand Corporation technical report. In 1962 he discussed the project at a meeting of scientists convened in Iowa by the National Academy of Sciences to develop plans for a space science program.

The response of the optical astronomers at the meeting to the possibility of launching telescopes of dimensions comparable to those used on the ground (a few meters in diameter) was lukewarm. However, a program of smaller ultraviolet telescopes (about a meter in diameter), the OAO, was established by NASA, mainly with the support of Nancy Roman, who had become the head of NASA's astronomy program soon after the creation of the agency in

1958. She had carried out research at the Yerkes Observatory in Chicago and the Naval Research Laboratory in Washington, D.C., but, like many optical astronomers at that time, had no experience in either instrument construction or management. She remained at NASA in that position for the next 21 years and had a substantial influence on the development of space astronomy in the United States, although in my opinion not always for the best. OAO was a mixed success: two successful launches, in 1968 and 1970, out of four attempts, produced interesting results in ultraviolet spectroscopy.

I first learned about the prospects for a Space Telescope in 1965 at the Woods Hole summer study sponsored by NAS, where the idea was received favorably. The academy commissioned a study by a group led by Spitzer, which published its findings in 1968 in the journal *Science,* followed by a full NAS report in 1969.

The 1970 decadal NAS *Report on Astronomy by the Greenstein Committee* was rather tepid in its recommendation to carry out the Space Telescope project, which it saw as a distant goal, and did not give it high priority. In fact, it did not appear among the top four new projects in astronomy. This ranking accurately represented the opposition of many optical astronomers to the project, because of its high estimated cost, even though it was initially quoted at a very optimistic $300 million. One can only imagine their reaction had they been told that the cost at launch would turn out to be more than a billion dollars.

Despite the lukewarm support offered by NAS, NASA initiated two feasibility studies at Goddard and Marshall in 1971. A small group of scientists headed by Roman provided the scientific guidance. In 1972 Marshall was designated lead center for the project. Bob O'Dell, director of the Yerkes Observatory, left his position there to become the project scientist at Marshall. Although the program was still poorly defined, in 1974 NASA decided to present it as a specific new line item in its budget proposal to Congress for the year 1975. This proposal came on the heels of the major overruns on the Viking program that had caused the drastic scaling back of the High Energy Astronomy Program, which I described in Chapter 9. Congress eliminated all funds for the Space Telescope (as it was called before launch), noting the low priority it had been given by the NAS Greenstein Committee.

There followed a well-orchestrated lobbying effort directed toward the astronomical community and Congress. Roman and O'Dell could not, as NASA employees, lobby Congress directly. Lyman Spitzer and John Bahcall, who had been invited to join a twelve-member working group headed by O'Dell, did not have this impediment, and when the project got into trouble, they began a vigorous campaign. Bahcall, according to his own statements, spent half his time over the next few years lobbying on Capitol Hill. Congress restored funds for a modest study in 1975. Although there were no development funds in the 1977 budget, in 1978 NASA was finally authorized to proceed with the development of the Space Telescope.

From the point of view of high-energy astronomy it was clear that this focus on the optical Space Telescope implied a long delay in getting started on a 1.2-m x-ray telescope, notwithstanding the high degree of community support. I was also concerned by the lack of experimental and operational know-how of optical astronomers, particularly for space projects. I thought it likely that the project would encounter considerable technical difficulties, which would result in substantial overruns and delays. In addition, the plan to have the Space Telescope serviced by astronauts carried aloft by the shuttle imposed severe demands on the program, because of the strict quality control requirements imposed by issues of human safety. Considering the scarce support by the astronomical community for an optical telescope, one could make a good case that the 1.2-m x-ray telescope should go first, and I was not shy in voicing this view at the Space Science Board of the National Academy of Sciences in the later 1970s.

However, the strong support by the NASA staff, the lure of utilizing the shuttle to do something useful, the very strong advocacy by Spitzer and Bahcall, and the lobbying efforts by industrial contractors made the Space Telescope program unbeatable. At the same time, the Advanced X-Ray Astrophysics Facility (AXAF), better known as Chandra after launch, received a modest start in 1977 for conceptual design studies and technological development.

As expected, the Space Telescope program experienced delays and overruns: its cost first rose from the early estimate of $300 million to $490 million in 1974, then to $700 million in 1981, and two years later to $1.2 billion. These overruns left a terrible impression in Congress, which by now had become skeptical of all new large projects in astronomy. Anything called a "telescope" was bad news, which is why AXAF was called a facility. Burt Edelson, NASA's associate administrator for science, is reported to have said in exasperation: "Why does AXAF look so much like Hubble? Can't you make it a cube, or at least a rectangle?"

The AXAF program could not be started until May 1988, thus leading to a 20-year hiatus in U.S. observational capabilities for high-resolution x-ray astronomy. Furthermore, Congress was wary of a repetition of the Hubble difficulties, and requested that the staff construct and demonstrate the performance of the x-ray mirror within the first three years of the program. This requirement was challenging, because no x-ray mirror of the size and quality required by AXAF had ever been built. This feat was achieved thanks mainly to the efforts of the CFA group at Harvard, led by Harvey Tananbaum, and their contractors. In the end Chandra was flown in 1999 and was an overwhelming success.

Even though I was ultimately proved correct in my technical and managerial evaluation of the Space Telescope and AXAF, during the debates of the late 1970s I had upset quite a few people at NASA and in the community—in particular, Nancy Roman. One day, at a meeting sponsored by the NAS and chaired by Gerald Tape of AUI, to study institutional arrangements for space observatories and the expertise required of their directors, Roman let me know that only over her dead body would I become director of a Space Telescope Institute.

Against NASA's Grain: The Founding of STScI

I have mentioned that, at the request of NASA, NAS had in 1976 undertaken a study of possible institutional arrangements for the scientific use of the Space Telescope. The study was carried out by a committee of the Space Science Board of NAS chaired by Donald F. Hornig. As mentioned in Chapter 11, its 1976 report recommended that an independent institute be established "to provide long term guidance and support for the scientific effort; to provide a mechanism for engaging the participation of astronomers throughout the world, and to provide a means for the dissemination and utilization of the data derived from ST."[1] This approach, while almost universally favored by the astronomical community, was strongly opposed within NASA and in particular by the Goddard staff. They envisaged a small organization run by NASA and staffed by civil servants, with a civil servant as its director. The argument that ultimately prevailed was that NASA had failed to, and probably could not, hire the caliber of astronomers that could be entrusted with the responsibility for the scientific success of the Space Telescope. The many arguments and debates surrounding this issue within NASA and the community are reported in detail in the book by Robert W. Smith.[2] I do not repeat them here, as I was unaware of these internal debates at the time. Finally, in December 1979, NASA issued a request for proposals for a Space Telescope Science Institute, with the responses due in March 1980.

Five different consortiums of universities submitted proposals, which included, as requested by NASA, the selection of a site where the institute would be located. The University Research Association proposed Fermilab as its site; the Association of Universities for Research in Astronomy (AURA) proposed a site on the Johns Hopkins University campus; University Space Research Associates, AUI, and Batelle Memorial Institute had picked a site near Princeton University.

The AURA proposal was ably shepherded by John Teem, president of AURA, and Arthur Davidsen, a professor in the Department of Physics and Astronomy at Johns Hopkins. It was written in collaboration with a designated subcontractor, Computer Sciences Corporation, and NASA judged it to be technically and managerially superior to the other proposals. An AURA contract with NASA was negotiated in early 1981, and an AURA search committee chaired by Margaret Burbidge of the University of California, San Diego, considered several potential candidates for director.

A call by John Teem in early 1981 to explore my interest in serving as director of STScI caught me completely by surprise. My first reaction was that such an appointment could not possibly be supported by optical astronomers and would be opposed by some people at NASA. However, I found the offer sufficiently intriguing that I asked for a week to think about it. I called Margaret and she allayed many of my concerns about the views of the optical community. I had long discussions with my wife, who suggested that I should think carefully before rejecting the offer.

As to NASA's consent to my appointment, a requirement of the AURA contract, Teem was able to obtain it, despite some concerns in upper management at NASA headquarters. Sam Keller, the most senior NASA manager closely involved with the Space Telescope, had developed a certain regard for my experience as a manager and inventiveness as a scientist in previous programs. In an interview with Robert Smith, he mused that "Riccardo's approach to a meeting is to first roll a hand grenade through the door, and after he has gotten your attention he gets down to the subject matter."[3] He also told Teem that in the execution of the program, I would use up three or four NASA program managers. John Naugle felt that "Riccardo's ego would not allow failure on this program," a characteristic he deemed good for all concerned.[4] Nancy Roman had by that time left the agency, so that she did not have to make good on her promise to commit hara-kiri.

As for me, I did not feel at ease in the academic setting of CFA at Harvard. The Einstein mission, though extremely successful, did not give me the deep emotional satisfaction that Uhuru had provided. The long delays to be expected in carrying out the AXAF mission, almost 20 years, made me feel that I was in suspended animation and was not allowed to work at my full capacity. I was quite conscious that a scientist has only one lifetime to achieve his goals, and that I was in danger of wasting mine.

During my work at AS&E and later at Harvard, I had also developed a vision of how a scientific organization should be run. It should be a place of clarity and scientific truth on the Uhuru model and dedicated to the pursuit of excellence in all aspects of the work. It should be an organization dedicated to a shared vision and a goal, in which freedom of research and the free exchange of ideas would be coupled with the internally imposed discipline necessary to maximize the results of our joint efforts. I did not know whether I was the man to achieve this lofty goal, but I was anxious to try. I believed that my scientific, technical, and managerial capabilities were at least equal to those of the people who had been involved in the program up to that point, and that I could make a real contribution to the success of the Space Telescope program. I also thought that I had a better vision than most professional astronomers of the future of observational astronomy both in space and on the ground. Its course would be dictated by the rapid development of optics, detector technology, and computer and information technology. I was being offered the opportunity to create, from the ground up, an institute that would fulfill that vision and become one of the premier astronomical institutions in the world.

I discussed the prospects and conditions with the AURA search committee and with the Johns Hopkins Physics and Astronomy faculty, particularly with regard to my personal research—which I intended to (and did) continue during my tenure as director—and the possibility of involving postdoctoral fellows in it. After these talks, I decided to accept the position of director of STScI and professor of astrophysics at the Johns Hopkins University. I received a two-year nonrenewable leave of absence from Harvard. Mirella and I looked for a house in Baltimore and moved there in September 1981.

Leaving the x-ray astronomy group at Harvard that I had formed during almost two decades of work was not easy. However, I felt that I had taught them—and learned from them—what I could, and that they would do very well on their own. Leaving a professorship at Harvard was an unusual decision, but some of the comments I made in the preceding section might explain it. Sometimes what one leaves is as important as one's destination.

Jumping into the Thick of Things: The Space Telescope in 1981

In 1981 the Space Telescope (I will henceforth use the name it was given after launch, the Hubble Space Telescope, or just Hubble) had been in development for four years. The official launch date was set for January 1985, and the program was by this time well into its construction phase. So much has been written about Hubble that here I give only a sketchy description of the hardware and the organization of the project, with some emphasis on how it appeared from my point of view. A comprehensive and scholarly description of the scientific, technical, managerial, and political issues that faced the Hubble program is given in Smith's book, *The Space Telescope*.[5] My only criticism of the book is that the conflicting visions of how to do science with Hubble are described as stemming from conflicts of personalities rather than from the profound changes that were occurring in observational astronomy. And although it is true that a great deal of politicking was going on, it ultimately had no real importance, and perhaps Bob Smith took it too seriously. I am reminded of a statement that Wernher von Braun made to me when we were discussing the curtailment of flight projects during the Nixon era: "Presidents come and presidents go, but we are still here." An article published in 1981 by Bob O'Dell gives some additional technical details.[6] A vast body of technical and scientific information is now made available by STScI directly on the web (www.stsci.edu/resources/).

Hubble consists of a 2.4-m f/24 mirror Ritchey-Chrétien telescope. This design yields coma-free images over a wide field, and thus permits the sharing of the focal plane by four instruments and the fine-guidance sensors (FGSs). A structure of graphite epoxy, called the optical telescope assembly, provides a rigid optical bench and attachments for the instruments and the FGSs. The optical telescope assembly is surrounded by the spacecraft (called the support system module), which provides structural support; the electrical power system (solar cells, batteries, and converter); the data recorder and transmission system; the command and control electronics; and finally the guidance and pointing system, which includes rate-control gyros, fixed-head star trackers, magnetometers, and the FGSs. The major components of the optical telescope assembly are the primary mirror, the secondary mirror, the optical bench structure, the focal plane structure, baffles, and the support for the FGSs and the scientific instruments (Figure 12.1).

The primary mirror is constructed as a light sandwich of glass with a low thermal expansion coefficient. Thin plates of glass are fused to the front and back of an egg-crate inner section. This design was adopted to ensure rigidity and stability of the mirror during fabrication and testing on the ground and operations in orbit. The use of actuators with such a rigid mirror, with a classical thickness-to-diameter ratio of 1 to 10, is restricted to very small corrections, and the twenty-four actuators provided for Hubble proved in-

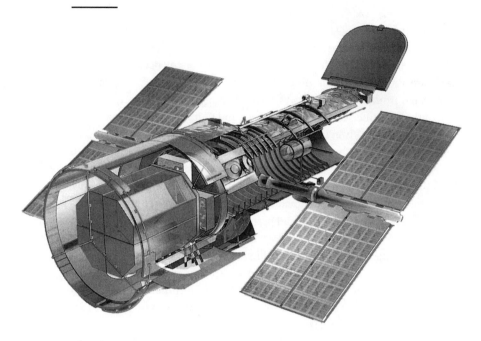

FIG. 12.1. *The Hubble configuration.*

effective in correcting the out-of-focus condition that was discovered after launch. (As a comparison, the actively controlled thin-meniscus mirrors of the VLTs of the European Southern Observatory, which rely on 180 actuators for obtaining an appropriate mirror configuration, are much thinner, with a ratio of 1 to 40.) The primary mirror had a diameter of 2.4 m, a focal ratio of 24. It was polished to better than 1/50 of a wavelength and vapor coated with aluminum and magnesium fluoride to improve ultraviolet reflectivity (Plate 8).

The secondary mirror has a diameter of 31 cm and is constructed of Zerodur, a very-low-thermal-expansion glass also used for Einstein, Chandra, and the VLTs. It was also coated, after polishing, with aluminum and magnesium fluoride and has six control actuators for fine adjustment and focusing in orbit. The secondary mirror reflects the light from the primary and focuses the images of celestial objects, through its central hole, onto the focal plane. The mirror assembly provides diffraction-limited images, at wavelengths longer than 1,150 Å, at the center of the field. This very high angular resolution (60 milliarcsec), the unique capability to observe the ultraviolet and infrared emissions from celestial objects above the earth's atmosphere (therefore unimpeded by its distortions and absorption), and the ability to obtain long uninterrupted exposures (on the order of days) are the primary factors that allow Hubble to compete effectively with 8- to 10-m telescopes on the ground.

The optical bench that keeps the precise alignment and spacing between the two mirrors is a rigid structure, built of graphite-epoxy, capable of maintaining the spacing to 2 μm and the co-alignment to 10 μm. The extremely low thermal expansion coefficient of the material maintains arcsecond stability of the image, as Hubble transits from the sunlit to the dark portions of its orbit, a displacement easily corrected with the FGSs.

The focal plane assembly provides mountings for four axial and four radial instruments and the three FGSs, for which the accuracy of the relative positioning is crucial. The installation and removal of these instruments is carefully controlled by guardrails, registration fittings, and latches. This system permits removal and repositioning of instruments by suited astronauts during servicing missions. The focal plane assembly also provides mountings for the fixed-head star trackers and the gyro rate sensors.

The Instrumentation of Hubble

The initial instrumentation of Hubble consisted of a wide field/planetary camera (WF/PC), a faint-object camera (FOC), a faint-object spectrograph (FOS), a Goddard high-resolution spectrometer (GHRS), a high-speed photometer (HSP), and the FGSs.

The WF/PC was developed at the Jet Propulsion Laboratories of the California Institute of Technology in Pasadena by a team led by Jim Westphal. It is a radial instrument that provides two fields of view: 2.57 arcmin^2 (f/12.9) and 77 arcsec^2 (f/30). The central beam of Hubble is intercepted by a flat pick-off mirror and split into four beams, which are focused on four CCD detectors of 800 × 800 pixels. The four resultant images are later reassembled on the ground in a mosaic image of 1,600 × 1,600 elements (Figure 12.2). The WF/PC has a wide spectral range (from 1,150 Å to 1.1 μm) and a wide dynamic range (from 8 to 28 visual magnitudes, a factor of ten thousand). The WF/PC was one of the most heavily used instruments on Hubble and obtained many of the breathtaking views of the heavens, such as the "pillars of creation" in the Eagle Nebula (see Plate 11), which have become Hubble icons.

The FOC is an axial instrument developed at the European Space Agency by a team led by Duccio Macchetto. Its primary role is to achieve the highest angular resolution imaging at f/48, f/96, and f/288 in very narrow fields of view of 22 × 22, 11 × 11, and 3.3 × 3.3 arcsec, respectively. The detector con-

FIG. 12.2. *The wide field/planetary camera.*

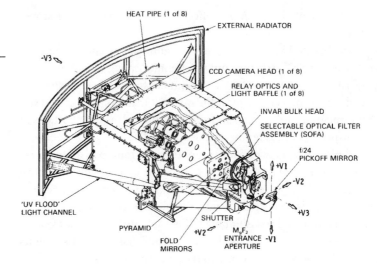

HEAT PIPE (1 of 8)

EXTERNAL RADIATOR

-V3

CCD CAMERA HEAD (1 of 8)

RELAY OPTICS AND
LIGHT BAFFLE (1 of 8)

INVAR BULK HEAD

SELECTABLE OPTICAL FILTER
ASSEMBLY (SOFA)

f/24
PICKOFF MIRROR

+V1

-V2

+V3

'UV FLOOD'
LIGHT CHANNEL

PYRAMID

SHUTTER

+V2

FOLD
MIRRORS

M₉F₂
ENTRANCE
APERTURE

-V1

sists of a three-stage image intensifier followed by an electron-activated silicon TV tube. This instrument allows the detection of single photons, with post facto image reconstruction on the ground. The FOC also has provisions for slit spectroscopy, objective prism spectroscopy, and polarimetry (Figure 12.3). It complements the WF/PC in giving better sampling of the Hubble point response, and better ultraviolet sensitivity, at the expense of infrared sensitivity.

The FOS was developed at the University of California, San Diego, by a team led by Richard J. Harms. It has the best sensitivity for faint objects, with a resolution of one hundred to one thousand for pointlike sources. It has two dispersive grating systems, which are viewed by two 512-channel digicons (Figure 12.4). These are silicon detectors bombarded with accelerated photoelectrons emitted by a tri-alkali or bi-alkali photocathode. The instrument can also perform polarization measurements. It has a wide wavelength bandwidth (from 1,150 to 7,000 Å). It is a very sensitive and therefore frequently used spectrometer for all problems, such as cosmological studies, that involve faint sources.

The GHRS was developed at Goddard by John C. Brandt and his team. Its purpose is to provide the highest spectral resolution achievable on Hubble (from two thousand to one hundred thousand). The wavelength range is shifted to the violet and ultraviolet in the range of 1,150 to 3,000 Å. The detectors are 512-channel digicons, as in the FOS (Figure 12.5). One of the detectors has a lithium fluoride faceplate to achieve the highest possible ultra-

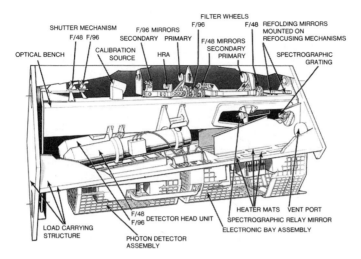

FIG. 12.3. *The faint-object camera.*

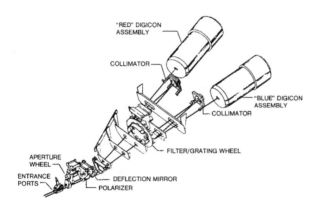

FIG. 12.4. *The faint-object spectrograph.*

violet sensitivity. The GHRS yields the highest resolution of emission and absorption line profiles for the study of luminous point sources, such as stars and quasars.

The HSP is the simplest instrument on board Hubble and consists of photomultiplier detectors covering the range from 1,150 to 6,500 Å with a maximum time resolution of 16 μsec. It was developed by Robert Bless of the University of Wisconsin and designed to detect time variability of celestial sources (Figure 12.6).

The FGSs were built at Perkin-Elmer and were intended for use both as instruments for astrometric measurements and for fine guidance. This dual use stems from the requirement that the pointing of Hubble must be accurate to astrometric standards in order to place the desired target within the slit of an

FIG. 12.5. *The high-resolution spectrograph.*

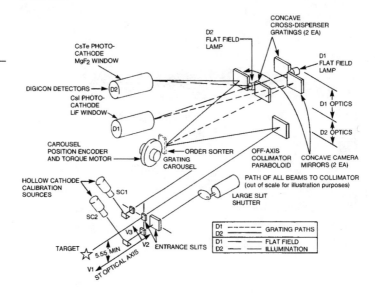

instrument and to maintain stability, during observations, of 0.007 arcsec. This value corresponds to a fraction of the slit width of a spectrometer or a fraction of a pixel in an imaging instrument. In each of the FGSs, the available field of view of 69 arcmin (Figure 12.7) is scanned with a Koester prism interferometer by means of rotating prisms (Figure 12.8). The fringes are detected by photomultipliers, which are sensitive to stars as faint as magnitudes 15–18. The FGSs can obtain accuracies of 0.003 arcsec at these magnitudes. This sensitivity to faint stars is required to ensure that at least a pair of guide stars should fall inside the aperture of each FGS. Bright stars (magnitude 10 or lower) are few and far between, and the 69-arcmin² field of view of the FGSs represents less than five parts in ten million of the sky; thus to have guide stars in the FGSs' field of view, faint stars (magnitude 15), which are much

FIG. 12.6. *The high-speed photometer.*

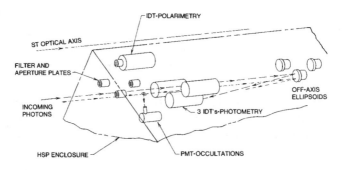

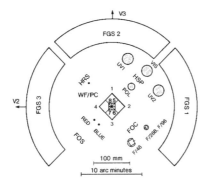

FIG. 12.7. *The disposition in the Hubble focal plane of the fine-guidance sensors and the instruments.*

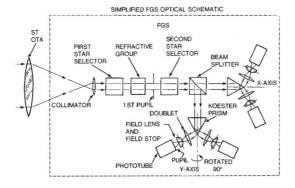

FIG. 12.8. *Principle of operation of the fine-guidance sensors.*

more numerous than brighter ones, must be used. An all-sky catalog of magnitude 15 stars did not exist at the time; therefore, STScI had to create it.

Hubble was carried into orbit by the shuttle, the only vehicle in the NASA fleet capable of lifting 17,000 pounds to 600 km above the earth's surface. Hubble orbits the earth every 100 min. This orbit is not ideal; from a scientific point of view, a much higher orbit would have resulted in unimpeded and more efficient observing for longer times. Later, in fact, higher orbits were chosen for Chandra and for the James Webb Space Telescope (the successor to Hubble), but in the early 1980s, the concept of shuttle servicing implied a low-Earth orbit that would be accessible to astronauts. Because Hubble is not visible from a fixed station on Earth except for a fraction of its orbit, it uses the Tracking and Data Relay Satellite System (TDRSS) for command reception and data transmission. TDRSS is a system of three communication satellites in geostationary orbits that relay data from and commands to Hubble through its ground station in New Mexico. From there the data are retransmitted to the Domsat satellite, which sends the data to the Hubble

Operation Control Center and Data Capture Facility, connected by ground line to STScI. Because the TDRSS system is not always available (because of higher-priority missions by the military or manned missions), the scheduling of data and command transmissions from and to Hubble requires a very sophisticated software system.

A Diffusion of Responsibility: Hubble's Management

The management of the Hubble program was dictated more by the interests of NASA's centers than by the requirements of the program. NASA had given Marshall the lead role in the program, but they had no authority over the portion of the program assigned by headquarters to Goddard. Lack of cooperation between the two NASA centers had caused a number of problems from the start of the program. As early as 1976, these problems prompted several astronomers on the Shuttle Astronomy Advisory Committee of NASA to advocate a reorganization of the project, a suggestion that headquarters did not follow. After prolonged negotiations, Marshall retained the lead role in the program (with responsibility for cost, schedule, and technical performance of the program as a whole) and the direct management of the contracts with Lockheed for the support system module and with Perkin-Elmer for the optical telescope assembly. Goddard was designated as the center responsible for the scientific instruments, the Hubble ground system, the mission and science operations, and the management of STScI. Lockheed and Perkin-Elmer were again associate contractors.

The worse aspect of this complex management approach was the lack of strong, coherent systems engineering. Marshall did not have the manpower to carry out this function in house and ultimately contracted with Lockheed to provide the systems engineering and the telescope assembly and verification functions. But Lockheed was not given prime-contractor responsibility for both their own and Perkin-Elmer's work, and so was not responsible for the ultimate performance.

To explain why systems engineering is important, one can take the guidance system as an excellent example. Perkin-Elmer had the responsibility to build the FGSs, whose signal would be fed to the guidance system constructed by Lockheed. The data from this system would be transmitted through the support system module, to TDRSS, Domsat, and finally to Goddard, which would use its command and data-handling subsystem to point the telescope in the desired direction. It is the function of a systems engineering group to ensure that all interfaces of this process are operational and compatible. Similarly, the assembly, test, and verification activities should be carried out under the control of a systems engineering group whose function is to determine which tests are required, analyze the results against predictions, and verify compliance with the scientific requirements. Ideally, this group should be part of the organization responsible for the success of the program. In practice, the fractured organizational structure of Hubble was the root cause of many of the inefficiencies of the program.

An additional problem was the large number of important NASA managerial positions associated with Hubble. Robert Smith lists twenty such positions held from 1977 to 1987 and occupied at various times by fifty-four individuals.[7] Such a plethora of managers is sure to hinder rather than help the work. Under these conditions, the competence, steadfastness, and dedication of the Marshall chief engineer for Hubble, Jean R. Oliver, was a great help in bringing cohesion to the program. Also important during the development phase was the creation of the Science Working Group under the chairmanship of Bob O'Dell as a forum for airing problem and addressing issues.

The Science Working Group was composed of the five principal investigators for the instruments (Robert Bless, John Brandt, Richard Harms, Hendrik Van den Hulst, and James Westphal), four interdisciplinary scientists (John Bahcall, John Caldwell, David Lambert, and Malcolm Longair), five NASA scientists (Lewis Hobbs, David Leckrone, Bob O'Dell, Nancy Roman, and Ed Weiler), one European Space Agency scientist (Duccio Macchetto), two Hubble scientists (William Fastie and Daniel Schroeder), and the team leaders of the Data and Operations Division (Edward Groth) and the Astrometry Science Division (William Jefferys). After the formation of STScI, the director (myself) and deputy director (Don Hall) became ex-officio members, and many of the senior members of the institute's scientific staff participated in the meetings by invitation. The main task of the working group was to bring issues raised by scientists to the attention of the project as a whole and ensure that the instruments were being developed on time and were being properly funded. The working group had less of an effect on project-wide issues, such as adequate testing, operations planning, and data handling.

STScI's Organization

The AURA management approach was modeled on that used successfully for ground-based observatories, modified as required by the observational experience with space observatories and through an ongoing process of learning. At STScI we considered the management structure as only a tool to permit the staff to work in a coordinated fashion, and it evolved significantly from 1981 to 1988.

The internal management of the institute was based on the recognition that technical leadership was the primary need in a task as complex as that assigned to STScI. A second important consideration was the need to facilitate communication throughout the organization both vertically and horizontally. Ease of communication permitted major decisions to be made by a consensus of informed people. Thus the process ensured that technical matters were openly and candidly discussed, and that the decisions, once taken, were fully understood and owned by the entire staff. It was part of management's responsibility to ensure that debate would come to a conclusion within the time allowed by programmatic needs, and that once a decision was made, all of the staff would work cooperatively to implement it.

To this day, AURA has the ultimate responsibility for management and is the legal entity that holds the contract with NASA. AURA is a nonprofit consortium of universities, founded in 1957, which at my last count included twenty-one institutions with major programs in astronomy. Its primary task is to establish, operate, and maintain national astronomical observatories for use by all qualified scientists. AURA is governed by a board of directors selected by representatives of the universities. The board elects the officers of the corporation and its several committees, including a specially chartered Space Telescope Institute Council. The governance model has changed through the years, but the AURA board has the sole authority to appoint the director and the deputy director; to grant tenure; and to approve personnel policies, salary scales, and contractual commitments.

The president of AURA is its chief executive officer and a member ex officio of the board. The STScI director reports to him. John Teem was president of AURA during the formative years of STScI; he brought to the job vast managerial experience, a belief in due process and fair dealing, and a great deal of personal strength.

The Space Telescope Institute Council is a special committee of the board, established at NASA's request, to oversee STScI operations. It has broad responsibilities, including the selection of candidates for director and deputy director (to be submitted for approval by the AURA board and NASA). It interacts with the director through quarterly meetings, and it can invite NASA representatives to attend its meetings. The council also appoints the Visiting

Committee of the institute and reviews and approves proposals by the director for tenured staff appointments. During the 1980s, the council had twelve members, and to recognize the international nature of Hubble, it included several from member countries of the European Space Agency.

Lyman Spitzer of Princeton University chaired the Space Telescope Institute Council during my tenure as director (1981–92). His technical competence, scientific reputation, and dedication to science were great assets to the Hubble program, STScI, AURA, and NASA. Spitzer and I were not always in agreement, particularly with regard to the degree of automation that I was striving to achieve for many of the operational and data-handling procedures. He was concerned that my plans were too bold and could not be successfully executed, and he therefore paid particular attention, at council meetings, to our progress in those areas. He always recognized, however, that it was my responsibility as director to make choices, and because of this fair and thoughtful behavior, I listened most carefully to his advice.

The Visiting Committee of STScI provides an independent review by the research community of the institute's performance. It is composed of members who have no connection with AURA, the European Space Agency, or NASA. During its yearly visit, it has access to any information it requires and can interview any member of the staff. Its report goes to the director, then, together with the director's response, to the Space Telescope Institute Council, to AURA, and finally to NASA.

Although AURA had the legal responsibility for the NASA contract, STScI interacted directly with NASA headquarters (at the level of the associate administrator for the Office of Space Science and Applications) to discuss policies and to represent the views of the community of Hubble users. STScI also maintained direct contact with the Hubble program scientist at NASA headquarters.

Goddard managed the STScI contract and provided technical direction to the institute. Smooth interaction between STScI and Goddard was essential to carry out the mission, and was in the main achieved, even though Goddard scientists had some difficulties in setting aside their resentment at the creation of an institute in the first place.

The European Space Agency participates in the program through a cooperative arrangement defined by a memorandum of understanding between NASA, the agency, and Hubble. The agency contributed the solar cell array, the FOC, and fifteen scientists and engineers to the staff of the STScI. The agency also established a European Coordinating Facility (hosted by the European

Southern Observatory in Garching, Germany), which worked cooperatively with STScI in creating and maintaining a full archive of the Hubble data.

The Johns Hopkins University is the host institution for STScI. It provides facilities and services, a variety of scientific support (including seminars and access to its library), and a great intellectual and cultural environment for the institute staff. As the first director of the institute, I was made a full professor in the Department of Physics and Astronomy. This appointment was quite important to me in the performance of my duties. In case of extreme controversy with NASA on an important point of principle, I had the option (and job security) to resign my position as STScI director without hurting my family; I should add that I am very glad I never had to resort to this option. The cooperation between STScI and Johns Hopkins went far beyond my appointment to the faculty: joint appointments of senior staff permitted the recruiting of some of the best astronomers in the community, and the initiation of an expanded graduate program run jointly by the university and STScI furnished research opportunities for faculty and staff. My personal relationship with Steve Muller, then president of Johns Hopkins, was extremely friendly and productive.

STScI's Relations with the Greater Research Community

Because of the early creation of the Hubble Science Working Group (preceding the establishment of STScI) and NASA's reluctance to yield control of the scientific program to an independent institute, the issue of who represented the research community's interests in the program and who would in fact lead the science was very much an item of conflict, even though the mandate by the community to STScI was quite clear.

For my part, being blissfully ignorant of optical astronomy, the answer was quite clear: I would rely as much as possible on the scientific views of the best optical astronomers and listen carefully to their advice. I would rely on my own experience on how to do science in space, how to manage large projects, and how to run an institute.

Scientific advice came to STScI through the Space Telescope Institute Council, chaired by Lyman Spitzer, whose members were senior scientists. However, because the council's functions included oversight and review of tenured appointments, I created a new committee, the Space Telescope Advisory Committee, whose members were outstanding astronomers and whose function was limited to providing scientific advice in the planning of the

Hubble program. This committee, for instance, gave us important guidance on fundamental scientific issues, such as the definition and creation of the large Hubble programs in the different subdisciplines of astronomy. Following their advice, I created, very early on, subgroups of outside astronomers for specific disciplines (for example, a subgroup of planetary astronomers), usually chaired by an STScI scientist in the same discipline (for example, Robert Brown for planetary astronomy). These groups gave us direct input on the requirements of the potential users, which, in turn, could influence the development of Hubble.

Because of the input from these committees and groups, we at STScI felt that we were the legitimate representative of the community's views. On the other hand, the Science Working Group, chaired by Marshall project scientist Bob O'Dell and composed mainly of the Hubble principal investigators, had developed a rather proprietary view of the project. The working group did not have established mechanisms to solicit community views, except through the NASA system, which was not at all what had been envisaged by the original 1976 NAS committee (chaired by Hornig and discussed above) that first recommended the institutional arrangements for Hubble. It did not help that the institute's reviews of different technical areas to assess what needed to be done appeared to working group members as a criticism of the latter's work in previous years. There were in fact clear deficiencies in the guidance they had given to the program since its start. The working group met only once every three months, and neither the principal investigators nor the project scientist had the staff required to conduct in-depth analysis of specific issues so as to propose solutions to problems as they arose.

Although the Science Working Group included (in addition to the five principal investigators), four interdisciplinary scientists, two telescope scientists, an operations team leader, and a leader of the instrumentation definition team for astrometry, many of the program reviews in later years revealed a lack of scientific guidance for the contractors developing essential pieces of the system. I discuss some of STScI's remedial actions in the next chapter. They included taking on a leadership role in the engineering process to ensure compliance with the scientific goals. This role was not understood by the working group.

However, as already stated, the most serious problem for the program was the lack of an overall system design for the scientific operation of Hubble, a task that the Science Working Group had not even attempted. This oversight can be explained in part by the lack of experience of ground-based astronomers in the design and use of space facilities, in part by NASA's reluc-

219

tance to give up authority and responsibility (even though they were not doing the job), and in part by the great changes in the style of doing astronomy brought about by advances in computer capabilities and software sophistication, of which both NASA and the working group seemed unaware.

The principal investigators conceived their task to be that of building the instruments for which they had been selected and producing good science with them. The idea that they were involved in building an international observatory that must allow the world's astronomical community easy access to Hubble and must provide them with quality data did not seem to be high among their priorities.

A Vision for STScI

I had carefully read the 1976 Hornig committee report, with which I was intellectually and emotionally in full accord. Its view that scientists must take responsibility for major facilities, for their development and their scientific use, resonated with what I had been attempting to bring about in x-ray astronomy. Along the way, however, some of the recommended roles, such as oversight of the hardware during development—which might have avoided Hubble's mirror problem—had been eliminated by NASA ukase. Because NASA had delayed the start-up of the institute, many of the development contracts for the software that the institute would eventually have to use were already in progress when the institute began to function. In other words, NASA's philosophy was that its contractors would deliver a turnkey system to the astronomers of the institute, which was conceived as a service facility. I was at total variance with this view, not only on scientific but also on practical grounds. In my experience, it was only through close cooperation between the technical developers and the ultimate users that a successful operational system could be achieved.

I had another fundamental objection to NASA's view. The major reason for creating an independent institute was to have the ability to recruit the very best scientists. They would be responsible for ensuring that Hubble, when completed, would be able to support a world-class observing program, with the full participation of the entire community. In thinking about the quality of the people necessary to fulfill these requirements, I felt there were two models to avoid. The first was that of NASA itself, which used a recruitment and promotion process different from that of the academic community. This process resulted in an in-house staff that was typically isolated from the rest of the scientists in their field and often not of top quality. I was convinced that the scientific staff should be—and be perceived as—very much a part of the community it served. I often used the image of a wide-mesh sieve with water flowing though it to describe how I saw the relationship of the institute's staff to the research community.

The second model to avoid was the approach used by some national facilities I had come to know, where the scientific staff providing services was separate from the staff conducting research. My own experience with the Einstein program was that the best scientists gave the best service; we should therefore ensure that all STScI scientists had a clear service assignment. The motto for tenure-track scientists at STScI was: Everybody works, everybody plays (does research).

As a result, appointments at STScI were limited to the number of scientists necessary to fill jobs in which scientific judgment was required. To retain their scientific credentials, the staff had to participate in active research. For scientists hired under the AURA tenure track, 50 percent of their time was set aside for research. This was an expensive requirement because, at least in principle, each such position was twice as expensive as a pure service appointment. However, my experience had been that the cost of a highly talented and motivated staff would be more than offset by their ability to fully and quickly understand what had to be done; reduce the requirements to those strictly necessary for science; and quickly complete the work, whether they did it personally or by leading a group. The concern to avoid overstaffing the institute was equivalent to the policy of setting a limit on faculty positions in a university.

To ensure the quality of the staff to be recruited, the senior scientific staff and I essentially followed university procedures: open advertisement, a senior selection committee for non-tenure-track appointments, and a full search (including outside letters of reference) for tenure-track appointments. For tenured appointments the STScI director presented his recommendations, together with the recommendations of the senior staff, to the Space Telescope Institute Council for approval. We were conscious that sheer scientific brilliance alone was not a sufficient criterion to hire somebody, because staff research at STScI was a derived rather than a primary requirement. We needed people with appropriate technical backgrounds who would and could take on service responsibilities, and who would be capable of leading or being an effective member of a group. These criteria played an essential role in hiring and promotions.

When I assumed the directorship of STScI in 1981, the launch of Hubble was still officially scheduled for January 1985. Unofficially, everybody involved in the program knew that this date was much too optimistic. In fact it was only the four years' delay caused by the tragic loss of the shuttle *Challenger* in 1986 that allowed all parts of the program (including the STScI) to be ready for launch in 1990.

Of course we at the institute did not know this in 1981. We were confronted with a huge amount of work. For one thing, we had to fully understand all aspects of Hubble operations in orbit. In this area I was very glad to have the support of Computer Sciences Corporation, a contractor that had the resources to provide us with qualified manpower in a short time. Of the approximately one hundred people ultimately provided to STScI by this corporation, twenty were trained as astronomers. They were judged strictly on the basis of their technical performance, but were treated as colleagues and encouraged to participate in the scientific life of the institute. They could apply for telescope time and could take time for research by obtaining grants or corporate research funding.

We also were blessed with a strong contingent of European astronomers sent by the European Space Agency as part of the latter's agreement with NASA. Other than for administrative purposes, no distinction was made in the assignment of tasks and responsibilities between staff from AURA, the agency, or Computer Sciences Corporation. Whoever could lead was given authority and responsibility. This open atmosphere was an attempt to recreate in changed circumstances the Uhuru spirit of giving preeminence to ideas and merit.

It was my great ambition to create an institute that would strive for excellence in all its work, but that would also provide a place were people liked to come to work, and where they would have ample opportunity for professional and personal growth. By adhering to high standards in hiring, we made sure we could tackle effectively almost any Hubble technical issues and that, after a stint at STScI, astronomers would be able to pursue successful academic careers at any major university.

We were in fact lucky in recruiting a brilliant scientific staff that only left STScI to accept academic appointments at major universities and observatories. In the initial selection of people, I was greatly helped by Don Hall, the deputy director chosen by AURA, who was a very good scientist and an excellent judge of people. He was knowledgeable about optical astronomy and astronomers, and I relied heavily on his advice.

Paradigm Shifts:
The Space Telescope Science Institute at Work

Assessing and Debating Institute Costs

When I assumed the directorship of STScI in 1981, I was determined that the institute would be successful and meet or exceed the expectations of the community. It seemed to me that we could do it, in a practical sense, by ensuring that after-launch scientific operations would be flawlessly executed. We should never be late in providing NASA with the operational and scheduling information they required, never be late in providing adequate guide stars, and never be late in anticipating hardware problems. In other words we should be the most efficient support contractor that NASA ever had, because only by achieving this kind of efficiency could we dispel the criticisms of the naysayers and have the freedom to carry out the research program according to our best scientific judgment. To achieve this goal, we needed to manage ourselves as tightly as I had ever seen it done in private industry. We needed to quickly assess the real cost of the institute and its staff, and analyze in depth all issues that could affect performance.

In my first few weeks on the job, together with some of the early senior staff, I wrote a plan for the internal organization of the institute that also described its interfaces with AURA and its committees (including the Space

Telescope Institute Council and the Visiting Committee), NASA headquarters, Goddard, Marshall, the European Space Agency, and our subcontractor, Computer Sciences Corporation. (This document, which was known in NASA parlance as MA-03, was one of a long list of documents required as deliverables by NASA for their approval.) It defined the scope of the institute and the manner in which it was to operate. It was approved, and notwithstanding many modifications, this blueprint remained in effect over the years, giving us a stable base from which to work. Even this early in the program, I included support for in-house research and outreach activities in the plan.

With the help of my colleagues (the professional staff at the time numbering about twelve) and using my best educated guesses (based on 20 years of successful space projects), I then estimated the cost of preparing the necessary hardware, software, facilities, and methods and procedures for operations three to four years hence, working from the ground up. The net result was a cost estimate that we considered correct but that unfortunately was larger, by a factor of two, than what had been anticipated by the 1976 Hornig committee and AURA. Unpopular as our assessment might be with NASA, I thought that it was important to obtain and provide a realistic idea of what it would take to do the job from the very beginning. This information is essential to the agency, because budgetary plans for congressional approval are drafted far in advance of their submission. It took a few years and many reviews to get NASA and the community to agree that the projections from STScI were realistic and appropriate. Although I did not later increase my estimate, except for additional tasks requested by NASA, each step I took to achieve what I had originally proposed was regarded by NASA as an overrun.

During the many debates that occurred concerning cost issues, NASA could always solicit and find scientists willing to write opinions inimical to the institute. NASA headquarters itself originated some of the misinformation on the projected size and cost of the institute, which raised the concerns of some astronomers. Negative remarks were also made by some of the principal investigators on Hubble as well as by Project Scientist Bob O'Dell.[1] Even the journal *Nature* waded into the debate in 1994 by showing a cartoon of the STScI as a baby chick of enormous size ready to devour all of NASA.

It has always baffled me that the same people did not express concerns over the huge overruns of the telescope itself. At launch, Hubble costs had risen to $1.2 billion (not including launch, at a cost estimated to be $600 million, or one-sixth of the shuttle program's cost), that is, some four times the initial estimate. The operational costs of Hubble were on the order of $250 million per year. At about $30 million per year, the budget of the STScI in the

1990s was comparable to that of a ground-based observatory; it was about one-eighth of the operations cost, which in turn was one-seventh of the construction and flight costs. Thus STScI made up about 2 percent of the known costs of the project and perhaps much less, since at that time NASA did not have a full cost-accounting system. I find it difficult to imagine that these simple calculations were not clear to everybody. Finally, by 1987, after many audits, reviews, and committee discussions, and with the strong support of the AURA board, these issues were set to rest and the STScI budget was accepted as being appropriate.

Grappling with Hubble's Technical Issues

The task of properly managing the institute paled, however, compared to the difficulty of fully understanding the technical issues that would make the mission a success or a failure. The basic problem was that neither NASA nor the core group of scientists originally involved in the program had realized the complexity of the task.

Hubble observing occurs on a 24-hours-a-day, 7-days-a-week schedule. The daily input of data into the Hubble archive today is between 3 and 5 gigabytes, and the data distributed by this archive amount to 10–15 gigabytes. Compared to the standards of the time, managing this data flow was akin to trying to drink from a fire hydrant. Thus the reception, calibration, and reduction of the data had to be highly automated if we were not to be swamped by the incoming data. The data distribution system required an equally high degree of automation to stay current and not be left hopelessly behind in getting the data to the astronomers. It also soon became clear that automation would be essential for both planning and scheduling operations and observation time. Similarly, the problem of providing guide stars with precise positions in a timely fashion for efficient observing had not been properly evaluated. Nor had the problem of data reduction and analysis after data reception been seriously considered.

The staff and I decided early on that we would dedicate our scientific and engineering talents to fully analyze each of these problems and assure ourselves that appropriate software and procedures had been developed; if these were not available, we would develop the needed tools ourselves. Since we were an unknown entity to the rest of the project, we decided to do it in gradual steps. We would consider a problem and try to understand it fully, better perhaps than anybody in the program so far, without regard for any com-

225

promises or misunderstandings that might have occurred in the past. To borrow a term from the military, the goal was to attain local air superiority. Once this was achieved on one subject, we would quickly move to the next one, until we had covered all aspects of Hubble that could affect its scientific performance. Given what appeared to be a very tight schedule, we decided we would not fish for minnows but only for great white sharks, that is, problems that could really damage the program if not fixed before launch.

Notwithstanding this prioritization, the amount of work to be done was daunting. It kept all of us at STScI fully busy, and all the staff had more to do than they thought they could accomplish. We were sustained by a feeling of camaraderie—there were few internal struggles for more responsibility, and considerable mutual support. It was, however, painful to do our work surrounded by the hostility and lack of understanding of some of our colleagues both inside and outside the Hubble program.

It would take a book to give a full account of all that happened between 1981 and 1990, the year of the launch; I will describe only some specific contributions made by the institute that I consider important.

Guide Star Catalogs I and II

Early studies of Hubble operations by the astrometry team led by William Jefferys resulted in the design concept of how to obtain the guide stars, which was incorporated into NASA's call for proposals for the STScI. The idea was to create an astrometry system capable of producing on demand the guide stars required by the Hubble observational schedule with an accuracy of 0.33 arcsec.

Schmidt plates of the sky would be used as the basic material. In the northern hemisphere the plates would be taken with the Oshin 1.25-m Schmidt Telescope at Palomar and in the southern with the U.K. 1.2-m telescope at the Anglo-Australian Observatory. The plates needed to be contemporary, so that the guide star locations would not be greatly affected by proper motion for ten years. Two microdensitometers would be used to scan specific fields for the necessary guide stars, and an astrometric plate solution (an analytical fit that yields a true sky position for each point of a plate) would give the required guide star list.

This approach sounded reasonable except for the following:

• A precision of 0.33 arcsec in stellar positions obtained from a Schmidt plate had been achieved only rarely. To our knowledge, only William van Altena together with his graduate students at Yale had achieved this precision during his astrometric research (in 1972–77) and under laboratory conditions.

• The planned scheme could produce a marginally adequate output of guide stars only if two scanning microdensitometers were kept fully operational on a continuous basis, with an operational staff of fifteen to twenty-five people.

• Original photographic glass plates (about 12 × 12 inches) would have to be handled on a daily basis by the operational staff. Any breakage could not be replaced faster than several weeks, during which time Hubble could not be pointed to that region of the sky.

Altogether it seemed to me to be a scheme doomed to failure. Perhaps these weaknesses were understood by the Science Working Group, but to accept such a risky approach seemed to me incompatible with the performance the community had the right to expect from Hubble.

Barry Lasker was a member of the staff of the National Optical Astronomy Observatory who had joined the AURA team in the preparation of the STScI proposal and had then become one of the first members of the STScI staff. He had clearly thought about this problem at great length, but it was not in his character to be confrontational, and he had preferred to await the arrival of the new director to express his misgivings. Together he and I developed a new solution that was quite simple but very demanding.

Our idea was to obtain all the necessary photographic materials before launch, and to scan all the plates, some 1,500 of them. We would then obtain astrometric solutions for all the plates and create a Guide Star Catalog containing all the stars in the sky to magnitude 15. This catalog would be maintained in the STScI computers in digital form and would be available on line to support the rest of the STScI software for proposal preparation, planning of observations, scheduling, target acquisition, and guidance control.

Should we not be able to complete the task before launch, we could accept operational limitations, provided the initial areas for survey were selected with some care. For instance, we knew we could complete scans before launch of the galactic center and the regions near the galactic poles, and of some of the perennial favorites, such as the Large and Small Magellanic Clouds and the regions centered on the Coma, Perseus, and the Virgo clusters. After launch, we could scan the rest of the plates to include the entire sky. Furthermore, even if we could not complete the task, we would be left in a much better position than if we had not started at all. If we could not scan the rest of the sky in real time, after having a substantial head start before launch, we could not complete the task at all without a head start.

The delay in the launch of Hubble from 1985 to 1990 enabled Barry and his group to nearly complete the scans of all 1,500 plates. He and his collaborators not only provided a perfect operational tool but also created a precious new resource for all astronomy.

The first hurried scans of the plates with a resolution of 1.7 arcsec produced Guide Star Catalog I, which contained 18,819,291 objects to magnitude 15.5, with errors in star positions between 0.5 and 1.2 arcsec. A second catalog (Guide Star Catalog II) was constructed by the year 2000 to magnitudes between 18.5 and 19.5 with scans of 1.0-arcsec resolution; it contained 435,457,355 objects, with errors in star positions between 0.3 and 0.75 arcsec. For each FGS field of 69 arcmin², some nine objects could be found on average in the first catalog and many more in the second one.

It turned out that the original requirement developed by the Science Working Group was much too strict for operational needs, and the approach developed by STScI worked beautifully. As a result, most of the requirements for pointing the telescope could be met automatically, using only 10 percent of the operators' time; the remainder of their time could be concentrated on the few really difficult cases of correcting observers' errors in the coordinates of study objects.

The results of this mammoth effort were of such quality that they were published by Lasker and his collaborators in the June 1990 issue of the *Astronomical Journal*. Details on the current status of this work are available on the STScI web site.

The scan of the plates for the construction of Guide Star Catalog II was done with such fine resolution that essentially all the information on the plate was transferred to the digital scan. Because the star images occupy only a small fraction of the field (most of it being empty of objects), data compression techniques could be used to fit the data onto commercially available disks for storage and distribution.

The inspiration for this next step came to me as the result of a visit in 1990 —after the fall of the Berlin Wall—to the Zentralinstitut für Astrophysik in Potsdam, Germany, as a member of a Max Planck Visiting Committee. Hilmar Lorenz and Gotthard Richter of the Zentralinstitut gave the committee a demonstration of a rapid algorithm based on wavelet transform, which had been developed by the astronomers under contract to the Soviet navy to reconstruct soundwave images obtained by submarines. It was also used for astronomical studies by Massimo Capaccioli, Enrico V. Held, Lorenz, Richter, and Rainer Ziener to perform surface photometry on NGC 3379.[2] I was struck by the fact that the method had to be very effective, given the modest computer capabilities available in Potsdam, and that it could be useful in several astronomy projects.

I inflicted the task of learning this program and modifying it for x-ray astronomy on Piero Rosati, who was doing his PhD at the Johns Hopkins University with Colin Norman and me. He would use the algorithm to distinguish between pointlike and extended sources in ROSAT images. Davide Lazzati continued this work for his Italian doctorate.

For optical observations, Richard White of STScI wrote a program based on the Haar transform (H Compress) generalized to two dimensions. This algorithm was used to compress the STScI digitized survey of the sky. Up to a compression factor of one hundred, there is little loss of information.[3] The all-sky map can be compressed to a few CDs and therefore made available to professional and amateur astronomers.

Thus STScI solved the operational problem; did it correctly and with great precision; removed the risk, the drudgery, and the cost of doing it on the fly; and produced a useful tool for astronomy that has had worldwide distribution. In this area, STScI had achieved local air superiority. It was a challenging job that was fun to do.

Revamping the Science Operations Ground System

As I mentioned previously, Goddard had decided to contract with an aerospace firm (TRW) for the development of the Science Operations Ground System as a turnkey system for STScI. This approach was less than desirable, because it did not allow the ultimate user (STScI) to have a say in the tools that would be available to carry out its responsibilities. NASA, however, has always behaved as though scientists were neither capable managers nor engineers, notwithstanding all evidence to the contrary. Furthermore, by starting the procurement of the ground system well before STScI was functional, Goddard retained control of the development of this important operational tool. This philosophy, wrong as it was, might have been workable had there been more input from astronomers or operations experts. But, as reported in all subsequent reviews by NASA itself, it is evident that neither the Science Working Group nor Goddard (with all the team leaders, project scientists, and project managers associated with the development of the ground system) gave the necessary guidance to TRW to provide software adequate to the task.

Enter Ethan Schreier, the scientist who became responsible for operations and data analysis at STScI when he joined the STScI effort in 1981. We had worked together as scientists and science systems engineers on Uhuru at AS&E and on Einstein at Harvard.

I should define here what was meant by science systems engineering, an expression we at STScI came to use with increasing frequency. It signifies the analysis of a scientific research problem in all its dimensions, even before developing the instrumentation. Starting with a clear definition of the problem, we would design instruments capable of obtaining the necessary data, then plan how these data would be analyzed, determine what errors might occur because of the intrinsic limitations of the instruments, and define the expected statistical weights of the observations. Only when it was clear that crucial results could be obtained would we proceed with the project. The principles of science systems engineering would be applied throughout the lifetime of the project to ensure that no changes occurred that would jeopardize its scientific success.

One of Ethan's most inspired actions was to recruit Rodger Doxsey, an x-ray astronomer from MIT who had worked on the HEAO program. Rodger became one of the few people in the world who understood Hubble as a whole, in all its complicated sophistication, and has been recognized by NASA and by the community to have played an essential role in the success of the project. He is an experimental scientist who could design and systems engineer experiments, lead a group in the development of sophisticated operational software, and, in a pinch, create in two weeks the management information system we needed to prepare our first budget at the institute.

In late 1981, Ethan gave Rodger the task of reviewing the ongoing development of the Science and Operations Ground System. Rodger found a grave situation: not only were there some seven hundred problems with the software, some of critical importance, but the overall design was fundamentally flawed in its technical and operational aspects. Rodger concentrated his efforts on aspects of scheduling and spacecraft control; Ethan himself pitched in on the software and data processing.

There was no concept of an overall command language in the software. No consideration had been given to the transportability of the system to computers other than the original VAX complement, and no study of throughput capabilities had been made. The ground system required operators to enter commands manually. Ethan and Rodger estimated that more than thirty-five people working shifts around the clock and typing commands at forty words a minute could not keep up with the required instruction flow. In addition, no application software had been envisaged, although it was required not only for data analysis but as an operational tool.

Ultimately the scheduling of operations and their command blocks had to be automated using very sophisticated systems. These included expert (or artificial intelligence) systems, which could take into account orbital constraints, instrument and telescope configurations and cycle times, availability of TDRSS for issuance of commands, and reception of data, all for the purpose of maximizing the effective observing time.

Apart from these technical problems there were also glaring omissions from the point of view of operations. For instance, notwithstanding NASA's oft-repeated commitment to use Hubble for planetary science, Hubble could not actually track planets with the system as it was. The planets move in the sky (because of their orbital motions around the sun) rapidly enough that, as Hubble tried to follow their motion, different guide stars would come in and out of the field of view of the FGSs. The Science and Operations Ground System as originally constituted could not cope with these changes in guide stars. What was needed was a software program that would permit the smooth hand-off of guidance from one set of stars to another. This deficiency was identified by STScI scientist Bob Brown, a planetary astronomer who was trying to understand how to use Hubble for his own research. A system was ultimately developed under STScI guidance that provided the necessary capability.

As another example, although a Ritchey-Chrétien telescope design allows sharing of the focal plane by several instruments without loss of angular resolution, no plan had been made to use more than one instrument at a time. STScI developed the software tools to permit this type of operation, which is now fairly routinely scheduled.

Ultimately the full system used by STScI for scientific operations grew to three million lines of software code, one of the more sophisticated systems in the world. It included capabilities to support proposal preparation, proposal selection, planning and scheduling, on-line data reduction and calibration, data analysis, data archiving, and distribution.

One of the key capabilities that STScI developed was the ability to reduce the incoming data using a pipeline processing mode. This mode applies calibrations to the data as they stream in and transforms them from numbers to measures of physical quantities ready for analysis by astronomers in a continuous flow at a rate two to three times faster than the input rate. Such high pipeline rates were necessary to permit off-line use of the system for data reprocessing.

The degree of automation needed to absorb the high data flow from Hubble had not been fully understood by TRW and its NASA sponsors. In particular, the need to calibrate the data in a pipeline mode seemed beyond the vision of the optical astronomers involved in the project. After heated discussions about the need for a Hubble Data Analysis System and for on-line calibration, we at STScI were considered dangerous visionaries who wanted to undertake an impossible task. Some Science Working Group members considered the idea of pipeline data processing unscientific and useless, and actually advised NASA to forbid STScI to even study the requirements for a data analysis system. Since Ethan, Rodger, and I had worked with a comparable data analysis system on the Einstein program, we knew we were not visionaries and only needed the rather modest resources necessary to do the job to be successful.

But our struggles to establish proper data processing systems were modest compared to the fights over data calibration.

Overhauling Hubble's Data Calibration

Data calibration was one of the most contentious scientific subjects in the Hubble program. The principal investigators of the various instruments were firmly convinced that they were the only scientists capable of properly calibrating the data from those instruments. They fully intended to receive their own data and calibrate them in a methodical albeit slow manner. They did not seem particularly concerned about guest observers, even though the latter would eventually be entitled to 70 percent of the observing time. A typical comment might be: "If the guest observers are not able to calibrate their own data they should not be allowed to observe."

This idea ran completely orthogonal to STScI's vision of its purpose. Our view was that it was the institute staff's duty not only to make observing time on Hubble available but to provide the observers with data of sufficient quality to meet basic scientific standards. Only in special cases should further data refinement be required.

I remember describing my methodology for data calibration to David Leckrone, the Goddard instrument scientist, in the fall of 1981. I started out with a model of the instrument, whose response to incoming radiation (which was determined by its design and by physical laws) could be described by a set of physical parameters. To calibrate the instrument meant to measure these parameters on the ground and verify them in orbit.

Thus, to perform a pipeline calibration, what we needed were models of the instruments and their parametric description; a method to measure these parameters on the ground; a method to periodically verify the values of these parameters in orbit; an observing program that obtained the necessary data from reference objects; a data reduction and analysis program that extracted the parameters from the data; and finally, a method to modify the input values of the pipeline calibration system to accommodate changes or improvements in our knowledge of the parameters.

In using this calibration approach, it was essential that the conditions of the entire Hubble system—for instance, temperature or electronic gain of the instruments and telescope—be kept constant (or at least be known), and the relationship between the measurements and their actual values also be well defined. I defined these requirements as science configuration management, something actually easier to achieve in space than on the ground, but something considered too restrictive and too disciplined by some romantic astronomers.

After my explanations, Dave Leckrone looked at me as if I had grown two heads. He was not alone in having doubts that this work could be done at all, and even Lyman Spitzer was concerned about our chances of success.

Robert Brown, the planetary astronomer who had joined the institute from the University of Arizona in 1982, took on the demanding task of conceptually defining the system, which he did successfully by 1983. It then took a considerable effort to complete the software development before launch, but that was also accomplished.

The calibration program at the institute was run by the Instrument Branch. Calibration observations take about 10 percent of Hubble's observing time and are themselves analyzed by the pipeline system. To my knowledge, the original principal investigators have never provided calibrations for their own instruments and instead used the reduced data provided by STScI. (Even more advanced systems were later developed by Michael Rosa of the European Coordinating Facility at the European Southern Observatory. Michael calls these procedures "forward calibrations" and has applied them to Hubble data to yield better accuracies than the standard system can supply.)

The program for data management that we implemented amounted to a paradigm shift in observational astronomy. First, huge quantities of astronomical data, such as those produced in orbit or on the ground by CCDs and infrared detectors, can now be analyzed and made available in real time. Second, the quality of the data has become the responsibility of the observatory staff, which was a profound change from the customary approach of ground-based observatories at the time. There the staff only provided access to the facility and some support to the observer, but the observer was responsible for the quality and the reduction of the data. This approach greatly limited the number of astronomers who could use specialized facilities. Third, data from the best telescopes in the world were immediately available to astronomers in all disciplines, without requiring specific expertise on their part in the idiosyncrasies of the instrumentation. The world of astronomy had become flat in the sense suggested by Thomas L. Friedman.[4]

The data from Hubble could then be used by different scientists for different purposes. This model of data use required the establishment of a vast archive of data readily accessible by astronomers, with the net result that research on archived Hubble data now exceeds that done on the data when it is first acquired. NASA once again had entrusted the development of the Hubble archives to an aerospace contractor, this time Ford Aerospace, Newport Beach, California, and it soon became apparent that this system would not be ready for launch. STScI was permitted by NASA to develop, together with scientists from the European Coordinating Facility and from the Dominion Astrophysical Observatory (now the Canadian Astronomy Data Center of the Herzberg Institute for Astrophysics), a quick-and-dirty prototype system to be used at launch. This system was developed in a very short time by two American, two Canadian, and two European postdoc fellows. With many improvements and modifications, it formed the basis of the archive still in use today by all three institutions.

Archives of this type are the basis of the National Virtual Observatory (an on-line archive set up in 2004 and sponsored by the National Science Foundation), where reduced and calibrated data from many observatories are made available to all astronomers. This new methodology has been transferred to the VLT of the European Southern Observatory, the largest ground-based array of telescopes in the world. It was also adopted in Chandra. The revolution in data handling may be a factor in explaining why the scientific productivity of Hubble, VLT, and Chandra is the highest in the world of astronomy. In my opinion, this methodological change is one of the important contributions of STScI to astronomy.

Links with the Research Community at Large: Academic Affairs at STScI

The creation of an Academic Affairs Division at STScI had not been envisioned by NASA or even by the Hornig committee. I felt, however, that we could maximize the benefit of being on a university campus by paying special attention to the quality of scientific life at STScI. Academic Affairs was intended as a home base for tenure-track astronomers. The staff actually belonging to this division was quite small, and their functional responsibility was to maximize support for research carried out by all astronomers at the institute. They created an excellent astronomy library, run by Sarah Stevens-Rayburn, one of the best librarians I have ever met. They organized seminars and symposia, as well as the publication by Cambridge University Press of monographs on astronomical subjects of relevance to the Hubble and the postdoc program.

We had distinguished scientists to head the division, including Len Cowie, George Miley, Colin Norman, and Nino Panagia. Cowie, Miley, and Norman took the initiative of organizing courses on astronomy at the Johns Hopkins University and supporting the graduate program of the university's Physics and Astronomy Department.

The Academic Affairs Division acted as the advocate and ombudsman to management in protecting the right to do research of the tenure-track astronomers. In truth, very little effort was required to persuade management of the necessity to allow scientists to remain scientists—after all, it was obvious that the STScI should be so organized that it could continue to do its work independently of any one individual's presence.

While the STScI deputy directors (Don Hall, Garth Illingworth, and Pete Stockman during my tenure) had the primary responsibility for search and recruitment activities for scientists, Academic Affairs supported the deputy director by organizing the process and being actively involved.

But perhaps the most important innovation of this division was the creation of the Hubble Space Telescope Fellowship program. The first suggestion that such a program could be beneficial came to me from Chris McKee while we were strolling in the English Garden in Munich, where we were attending a meeting in 1982. His suggestion was general, but it started me thinking about what kind of a program we could have at STScI.

It was important to me to create a fellowship program that would attract the very best young people to work with Hubble. I thought we could distinguish the program from those that would be carried out under the principal investigators' grants (in which the postdocs were treated in essence as research assistants to the principal investigators) by giving unparalleled freedom to the fellows. So we created a program in which the fellows could engage in independent research on subjects related to the Hubble program at any institution in the United States (including the STScI), by previous agreement with that institution. The only limitation was that no more than two fellows could be at a particular institution in any given year. The fellowship was for a period of two years, renewable for one more, at a competitive salary. We also provided a small purse for travel and incidentals. The fellows had no duties except to come to STScI once a year to give a seminar on their work. This freedom and independence (as well as the salary) attracted to the Hubble program, over the years, some of the best young scientists in the world. Receiving a Hubble Fellowship became a coveted honor, and the few applicants who refused the fellowship or left before it expired did so because they had accepted attractive academic positions at major institutions.

The program was started with fifteen fellowships per year, so that at any one time we had to support some thirty to forty-five fellows at a total cost, including benefits, of about $3 million per year (10% of the STScI budget and 1% of the operations budget).

The program has been very successful, and its style has been adopted in other NASA programs, most notably Chandra and the Spitzer Space Telescope. Its creation was for me a statement of principle for the pursuit of excellence and against conformism and mediocrity.

Observer Support: The Struggle to Develop a Coherent Approach

As part of our science systems engineering approach, the scientific staff at STScI and I decided to study what would be needed by the astronomical community to absorb the flow of data originating from Hubble. Of the 3 to 5 gigabytes of data received daily, some 10 percent would have to be analyzed by the STScI for the purposes of calibration. All the rest would be reduced and calibrated by the pipeline processing system and distributed to the community for further analysis. The question in our minds was whether the community had the computer capability to digest these data and the number of scientists required to effectively use the data.

In the mid-1980s, Neta Bahcall, who was at STScI at the time (prior to accepting a full professorship at Princeton), led a NASA-STScI study group that undertook to answer this question. Ed Weiler, the NASA headquarters program scientist, participated fully in this study. Members of the group consulted widely with the community and found that U.S. universities did not possess sufficient computer capabilities available for Hubble astronomy to fully utilize the data. It was also clear that scientists needed technical support in the handling of the data and the development of specialized software. The committee wrote a report on their findings in the late 1980s that was well received at NASA headquarters. Charlie Pellerin, the director of NASA's Astrophysics Division, agreed to the program proposed by STScI, which would make available grant funds to winners of observing time, so that they could analyze their results in a timely fashion. The program included STScI-university cooperative agreements to share the cost of additional computer capabilities, which would remain at the university at the end of the program. Although funding for data analysis by guest observers had been provided by NASA in some previous missions (for example, the International Ultraviolet Explorer and ROSAT), the Hubble program was much larger in scale and was managed by the research community itself.

This program was received favorably by the astronomical community, and the idea of associating financial grants with observing time became the norm for other NASA observatories, including Chandra and Spitzer. The program was so successful that NASA soon provided more grant funding to astronomy than did the National Science Foundation, the agency that had traditionally supported astronomical research. The foundation has not yet adopted this concept for its ground-based observatories, but it might well be forced to do so by the large quantity of data expected from the Atacama Large Millimeter Array and from optical telescopes of the 30-m class and larger, now under study.

The growth of NASA's role in funding astronomical research eventually led to the proposal that NASA should take sole responsibility for all astronomical research. A blue ribbon panel was formed in 2000 by the National Academy to investigate this issue. When asked my opinion, I advised against this action. My reasons were the importance of retaining multiple funding sources in any scientific discipline; NASA's increasing dominance in setting the research agenda; the growth of in-house research at NASA centers; and the fact that, for all its faults, the National Science Foundation still allowed the community to develop and operate its facilities. I was pleased that the decision was made to preserve a role in astronomy for both agencies.

But it is quite clear that NASA grants have been vital to the progress of U.S. astronomy. In Europe and Japan, scientific research is mainly funded through institutional support. The universities and various societies, such as the Max Planck Institute in Germany, provide the salaries of the researchers and rely less on grants. In the United States, research is much more dependent on "soft money," which disappears at the end of a project.

The European system has the great advantage of stability, but it is less capable of responding to scientific opportunities. The U.S. system is more agile in this sense, but it lacks long-term stability for institutions not supported by private money, such as universities. Even for universities, reliance on soft money has grown to the point that it affects their independence in selecting the direction of research and their ability to take positions that differ from the government's in matters of national interest.

It is notable, for instance, that the start of the Space Shuttle and Space Station programs and their many technical, managerial, and financial problems were not forcefully debated throughout the scientific community. Nor were the serious mistakes being made brought to the attention of Congress by the National Academy of Sciences. It took 20 years and the administration of George W. Bush to recognize the folly of these NASA programs and to cancel the shuttle program and decrease U.S. participation in the International Space Station as a prerequisite to a rational manned and robotic program of exploration of the planets.

Sharing the Joy and Building Trust: STScI Education and Outreach Programs

I have always believed that the results—and the joy—of doing science should be shared with the public at large in the conviction, perhaps misplaced, that it is important to spread the light of reason in a world that is becoming more fundamentalist rather than more rational. But to reach out, it is necessary to do more than publish beautiful pictures or astonish the public with pronouncements about black holes, especially when most people are not even aware of the source of energy in our own sun (nor do many of them think to wonder about it). I felt it was all right to dazzle if we also entertained and taught.

There were many aspects to STScI's educational program. The most basic one was directed toward the Baltimore schools and the Baltimore Science Museum. Every day, on average, one of the STScI staff members would be giving a talk at a local school. We obtained some private funds to start a collaborative program with the Baltimore Planetarium, which reached a different segment of the public.

One idea I particularly liked was to work with amateur astronomers. When at Harvard, I had worked with the American Association of Variable Star Observers, based in Cambridge, Massachusetts, to study x-ray sources. The idea was to make some of the Hubble time available to amateurs. There are about 250,000 amateur astronomers in seven different associations in the United States. We asked the presidents of those associations to work together to create a unified amateur program and a selection procedure, to be run by them, to choose among competing proposals. Although the Hubble staff would give them technical support during observations and data analysis, we offered no financial support.

The program was quite successful: some discoveries were made, and the selected observers—students, teachers, or simply amateurs—became celebrities within their communities because of their association with Hubble. Many of the winners told their stories on local radio, on television, and in newspapers; some wrote books about their experiences. The best features of the program were that most of the work was done outside the institute—thus our small resources were expended with great leverage—and that pedagogically it was done correctly. The program was restricted to less than 1 percent of the Hubble time, but it still received criticism from some members of the community as "a waste of taxpayers' money," and it was stopped after I left STScI.

We were lucky in having Eric J. Chaisson as the head of the Education and Outreach program at STScI during 1987–92. Eric was a radio astronomer who had given the best-attended undergraduate lectures at Harvard (five hundred to a thousand students!) and had written several astronomy books intended for the general public. An updated edition of his 1993 book *Astronomy Today* is still the most widely used text in introductory astronomy classes. He left STScI—after becoming embroiled in a feud with NASA and with many astronomers by publishing *The Hubble Wars,* a highly critical view of the Hubble program—to become head of the Wright Center for Science Education at Tufts University in Medford, Massachusetts, a position I believe he still holds. The book is flawed because Chaisson wrote it in the conviction that Hubble would fail, and gave a slightly exaggerated view of his own influence on events, but it portrays with some authenticity and verve the squabbles of Science Working Group scientists among themselves and with STScI, as well as some of the interactions with NASA.

Chaisson ran the amateur program with great enthusiasm. He was also able to set up collaborative programs with local and public television stations. The typical video release from STScI contained three elements: the announcement of a discovery, the story by a staff member of how he or she had became involved with Hubble (to present a role model à la Jean Piaget), and the explanation of some specific physical concepts.

Perhaps one of Chaisson's most persistent influences was his creation of science writers' workshops well before Hubble's launch. He invited science journalists to day-long workshops in which the science behind the anticipated Hubble observations was discussed with institute staff. The honest exchanges laid a foundation of trust, which made the institute a valued resource for expert commentary on new astronomy results, whether from Hubble or from other observatories. The stories emanating from STScI attempted to be as accurate as possible scientifically and factually and were often in conflict with the more facile NASA style. To give an example, NASA always claimed that Hubble would look farther in space and farther back in time than ever before. This statement is clearly untrue, given the discovery by Arno Penzias and Kenneth Wilson in the 1960s of the 3-K cosmic background radiation, the first radiation that escaped the original Big Bang fireball, when the universe was only half a billion years old. Some large ground-based telescopes can also see objects at greater distances than can Hubble. The real advantage of Hubble is its exquisite resolution, which makes it possible to do much unique science, and we saw no need for misleading statements. These arguments did not seem to placate NASA, because we were inadvertently challenging the agency's control of the information going to the public.

Ultimately NASA's viewpoint did not prevail, and in later years it recognized the strong scientific contribution that STScI could make even to its outreach programs, the professional quality of the institute's products, and the careful attention given by STScI to reaching the general public with appealing and exciting images. The work was done by five or six dedicated members of the staff when I left the institute in 1992, but NASA's recognition of the institute's value eventually led to a substantial increase in its outreach staff. I believe it is fair to say that STScI set the standards of public information activities from a live observatory.

This chapter has told the story of some of the intellectual and methodological contributions made by the new institute to modernize astronomical research. Next I turn to how it performed at launch and when confronted with calamity.

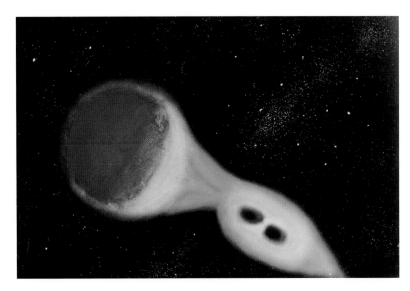

PLATE 1. *Artist's conception of Her X-1. The gas from the normal star falls onto an accretion disk around the neutron star. As it nears the neutron star, the gas becomes extremely hot and fully ionized, and thus it is guided toward the magnetic poles of the neutron star.*

PLATE 2. *Artist's conception of Cyg X-1, showing the accretion disk but (in contrast to a neutron star's accretion disk) the lack of features in the accretion flow.*

PLATE 3. *Skylab in orbit.*

PLATE 4. *X-ray image of the sun, obtained with Skylab.*

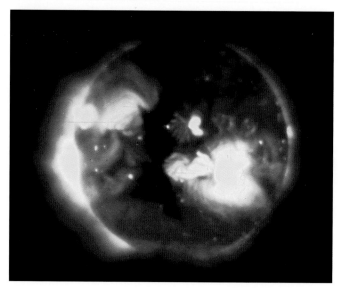

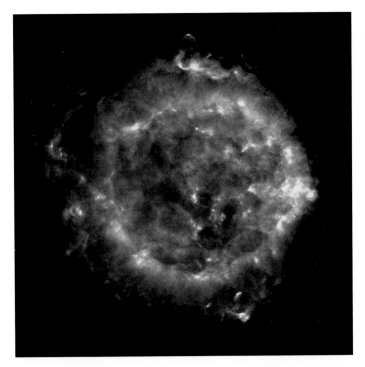

PLATE 5. *An x-ray picture of the Cas A supernova remnant, obtained with the high-resolution imager on Einstein and showing the hot plasma heated by the shock from the supernova explosion.*

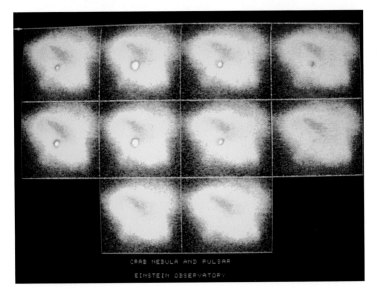

PLATE 6. *Time-lapse photography of the x-ray emission from the Crab Nebula and its pulsar obtained with the high-resolution imager. The diffuse emission is produced in the nebula, whereas the point source is the pulsar.*

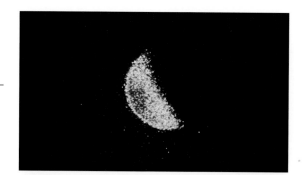

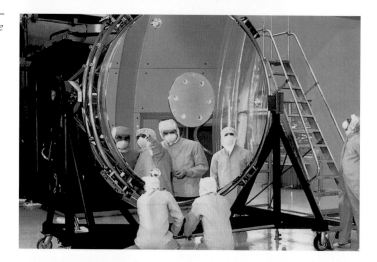

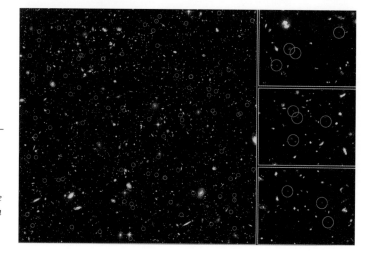

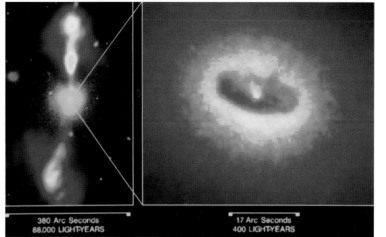

380 Arc Seconds
88,000 LIGHT-YEARS

17 Arc Seconds
400 LIGHT-YEARS

PLATE 10. *Radio and optical images of the Galaxy NGC 4261 obtained from the ground* (left), *and a picture of the inner region obtained with Hubble* (right; note the change in scale).

PLATE 11. *Star-forming regions in the Eagle Nebula.*

PLATE 12. *Dust disks surrounding young stars and the jets ejected by them.*

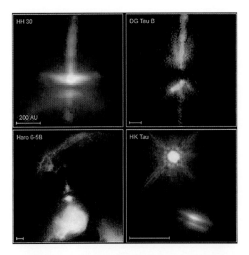

PLATE 13. *Pictures of the Crab Nebula obtained in the x-ray domain with* Chandra (left column) *and in the optical domain with* Hubble (right column).

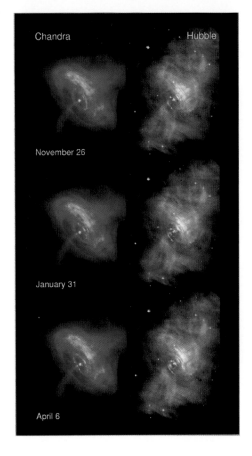

PLATE 14. *Aerial view of the European Southern Observatory's La Silla Observatory, Chile.*

PLATE 15. *Aerial view of Cerro Paranal in 1993. Excavation for the four unit telescope foundations and for the access road can be seen.*

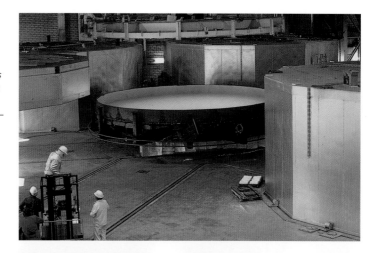

PLATE 16. *An 8-m blank made of Zerodur for one of the unit telescopes, shown in the process of ceramicization at Schott.*

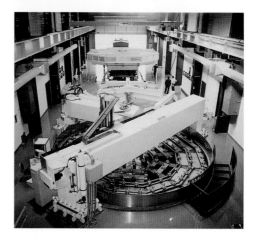

PLATE 17. *Two views of one of the 8-m mirrors at REOSC during polishing operations.*

  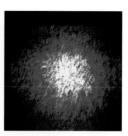

PLATE 18. *Simulated distortions of the atmosphere* (left), *Shack-Hartmann pattern in the wave-front sensor* (middle), *and final star image obtained with the model* (right).

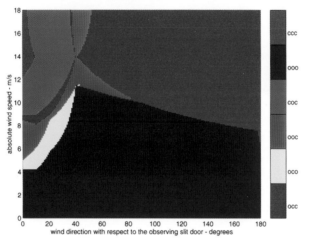

PLATE 19. *Recipe for optimal setting of the ventilation doors, louvers, and windscreen as a function of wind speed and direction, computed using a finite element model.*

PLATE 20. *One of the early images (November 1998) obtained with the Focal Reducer Low-Dispersion Spectrograph on Antu (Unit Telescope 1).*

PLATE 21. *A picture of the Hydrogen-alpha emission from SN 1987A obtained with the Ultraviolet High-Resolution Spectrograph at the Kueyen (Unit Telescope 2) focus, showing the high resolution of the spectrometer.*

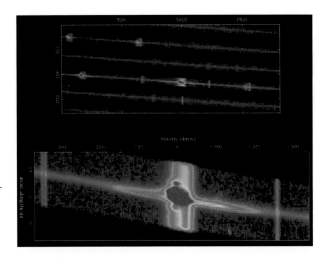

PLATE 22. *An aerial picture of the Paranal platform, showing the four unit telescopes, the stations for the auxiliary telescopes for interferometry, and the location of the interferometric tunnel.*

PLATE 23. *The residence at Paranal houses offices, a restaurant, and sleeping facilities (120 single rooms). It is designed to be as unobtrusive as possible in the landscape. The Very Large Telescope/ Very Large Telescope Interferometer observatory is visible in the background.*

PLATE 24. *The interior of the residence is designed to provide relief from the harsh desert conditions at Paranal. The automatic curtains for night blackout of the 30-m dome can be seen in the center. The pool is part of the humidifying system but can also be used for recreation.*

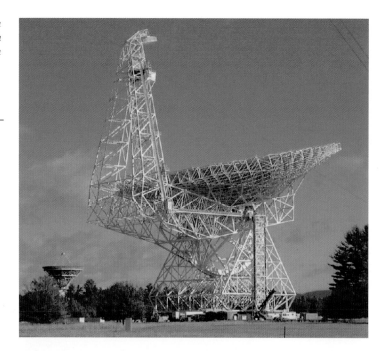

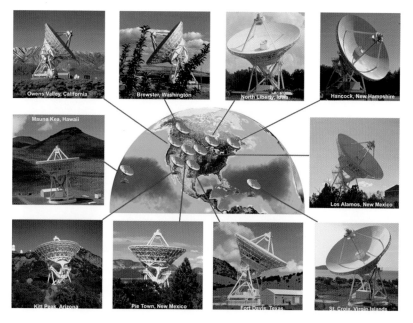

PLATE 27. *A composite of photographs of the Very Long Baseline Array antennas and their locations around the world.*

Owens Valley, California
Brewster, Washington
North Liberty, Iowa
Hancock, New Hampshire
Mauna Kea, Hawaii
Los Alamos, New Mexico
Kitt Peak, Arizona
Pie Town, New Mexico
Fort Davis, Texas
St. Croix, Virgin Islands

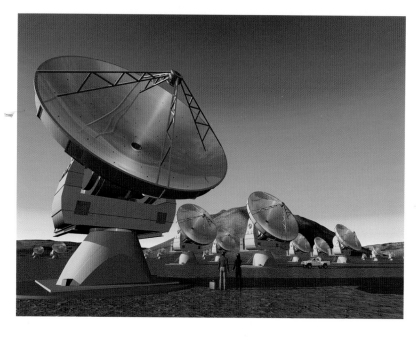

PLATE 28. *Artist's conception of the Atacama Large Millimeter and Submillimeter Array.*

PLATE 29. *Artist's conception of Chandra in orbit.*

PLATE 30. *A stunning x-ray image of the Crab Nebula, showing the pulsar, and the jets of particles and shocks produced by it.*

PLATE 31. *The jet originating from the supermassive black hole in the center of the radio galaxy Cen A.*

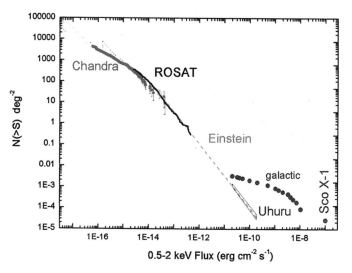

PLATE 32. *Plot of the number of sources observed in a square degree of the sky as a function of x-ray flux. Shown are the results of the Sco X-1 discovery flight, Uhuru (for galactic and extragalactic sources), Einstein, ROSAT, and the 10^6-sec exposure of Chandra.*

PLATE 33. *The 10⁶-sec exposure in the Chandra Deep Field South. The colors correspond to the energy of the radiation, red being the softest and blue the hardest.*

PLATE 34. *The author in front of the Green Bank Telescope, 2002.*

The Space Telescope Science Institute: Launch Readiness and Its Finest Hour

Coming of Age: STScI at Launch

While paying great attention to the technical and scientific issues of the Hubble program, the institute worked hard to prepare itself for its service role in science operations, which included proposal selection, study of the proposed observation's feasibility, assignment of guide stars, planning and scheduling, data reception, data reduction and calibration, data archiving and distribution, public dissemination of information, and grant support for the observers.

The March 1992 report on STScI was the tenth in the series of annual reports to the AURA board of directors.[1] It describes a mature institute with a well-defined, though still flexible, organizational structure that had been tempered by having gone through launch (on April 24, 1990), orbital verification, science verification, the discovery in orbit of the out-of-focus telescope optics, the studies and proposals for recovery of focus, and the beginning of scientific observations.

I choose to describe STScI at this time because by March 1992, STScI had proven its value as a major new institution devoted to maximizing the scientific returns from Hubble and because a new way of doing astronomy had been established. From a personal point of view, it defines what I left to my successor, because by the end of that year I had resigned the directorship of STScI.

Figure 14.1 shows the organizational chart of the institute in March 1992. In the director's office, the deputy director was Pete Stockman, the associate director for operations was Ethan Schreier, and the associate director for program management was Robert Milkey. During the four years of delay in the launch of Hubble (because of the January 1986 *Challenger* tragedy and the subsequent halt of shuttle operations), STScI had taken on more functions in software development and systems engineering.

All aspects of the old and the new work were brought to successful conclusion by the institute staff, which had grown to 412 (including AURA, European Space Agency, and Computer Sciences Corporation personnel), a

FIG. 14.1. *The organization of STScI in March 1992.*

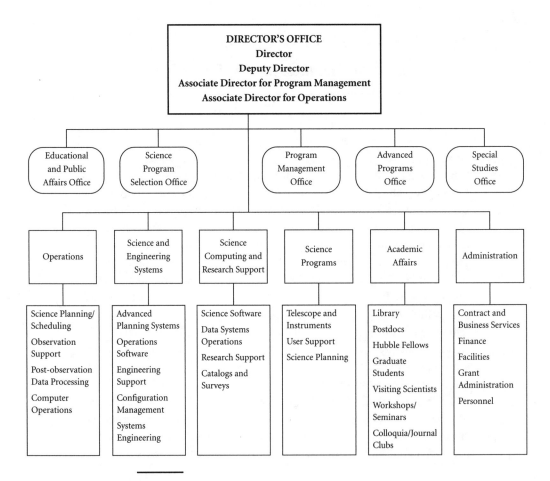

staff level comparable to that of ground-based National Science Foundation observatories, such as the National Optical Astronomy Observatory and the National Radio Astronomical Observatory. The 1992 estimated budget was $35.6 million, $33.8 million for operations and $1.8 million for new hardware and software projects.

Of great importance in providing for the recruitment, relocation, housing, and support of this growing staff was the flawless performance of the Administration Division. We had the considerable advantage that, because STScI had started from zero, no bad habits or false complacency had yet set in, and we were able to motivate the Administrative Division with the same drive for excellence that pervaded STScI.

Vincent Severo, who headed the division in its formative years, was a hidalgo of Filipino origin, and his hard work and enthusiasm contributed greatly to this effort. His most disparaging remark about a person was that he had "no fire in his belly." This fire was certainly needed as we moved from temporary quarters to a new building (of only half the size ultimately required) on the Johns Hopkins University campus in Baltimore, rented space outside campus, converted a floor of the garage to temporary office space, and finally doubled the space of the original building by adding two wings. All of this naturally had to occur without impeding the mission schedule. His successor was George Curran, who was extremely competent, if perhaps not quite as fiery.

I do not intend to describe all aspects of STScI's work, as that certainly merits a book of its own (which I hope will be written one day), but only outline the great effort by the staff to develop a full scientific and engineering understanding of the telescope and its instruments, the success of the software and procedures developed for operations, the discovery in orbit that the telescope was out of focus, and the contributions of STScI to the recovery of Hubble's design performance.

Science Programs Division

The Science Programs Division had oversight of the development and completion of the science programs with Hubble. The division was responsible for the solicitation, processing, and organization of the peer review of proposals; development and execution of the calibration plan to support the users; development of a long-range observing plan; and coordination of the on-site support for general observers and guaranteed-time observers. The division prepared the calls for proposals and the *Users' Instruments* handbooks. The division was composed of a division office and three branches.

The User Support Branch was the main contact with the community throughout the proposal cycle. The support branch provided services to the guaranteed-time observers, the general observers, and the archival researchers by supplying information, receiving and processing proposals, supporting the Time Allocation Committee, maintaining the observing program databases, supporting on-site users, and receiving and responding to feedback from users. In 1992 there were already more than five hundred guest observers using Hubble.

The Science Planning Branch developed the yearly observing plan for Hubble. It provided proposal assistance to the observers and technical support to the Time Allocation Committee on scheduling matters and observing feasibility. The planning process included scheduling feasibility, preparing ephemerides and determining available windows for solar system targets, and integration of all programs into the yearly plan. The planning branch monitored and reported progress in the execution of the proposals, repaired failed observations, accounted for time utilization, and developed software and procedures to improve science efficiency.

The Telescope and Instrument Branch had the task of evaluating the in-orbit performance of the optical telescope assembly and the instruments and passing this information on to NASA and the astronomical community. The scientific calibration performed by the branch staff was used to update the Project Database and the pipeline calibration processing system. The staff was also charged with maintaining expertise on the telescope and on each of the instruments, existing or planned. The branch scientists worked with branch engineers to maintain and improve the scientific returns from the observatory.

During the 1980s the Science Programs Division—and in particular the Telescope and Instrument Branch—underwent various structural changes to respond to changing requirements of the program. Many excellent people contributed to the success of these adaptations. Duccio Macchetto played the major role through all of the transformations. He was the most senior scientist on the European Space Agency staff, and he became the chief of the Instrument Support Branch in 1983 and head of the Science Programs Division in 1988. His experience in hardware and mission operation on the International Ultraviolet Explorer satellite (launched in 1978) and his scientific judgment contributed greatly to Hubble's success. The success of Hubble and of its new approach to astronomy was not due, however, to any single person but to the creative contributions of many members of the staff, particularly in this division. I mention only a few who provided leadership when it was needed: Robert Brown, Holland Ford, Don Hall, Larry Petro, and Pete Stockman (before and after his appointment by AURA as deputy director).

Of great utility were the meetings of the Telescope Time Review Board chaired by Duccio and attended by staff of the Science Programs Division, Operations, and the Science and Engineering Systems Division. This board provided an important forum to discuss how to use the telescope more efficiently and to plan for the future, including the servicing missions. The recommendations of this board came to me for concurrence and action.

I cannot overemphasize the importance of the ability of scientists and engineers at STScI to communicate with one another, thereby speeding up the learning process and assessing problems from different points of view. At times we went to the extreme of switching staff members around to carry out, for a while, one another's jobs. It was this open atmosphere that, in my opinion, created a feeling of unity and shared goals among the staff and made the work so successful.

STScI Preparations for Operations

From the start of STScI, Ethan Schreier had responsibility for all science operations and development as well as computing support. The organization chart (Figure 14.1) shows a later refinement: as the scope of the work rapidly expanded and we approached launch, it was thought useful to create three separate divisions responsible for different aspects of the work. However, even after he became associate director for operations, Ethan maintained oversight of these three groups and continued, as he had done throughout the program, to make the major decisions on operations programs and computer issues. The three divisions were the Operations Division, the Science and Engineering Systems Division, and the Science Computing and Research Support Division.

The Operations Division had responsibility for science planning and scheduling on a day-by-day basis, real-time support of observations, post-observation on-line data processing, and computer operations. James Crocker and later Pat Parker headed the group.

The Science and Engineering Systems Division was responsible for the maintenance and improvement of operational software systems and for understanding at an engineering level how both instruments and spacecraft operate and how they are controlled from the ground. Rodger Doxsey was then and remained for many years the head of this group. One of the difficulties

Rodger had to face was that Lockheed had not planned to transmit any engineering data to the institute. Their view was that we did not need to know with what precision the telescope was being pointed during the actual observations, because the Lockheed system would always point well enough. This attitude was too much even for NASA, and STScI did in fact receive the data. The further difficulty Rodger had to face was that, naturally, there was no software system to transform the engineering data stream into useful information.

The Science Computing and Research Support Division was headed by Ron Allen, a radio astronomer, and was responsible for the maintenance and improvement of the Guide Star Catalog, the development of the data analysis and distribution system, the IRAF (Image Reduction and Analysis Facility) / Hubble Data Analysis System, and many algorithms for data enhancement. This group provided research support to Hubble users and the staff. The Science Software group also maintained 120 Sun workstations and 38 VAX minicomputer systems.

Among the many decisions made under Ethan's direction, one of the wisest was the choice by STScI not to develop its own high-level command language but to adopt the command language being developed by Doug Tody's group at the National Optical Astronomy Observatory. Our European colleagues grumbled a little, because they had started development at the European Southern Observatory of the Munich Image Data Analysis System, but STScI's adoption of the more general language made IRAF the language of choice in U.S. space- and ground-based systems. I believe that Chandra has also adopted IRAF. This adoption was a good idea not only because none of the observatories had the resources to develop and maintain independent, full-scale systems of their own, but also because it saved the users the annoyance of having to learn different languages.

It was also Ethan and his staff who steered STScI away from mainframe computing toward workstations, which provided more output for the cost. He thus had a profound influence on how computer and software capabilities developed at the institute.

Operations Division

While engaged in all these technical and intellectual issues, we certainly could not neglect the operational side, because our success as an institution depended greatly on keeping the users and NASA happy. Ethan Schreier and I had discussed often what our approach should be: I was very much in favor of a disciplined, military-like organization for the Operations Division, whose task it was to produce a continuous flow of operational schedules and commands for NASA and to receive and process the Hubble data. This was the aspect of the institute that we presented to NASA. A failure in operations would damage Hubble by curtailing its scientific potential and would damage STScI by endangering its chances to renew its AURA contract.

Ethan, on the other hand, saw the telescope operators as having a greater role in contributing to the overall success by understanding the meaning of the observations, assisting while they were being carried out, and at times correcting errors in pointing or in commands to the instruments. This approach also allowed rapid fixes to problems in the operations systems. We ended up with a successful compromise.

In fact, many of the operators (who came mostly from Computer Sciences Corporation) had degrees in astronomy and were themselves active astronomers. Because the science operations were conducted without the presence of the observers 24 hours a day, 7 days a week (in three daily shifts), it was the operators who monitored the observations in real time. The operators were therefore the only people who could intervene at a moment's notice and save an observation that might be pointed toward the wrong object. They would know that it was wrong because they had actually read and understood the proposal for the program that was being executed. They fixed such problems routinely during the program.

We were lucky in hiring as head of the Operations Division a brilliant engineer, James Crocker, who was not only an outstanding manager but a highly creative person. He in turn was able to hire and motivate an experienced and competent staff, including Pat Parker, who had been hired by Ethan and initially served as Jim's deputy and then his successor.

The problem faced by Jim and Pat was training the staff in the use of the fairly sophisticated software and operational procedures that STScI had developed. It was also important that this staff manage the highly complex interfaces between their own and other divisions.

Taking a leaf from Jean Martinet, the French general under Louis XIV who invented close-order drilling, we had all members of the staff go through three readiness reviews conducted in front of three seasoned consultants, who happened to be some of the best operations specialists at NASA but were not involved in Hubble at the time: Joe Rothenberg, Tom Ratio, and Fletcher Kurtz. It is fair to say that the first review (in 1983 or 1984) was a disaster, the second had obviously benefited from the criticism and the advice we received, and the third one (around 1989) was good enough to go to launch, even though we were still a little nervous. I know of no other group in the program that put itself through such exhausting paces.

The fact that Crocker's and Doxsey's groups had been working together and helping each other in all aspects of operations was very important. There were differences in approach but no turf battles, just a lot of hard work. Both Jim and Rodger were totally dedicated to Hubble's success, and being smarter than most people I have known, they played an essential role in the success of the entire project.

In September 1989 all of the TRW software had been delivered. This was a deliberate decision by STScI. Although a number of systems were known by the institute staff not to be fully adequate to support operations, we agreed with NASA that we would accept delivery, confident that we could oversee the critical fixes necessary for launch. Several TRW engineers later joined the institute to implement these fixes. Some 6,060 software problem reports had been filed—some of critical importance, others of moderate or low impact. The groups led by Doxsey and Crocker had solved about 5,300 of these problems by launch, leaving 760 still to do. (After launch, this number kept increasing as new issues were uncovered during in-orbit operations, and a constant and successful effort was required to maintain the system. We had opened about 10,000 software problem reports and had closed 9,000 of them by 1992.) Despite these problems, in 1989 the necessary software for command of the scientific instruments and analysis of the data was ready for the verification tests at Lockheed.

At launch the postoperation data processing system was fully capable of keeping up with the data flow and digesting 99 percent of the data within 48 hours. By November 1992 data from some 30,580 scientific observations had been received, and for only 3 percent of them did the processing time exceed 48 hours. Each month 20,000 datasets were archived, of which 5,000 were calibration data. The archiving and reprocessing volume during October 1992 had reached 102 gigabytes of data per month. This rate has continued to increase after each of the servicing missions, increasing by a factor of sixty in 1994 and by a factor of thirty-three in 2000. These are very large numbers by astronomical research standards and show the necessity of the high degree of automation that we had been able to achieve in data handling.

Also in the month of October 1992, 255 FITS (an international standard digital format) tapes were distributed to observers and 50 optical disks of archived data were transferred to the Space Telescope European Coordinating Facility. Hubble's observing efficiency in 1992 was 35 percent—the relatively low percentage was due mainly to seeing constraints imposed by the orbit—

but its efficiency was continually growing as the most important impediments were eliminated one by one.

Notwithstanding the skepticism of our colleagues, in 1990 we had brought to successful conclusion a very complex system of sophisticated software that allowed us to plan and schedule observations, provide the guide stars from a catalog, retrieve the Hubble data, reduce and calibrate the scientific data on line, and finally archive and distribute them within 48 hours. At the same time STScI was involved in planning and developing software for the new instruments for the first servicing mission and in developing image restoration techniques to cope with Hubble initial image degradation. For this work in 1991 Rodger Doxsey received the highest NASA award, the Distinguished Public Service Medal, for "outstanding leadership in developing concepts and implementing systems for science operations of HST." In the same year James Crocker received the NASA Public Service Medal for leadership and management in the preparation and execution of science operations for Hubble. NASA also recognized with a Public Service Group Achievement Award the STScI team that implemented the orbital verification program. Finally in 1993 I received the NASA Distinguished Public Service Medal for "outstanding leadership for the development of STScI."

Flirting with Disaster: Clashes over Tests and Calibrations

A basic problem in the Hubble program was the lack of experience of most of the scientists involved in the early stages of the program with the development and operations of space experiments. None of them had carried out a project as challenging as Hubble or fully understood what it implied. They were timid in presenting the program needs to NASA, accepted risky compromises, and chose to rely on the supposedly greater wisdom and experience of the contractors and of Lockheed in particular, because of the latter's reputation for doing classified satellite work for the government. This tendency to defer was reinforced by a climate of what was, in my opinion, unnecessary secrecy and the fear that an open discussion of technical problems would reveal significant military secrets to the Soviet Union. I have no way of knowing whether these fears were genuine or simply a way to claim a closer connection to previous developments that in fact existed, and thus convince Congress that Hubble was a simple step from classified surveillance satellites. The Hubble program started before glasnost and the fall of the Berlin Wall, and such security concerns were perhaps justifiable.

I had held a clearance for classified work since 1961 and could understand the concerns, but my practical experience had been that technical secrets were often not secrets at all. The performance of some U.S. classified systems was in fact met or surpassed in private industry elsewhere in the world, and comparable technology was available on the open market. I also did not think that we could enter into the international development and use of Hubble with its technical specifications shrouded in mystery. I gave up my clearance and insisted that no classified documents be retained at STScI: whatever information we needed for our work had to be unclassified. In practice there was no problem, because the astronomical applications of high-precision optical and guidance systems are so much more demanding than military applications that any significant transfer of technology would be in the other direction.

However, the total reliance on Lockheed to integrate the telescopes and instruments because of their experience in classified missions had the unfortunate effect of lulling the Science Working Group and NASA into unjustified complacency.

Peter Stockman, who had been the head of the STScI telescope branch of the Instrumentation Division prior to his appointment as deputy director of STScI in 1988, took on the issue of calibration and verification of Hubble as a whole. Pete was an experienced experimentalist from Robert Novick's group at Columbia University, which had been involved in space experiments. He was one of the most technically competent of all the astronomers at STScI and was a great support to me in all my decisions. I used to joke that when we disagreed on something, I later found that his views on the subject were those I hoped I could reach myself after giving it considerable thought.

Both Pete and I were quite concerned that there would be no end-to-end testing of Hubble on the ground. By end-to-end testing, I mean a "dry run": placing the integrated spacecraft or the optical telescope assembly in a vacuum chamber, illuminating the telescope with a suitable light source, detecting the images or spectra with the actual instruments, and receiving and analyzing the data with the ground software system.

This end-to-end test approach was what the Marshall-CFA team had used for Einstein in 1977 and would use for Chandra in 1997. Stockman and I thought that a collimated x-ray beam would certainly be harder to achieve than an optical parallel beam. Therefore, errors in the instrument's focus should have been easier to detect during end-to-end testing of Hubble than during testing of Einstein and Chandra.

I remember a discussion that took place at Marshall during one of the program reviews in the late 1990s. At these NASA reviews, one was not expected to bring up real issues, because they might embarrass some of the participants; we chose, however, to ignore this etiquette, because we thought that the problem of testing was urgent, and Pete spoke for the STScI in expressing our concerns. He insisted that we wished to participate in the detailed planning of the calibrations of the entire Hubble system, and that we wanted to have STScI personnel (in addition to the principal investigators) at Lockheed during the tests.

These requests provoked the strongest reaction on the part of one of the Lockheed managers, Dominick Tenerelli. He insisted that Lockheed was an experienced contractor that knew perfectly well how to do thermal vacuum systems testing. He made clear that the fewer scientists around the better, and then attacked Pete personally, questioning the latter's experience in space experiments. In fact, Tenerelli had experience in military, not scientific, programs. None of the NASA people, including the Marshall project scientist or the program manager, intervened. None of the scientists of the Science Working Group had any objections to Lockheed's position, so that I found myself alone in objecting as strenuously as I could. This encounter changed nothing in the rather arrogant stance that Lockheed was taking and continued to take toward the scientists: Lockheed knew best, and the STScI contributions were tolerated but not welcomed. Unfortunately, neither Lockheed nor the scientists involved in the program had any experience of the close working relationship between scientists and engineers that prevails in successful scientific space programs.

But of course we could not give up. Because NASA had decided against end-to-end testing, the verifications at Lockheed were not representative of the integrated system. The measurements could only verify, with test light sources for each instrument, that the quantum efficiency of the detectors or the gain of the on-board electronics had not changed when the instruments were integrated into the test facility.

Stockman and I were still very concerned about the lack of information regarding the performance of the telescope and the system as a whole, so we invented another approach to the problem. Beginning in 1988 Stockman and a team of STScI scientists designed a set of typical observations that Hubble would carry out in orbit and which would test different aspects of the system. In a typical gedanken (thought) experiment, the team planned to follow, by simulation, the photon beam from its entry into the baffled aperture of the telescope, through the telescope and instrument optics, and onto the detector. At each step of the way, they would ask for mirror efficiencies; scattering constants; baffle shadowing; and detector efficiencies, gains, and all other parameters necessary to compute output. At each point they would ask for proof that there existed either direct and certified laboratory measurements or calculated values. Clearly, given the available time, this effort could not be prolonged, but it could bring some systems engineering to bear on the program.

NASA refused STScI both the funds and the access to the documentation necessary to carry out this work. In view of what happened after launch, I still ask myself whether it would have made any difference had STScI been allowed to proceed. Perhaps we could have seen the test results on the telescope that the responsible engineer had refused to sign. Had we found the problem, could the project have fixed it? I do not know, but so many other problems had been fixed that most likely we could have. In any case, we proceeded to launch without end-to-end testing or verification.

Hubble's Blurry Vision: Reactions and Recovery

Discovery and Initial Reactions

The Hubble Space Telescope was launched on April 24, 1990. It was soon clear to astronomers that the telescope suffered from spherical aberration, a condition not uncommon in ground-based telescopes at first light. Spherical aberration is an optical condition in which light beams impinging at different radii on the primary mirror come to focus at different distances. Because most of the light comes from the outer rim of the mirror, the central image contains little of the total beam, which is spread out as shown in Figure 14.2. I believe the situation was first clearly understood by Chris Burrows, a young physicist who held a dual appointment at STScI and the European Space Agency. He quickly produced a computer model in May 1990 that reproduced in great detail the observations of stars from Hubble's WF/PC and FOC, showing the spread of the light in the focal plane.

NASA reached the same conclusions in June 1990. In July it convened an Investigatory Board, chaired by Jet Propulsion Lab Director Lew Allen, which reported its findings on the probable cause of the error in November of the same year. As it turned out, in late 1980 or early 1981, a technician at Perkin-Elmer had improperly assembled a device used to measure the configuration of the primary mirror. As a consequence, the mirror had been built to the wrong shape. Tests performed at the time showed that there was a problem, and the responsible engineer refused to certify the results. This warning was

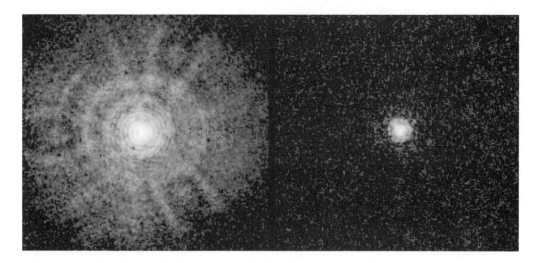

not heeded and, in the absence of a full end-to-end test, the problem could not be discovered before launch.

In the article submitted to the *Astrophysical Journal* by Burrows and others in October 1990,[2] a computer simulation perfectly mimics the observed behavior measured with WF/PC and FOC, which suggests that the thought experiment Stockman and I had proposed before launch might have worked.

What I found rather sad was the reaction of scientists who had been involved for years in the program. Jim Gunn of Caltech, for instance, refused to have anything to do with the program at all. Senator Barbara Mikulski of Maryland called it a techno-turkey. And in a climate of defeatism and disappointment, Hubble was jeered at as a lemon in the press.

NASA, on the other hand, reacted constructively. The shuttle servicing mission that had already been planned gave us the opportunity to modify the instrument optics (for instance, provide the WF/PC II with relay optics designed to correct the aberration). The solutions for the two spectrometers and the FOC were not, however, immediately obvious, because no new versions of these instruments were being planned.

In addition, there was a grave problem in the pointing of Hubble, which also degraded its performance but did not receive as much public attention. Contrary to Lockheed's assertions prior to launch, Hubble was not pointing very well. Its behavior could be interpreted thanks to the engineering data.

In part these pointing difficulties were a consequence of the out-of-focus optics, which prevented the FGSs from achieving their highest accuracy.

FIG. 14.2. *Faint object camera picture of a star before* (left) *and after* (right) *the optical correction of the instrument by the Corrective Optics Space Telescope Axial Replacement (COSTAR).*

But much more serious was the loss of lock (failure of the telescope to lock on to guide stars and thus to track its target), which stemmed from the design chosen by the European Space Agency of floppy supports for the solar cells. This design caused oscillations in the large, winglike solar panels produced by thermal shock every time Hubble passed from the sunlit to the dark side of its orbit. As a result, there was a loss of lock during the transition; but even during quiet periods, the design reduced the pointing precision to 7 milliarcsec in less than an hour rather than the desired 3 milliarcsec over 24 hours.

Such design problems clearly show the lack of systems engineering analysis and design in the construction of Hubble. Is it possible that the stiffness of the solar cells had never been specified? There were existing designs that provided the necessary rigidity; why were they not adopted? I do not believe these questions will ever be answered, because they are entangled in denials, international and institutional sensitivities, and politics.

Solutions: STScI's Finest Hour

At STScI, we had already managed to rally after the 1986 *Challenger* tragedy. Now we had to recommit ourselves to bring the preparations for science operations to the high degree of sophistication required, which they achieved before the 1990 launch. By the launch date, our confidence in NASA and its contactors had plummeted. However, we had acquired a great deal of self-confidence by succeeding in the tasks that were under our control, and by performing well at launch and during orbital verification and science verification. The institute had stood as a rock of competence, self-confidence, and achievement amid the pervasive confusion generated by the lack of readiness by Marshall to conduct orbital verifications and the difficult handoff of responsibility to Goddard and STScI. STScI command and scheduling software was up and running within 24 hours of launch.

My early mountain training as a climber in the Dolomites, which had taught me never to give up, came to my help now. In the early spring of 1990 I told the STScI staff that it was time that the scientists take their future into their own hands and come up with suggestions on how to fix the problems, rather than moaning and groaning. The response by the staff was immediately positive and gave STScI a great boost in morale.

STScI scientists Robert Brown and Holland Ford approached me within days with the suggestion that we should convene a panel of experts from all over the world to study the technical problem and propose solutions without pre-established constraints. I agreed. The panel started its work in August 1990. The charter of the panel reads:

> The Panel will identify and assess strategies for recovering HST [Hubble Space Telescope] capabilities degraded by spherical aberration. It will review the current state of the Observatory, the Allen Board findings, the scientific potential of the ideal HST, and the tentative science program of the unimproved HST. It will develop a comprehensive framework for identifying possible improvements, including OTA [optical telescope assembly], instrument, spacecraft, and operations-level changes, and including hybrid combinations. Within this framework the Panel will develop and debate the technical and scientific merits of particular improvements. On the basis of their findings, the Panel will formulate a set of recommendations and conclusions.
>
> The Panel will cast its net widely, especially seeking the ideas and appropriate involvement from STScI staff. It is expected that the whole Institute will be informed regularly of the Panel's thinking and process. As necessary in the course of its work, the Panel can request STScI support for short studies of specific technical or scientific issues that may arise. They may also request the support of outside experts including but not limited to NASA and NASA contractor personnel.
>
> The Panel is appointed by and reports to the STScI Director, who will take the Panel's findings to NASA.[3]

Brown and Ford co-chaired the panel, which included Roger Angel of the University of Arizona; Jacques Beckers of the European Southern Observatory; Pierre Bely of STScI; Piero Benvenuti of the Space Telescope European Coordinating Facility; Murk Bottema of Ball Brothers Research Corporation; Chris Burrows of STScI; James Crocker of STScI; Rodger Doxsey of STScI; Sandra Faber of the Lick Observatory; Edward Growth of Princeton University; Shrinivas Kulkarni of Caltech; Bruce McCandless of the Johnson Space Flight Center (an astronaut); Francesco Paresce of STScI; Lyman Spitzer of Princeton (chairman of the Space Telescope Institute Council); and Raymond Wilson of the European Southern Observatory. The panel started its work on August 17–18, 1990, and had meetings at STScI and the European Southern Observatory through October 17–18, 1990.

Meanwhile I was considering the problem and its ramifications. Given the cost and risk of attempting a servicing mission to Hubble, I was afraid that science might not have high enough priority at NASA to warrant the gamble.

Should NASA decide to proceed, it would require a high degree of commitment from all involved. If successful, the repair would prove the value of man in space; if unsuccessful, it might pose grave political problems in Congress for the entire Space Shuttle and Space Station programs.

I started thinking up alternative solutions. I knew that Kodak had also been granted a NASA contract to build a twin of the Perkin-Elmer mirror (just in case), and that that mirror had been completed and was sitting at Kodak. To my knowledge, that mirror had no known defects. It occurred to me that we could build a spacecraft and improved instruments for much less than Hubble had cost. I proposed to solve the issue of launching the alternate telescope by making a deal with the Soviet space program. Hubble weighed some 22,000 pounds at launch; it therefore needed a very big booster, like that for the shuttle. However, even the shuttle was insufficient to place the spacecraft into a high, geosynchronous orbit. Yet such a high orbit would have more than doubled the effectiveness of Hubble, because the telescope would not be regularly eclipsed by the earth. This additional observing time would be given to the Soviets in return for the use of *Energia,* the most powerful space transport ever built (which had never again flown after its maiden voyage in 1987). In addition, line-of-site communications would be possible if the telescope were in geosynchronous high orbit, which would greatly simplify command communications and data retrieval.

I discussed this possibility with Rashid Sunyaev (head of the High Energy Astrophysics Department of the Academy of Sciences of the USSR) and Roald Sagdeev (director of the Soviet space program), who were enthusiastic about the idea. Politically the idea reached as high as the Soviet foreign minister, Eduard Shevardnadze, whom I met in Washington, D.C., at the Soviet embassy in late 1990. Glasnost was in the air and the minister seemed quite positive. When I discussed this possibility with NASA headquarters, however, I received a glacial reaction. Nothing ever came of it, presumably for security or political reasons.

This idea was not a crackpot suggestion, even for the times. Geostationary orbits for astronomical satellites had been considered previously and used for the International Ultraviolet Explorer (launched in 1978). But in the early 1980s, a complete dependence on the shuttle for launch had restricted us to low-Earth orbit (which, however, did make servicing the telescope possible after launch). The *Challenger* disaster in January 1986 threw this approach into disarray. The shuttle was less safe than had been claimed, its availability was uncertain, and the servicing costs were much greater than had been anticipated. There followed a period of great uncertainty about the future of space astronomy missions.

When invited to give a talk on this subject at the American Association for the Advancement of Science annual conference in Chicago (February 14–18, 1987), I decided to take a positive approach. Rather than lamenting the errors of the past, I developed the theme "New directions for space astronomy." I suggested that there was a viable alternative to servicing, consisting of complete replacement of an ailing or aging satellite with a new one, at less cost than servicing. This approach would channel more resources into the development of new technology rather than into operations and would permit the use of highly eccentric or, in general, high orbits, which would greatly enhance observing efficiency, because the earth would not block the satellite's view for half the time, as happens in low-Earth orbits. This alternative program would consist of continuous development, construction, and launch while observing with existing spacecraft.

In 1992 the managers of the AXAF program (now known as Chandra) reached the same conclusions. While the shuttle was still used as a heavy-lift vehicle to low-Earth orbit, the payload also included an inertial upper-stage engine, which inserted the satellite into a high-Earth orbit that extended one-third of the distance to the moon. It has now operated continuously for six years without servicing. The Spitzer Space Telescope is in solar orbit and the James Webb Space Telescope will be placed at Lagrange point L2. (The Lagrange points, or libration points, were discovered by the Italian-French mathematician Joseph-Louis Lagrange. They are the five special points in the vicinity of two orbiting masses where a third smaller mass can orbit at a fixed distance from the larger masses. L2 is the earth-moon libration point.) These changes in orbit represent a massive shift in strategy by NASA, to which my talk, followed by an article with Bob Brown in *Science* in 1987,[4] may have contributed (though this was never acknowledged).

Thankfully the STScI Strategy Group came up with excellent suggestions and a great new idea. The panel decided that the optical problem was thoroughly understood and the solution straightforward: each of the science instruments required two corrective mirrors, one of them built with an aberration identical to, but of opposite sign to, the aberration in the primary, which exactly canceled the spherical aberration. The implementation of this correction was immediately obvious for WF/PC by changing out the entire instrument, as had already been planned for the first servicing mission. The group proposed a new device called the Corrective Optics Space Telescope Axial Replacement (COSTAR), which would fully restore the high-resolution imaging capability of the FOC and also the spectroscopic capabilities of GHRS.

COSTAR consisted of a system of pairs of corrective lenses that could be deployed in orbit in front of the appropriate axial instrument on command. COSTAR would replace the high-speed photometer, the least used of the instruments, in its bay during the servicing mission. James Crocker of STScI came up with this solution, which was adopted by the entire panel. He told me he got the inspiration for this elegant mechanism, which permits the positioning and retraction of the correctors, from the design of the shower head in his hotel room in Garching, where the panel met in September 1990.

The panel also recommended that the solar-cell panels, which had caused pointing problems, be replaced with a set of rigid panels provided by TRW. I fully agreed with these findings when the panel presented them to me orally, and I took the recommendations to NASA headquarters in mid-October 1990. The STScI suggestion was well received by Charlie Pellerin, director of the Astrophysics Division, and the panel was invited to give a presentation at headquarters on October 26. In the weeks that followed, NASA carried out an intensive study of the feasibility and cost of COSTAR. In December 1990 NASA authorized the COSTAR program with Jim Crocker as the principal investigator of the project. He succeeded in having COSTAR ready for launch in two and a half years, by working closely with the engineering group at Ball Brothers, so that it could be carried aloft in the first servicing mission in 1993.

I think this performance established STScI's competence as an instrumentation builder as well as a science operator. The successful proposal by Holland Ford and his team for the Hubble advanced camera system resulted in a camera much superior in performance to the WF/PC-2. The advanced camera system was designed and built by scientists and engineers at the Johns Hopkins University, Ball, and STScI between 1996 and 1999 and flown in servicing mission 3B in March 2002. Because of its larger field of view and broader wavelength, the coverage it provided is ten times the discovery space of WF/PC-2.

The commitment of NASA to the success of the servicing mission had many important consequences. Success would be of great scientific value, and NASA would give a clear demonstration of the utility of the shuttle and of manned servicing. It would also provide a much-needed boost to the spirits of the public and of Congress, not to mention renewed confidence in NASA (which was already in trouble because of the great difficulties encountered in the implementation of the International Space Station). The full commitment by NASA also made work much easier, eliminating many bureaucratic barriers and delays.

Joe Rothenberg was the Goddard associate administrator for space flight for Hubble (1990–94) and subsequently Goddard director from 1995. He was uniquely helpful in advocating and ensuring the success of the Hubble servicing mission. Under his leadership, Frank Cepollina, an excellent mechanical and systems engineer, and his team performed wonders. A new relationship soon developed between STScI and Goddard, each institution recognizing the contributions of the other, and both working together toward a common goal. Rothenberg recognized the quality of the scientific leadership given to the program by STScI. As the Hornig committee had advocated back in 1976, the institute had taken on the responsibility of principal investigator for the entire program on behalf of the research community.

We at STScI worked enthusiastically with the engineers at Marshall and Goddard and the astronauts at Johnson to achieve success in restoring Hubble's vision. One of the astronauts on the mission, Jeffrey Hoffman, had worked as an x-ray astronomer at MIT and was known to many of us as a colleague, thus extending the sense of trust and collaboration from the scientists to all members of the team.

(This great spirit of cooperation lasted only as long as Rothenberg was at Goddard, and did not unfortunately extend to the preparations for Hubble's successor, the James Webb Space Telescope, except in its early phases. In this new program Goddard once again imposed an arm's-length relationship with STScI, in an evident effort to reestablish managerial, technical, and scientific control, with poor results for the program.)

The first servicing mission, 3A, was launched in December 1993. It was entirely successful, and by restoring Hubble to its full capability it enabled us to carry out the scientific mission originally intended. Hubble's spectacular data have resulted in more refereed publications in astronomy per year than for any other telescope in the world. The only challengers are Chandra and two ground-based observatories: the European Southern Observatory's Very Large Telescope and W. M. Keck Observatory, in that order. Ten years after the formation of the institute, the scientific staff of STScI, with no privileged access to Hubble, was producing results that placed it fifth in scientific productivity among astronomical institutions in the world.

FIFTEEN

Meritocracy: Selection of Observers from the Research Community—
Balancing Service and the Pursuit of Science: Research by STScI Staff—
My Forays into Research and the Politics of Science—Hubble in Its Splendor—
Getting Away from the Past: From STScI to the European Southern Observatory

Science at the Space Telescope Science Institute

Meritocracy: Selection of Observers from the Research Community

In the previous two chapters I have emphasized the aspects of STScI that had to do with service to the research community. Just as important to the scientific staff was the opportunity to work in an environment where science was actually being done and that provided the intellectual stimulation conducive to learning.

We were very much helped by the institute's having become a nexus for scientific encounters. Time allocation for Hubble was the prerogative of the STScI director, but the director in turn followed the advice of the Time Allocation Committee with few exceptions. Use of the director's discretionary time ultimately had to be justified to the Space Telescope Institute Council and never amounted to more than 1 percent of the total available observing time. A large number of scientists were involved in the work of the Time Allocation Committee and its several subcommittees, so that the committee did not take on the aura of an upper-caste clique.

The composition of the committees involved in reviewing the proposals for observations read like a who's who of astronomy. We made sure that the best scientists in the world in each subfield would act as reviewers, without regard to nationality or institutional affiliation. This approach resulted in an extremely fair selection system, which was essential, given that only one out of every seven proposals could be granted observing time. The lack of bias on the part of the Hubble Time Allocation Committee is demonstrated by its allocation of up to 25 percent of the available observing time to European astronomers, even though the European Space Agency's contribution "entitled" them to only 15 percent of the viewing time. This drive for scientific excellence rather than "juste retour" was an important factor in the success of Hubble.

Literally hundreds of astronomers came to STScI because of their work on various committees and study groups or for purely scientific reasons. Many of them gave formal or informal seminars, which were organized by Academic Affairs. The STScI staff flourished in a setting alive with activity and intellectual stimulation.

Little direction was given to the Hubble Time Allocation Committee except in a broad policy sense. With the advice of the STScI scientific staff, the deputy director and I had organized the committee into several subdiscipline groups, so that the considerable number of proposals could be handled more efficiently. Each group had at its disposal a fraction of the available observing time. At the beginning of the proposal process, this fraction was set in rough proportion to the number of proposals received in a given subdiscipline. However, we wanted to allow for the ratios to change with time as scientific interests shifted. We adopted a suggestion by Dick McCray of Colorado University on how to compare apples and oranges. For most proposals, it was immediately obvious which were very good and which very poor. The good ones were accepted and the poor ones rejected quickly. Then there was always a zone in the middle, from which proposals were chosen according to a cutoff line. The total number of proposals accepted could be affected by changing the cutoff line. This was done by a vote from the chairs of the subdiscipline groups during each proposal cycle, on a program-by-program basis. It turned out that the votes would easily converge: a good planetary program, for instance, was preferable to a bad stellar astronomy one—even in the opinion of a stellar astronomer.

Another policy issue was the ratio among long, medium, and short observations. In the Einstein program, we had decided that an equal-time distribution among the three categories was a good compromise between elitism and populism, provided the ratio of the observing times was set to 10. Thus for that program there were three observations lasting 10^6 sec, thirty of 10^5 sec, and three hundred of 10^4 sec. After discussion with the Space Telescope Advisory Committee, a similar approach was adopted for Hubble.

Finally, the Time Allocation Committee set aside time for key programs, such as the measurement of the Hubble constant (a cosmological parameter of interest in determining the age of the universe). The guidelines were that these programs had to be few in number, should be done quickly to allow time for others, and should not be allowed to proliferate. They should be done only in those fields where the community thought they would be appropriate. For instance, the ultraviolet spectroscopists did not want key programs, and planetary astronomers wanted observing campaigns instead of key programs (which would allow them to maximize observing time when the conditions were most favorable, for instance, at the closest approach of a planet).

Thus the policy directions were very few and were supported by the community. The policy was built on a spirit of cooperation in the use of this marvelous and unique facility.

Balancing Service and the Pursuit of Science: Research by STScI Staff

Early in the life of the institute, Len Cowie, Don Hall, and I had envisioned a limit to the support that NASA would be willing to give to in-house scientific research. We thought that between 5 and 6 percent of the institute's costs should be devoted to the support of research done by the in-house staff. This amount was not much different from the fees NASA would pay a good contractor and could be justified on those grounds alone (although I think that in-house research was one of the reasons for the high performance of STScI and thus needed no further justification). In our budget we included 50 percent of the salaries of the AURA tenure-track staff members (about twenty-nine individuals), 20 percent of the salaries of the non-tenure-track scientists and the director, and other items, such as travel to meetings and director's discretionary funds. Because these amounts exceeded the 5–6 percent limit we had set ourselves, we sought additional support by competing for grants. In fiscal year 1991, we had a total of fifty active grants or contracts adding up to $4,839,660. Most of the grants came from NASA headquarters and NASA centers (including Goddard, Marshall, NASA Ames Research Center, and the Jet Propulsion Laboratory), and a few were from universities. Some grants were given to the STScI staff that had been selected as guest observers. The National Science Foundation refused to consider any grant request from STScI staff for reasons never clear to me.

The STScI staff received no preferential treatment from the Time Allocation Committee. When this possibility was suggested by a member of the Visiting Committee as an aid for young scientists, the staff rejected it as demeaning and offensive. Over the years scientists at STScI won competitively, through the peer review system, a sufficiently large amount of time to annoy some scientists in the community. A committee set up in the late 1990s to study whether any impropriety had occurred concluded that the staff was winning its time through good scientific proposals. This result then led to the proposal, by a senior astronomer in the community, that STScI staff members should be judged more severely than other astronomers, so that their share of observing time could be reduced. I must say that I find such a suggestion appalling.

Another committee was set up years later (2002) to investigate whether STScI's hiring or promotion practices discriminated against women astronomers. STScI had always been at the forefront of affirmative action. In September 1992, at the suggestion of Megan Ury and Ethan Schreier, STScI organized and hosted "Women at Work: A Meeting on the Status of Women in Astronomy." At this meeting, the participants drafted a document that became known as the "Baltimore Charter," listing the improvements deemed necessary to the resolution of outstanding issues. This charter was later adopted by the American Astronomical Society and other scientific societies in the United States. I believe I was actually the one who suggested that the meeting would have more of an impact if it produced a tangible result.

I was somewhat taken aback, therefore, by the findings of the 2002 committee, which judged that STScI did not, in fact, discriminate except on the basis of excellence, but that this very fact created too harsh and competitive an environment. I must say that I find this point of view the epitome of antifeminism, because it assumes that women cannot compete on a par with men.

In any case, by 1992 the STScI preprint series had already produced some six hundred publications by or with the participation of the STScI staff. Despite their heavy load of functional work, the scientific productivity of the staff was comparable to or greater than the average in the astronomical community.

My Forays into Research and the Politics of Science

While at the institute I pursued the two lines of research in which I had been interested for several years: the nature of the XRB and the study of x-ray emissions from clusters of galaxies.

The X-Ray Background

Ever since my colleagues and I at AS&E discovered the XRB in the rocket flight of 1962, there had been a lively debate on its origin. I was convinced that it was due to the superposition of individual extragalactic sources so numerous that they could not be distinguished from one another because of the poor angular resolution available prior to Einstein's launch. Using results from Einstein, I was able in 1980 to resolve 25 percent of the XRB into individual sources in the energy range of 0.5–3.0 keV. But the chance coincidence of the XRB spectrum measured by HEAO-1 with that of a thermal bremsstrahlung source at a temperature of 40 keV convinced some astronomers that the emission was from a hot plasma pervading the universe. As I discussed in Chapter 10, this explanation made little sense in terms of known cosmological density and energy budgets, but the debate continued.

In 1987 I wrote a paper with Gianni Zamorani, who was visiting the institute, in which we gave a demonstration by "reductio ad absurdum" that the diffuse hypothesis could not be true. By that time it was known that 50 percent of the background was due to sources that had steep (soft) spectra. It followed that if the remainder was due to diffuse emission, it had to have a spectrum much harder than a bremsstrahlung spectrum, so that the sum of the two could reproduce the observed XRB. The only way to obtain such a spectrum was for the hypothesized plasma to be contained in dense clouds that would cut off the soft emission. We knew that we needed very many of these clouds to give the appearance of uniformity. By summing the total mass of matter contained in the clouds and requiring that it not exceed the total mass in the universe, we could set a bound on the size of each cloud. It turned out that this bound was the size of a galaxy, thus negating the diffuse background hypothesis.

It is interesting that this simple but powerful argument was not generally appreciated except by a few astronomers, such as Len Cowie, possibly because of its Jesuitical sophistication. The referee of the paper himself recommended that we further discuss the nature of these discrete sources, as if for our argument they needed to be real objects.

There seemed to be no other way to demonstrate the case for discrete sources except to detect them all directly at fluxes much fainter than could be detected by Einstein. With the 1990 launch of the ROSAT satellite, we would be able to resolve 80 percent of the background, but again at relatively low energies (0.4–3.0 keV), leaving the issue of the higher-energy background still unresolved. We had to wait for Chandra to be launched in 1999 for the high-energy response, sensitivity, and angular resolution necessary to fully resolve the problem.

In 1984 I responded to a NASA call for proposals to become an interdisciplinary scientist for the Advanced X-Ray Astrophysics Facility (renamed Chandra after launch) on the basis of continuing my research on the XRB. I won the proposal, became a member of the AXAF Science Working Group, and participated in its meetings and deliberations. The working relationship between Martin Weiskopf (the AXAF project scientist at Marshall), the Science Working Group, and the community seemed to me much healthier and productive than had been the case for Hubble. Most of the working group members were skilled and experienced experimentalists and so was Martin. In addition, he was supported by a small staff that was highly effective in facilitating the full study of issues as they arose. TRW was also an easier and less arrogant contractor than Lockheed had been, possibly because there was no issue of national security. It is also possible that this better working atmosphere was due to better delineation of the responsibility and control that Marshall had on this program; perhaps the Hubble experience had shown what not to do. I finally got to use Chandra to do science after its launch in 1999.

Clusters of Galaxies

I turn now to the second line of inquiry in which I was interested: clusters of galaxies. Clusters of galaxies are the largest aggregates of matter in the universe, consisting of thousands of gravitationally bound galaxies. They were discovered in optical survey plates obtained with the 48-inch Palomar Schmidt by George Abell while working on his PhD thesis. He published a list of about four thousand clusters in 1948, when he was 31 years old. The importance of his work was not recognized at the time, and it is only in the past few decades that the study of these objects has received increasing attention. The understanding of their formation and evolution is essential to the unraveling of some of the most important questions in cosmology.

X-ray observation of the intercluster medium in clusters of galaxies was one of the most significant discoveries of Uhuru in 1972 (see Chapter 7). It revealed the existence of a low-density, high-temperature plasma pervading the space between galaxies whose existence had not been previously known. The volume filled by this gas is so large that even though its density is very low, its mass is greater than that of all the galaxies and stars in the cluster.

After having been enriched in heavy elements by nuclear fusion inside stars, this gas is injected into the cluster by supernova explosions.

As discussed in Chapter 10, the individual member galaxies themselves do not have sufficiently deep potential wells to gravitationally trap this gas, so that the gas escapes the confines of each galaxy and is only retained in the cluster by the combined gravitational effect of all galaxies and their attendant dark matter (see Figure 10.6). It is the dark matter, which exceeds normal baryonic matter by a factor of ten, which binds the system of gas and galaxies. As time passes, clusters shrink on themselves (or more correctly, undergo gravitational collapse). The changing potential accelerates all particles within the cluster and increases their energy, thus heating the gas to high temperatures.

The net result is that the x-ray luminosity of a cluster is not simply the sum of the luminosity of the galaxies; added to this contribution is the much greater luminosity of the hot intergalactic gas, which can be five hundred times more luminous than the galaxies themselves. A cluster of galaxies held together by its own gravity is not therefore simply a local clumping of objects but a well-defined dynamic system. As it collapses, the high x-ray luminosity flags the collapse, and the extended nature of this luminosity distinguishes it from that of quasars.

Richard Burg and I had worked with William Forman and Christine Jones of STScI on the x-ray luminosity function of a complete sample of 200 Abell clusters from the Einstein data. While I was all for publishing our observational results as they stood, it was Richard who insisted that we should provide a physical explanation of what we were observing. This decision of course delayed publication by a few years, but we gradually came to realize that x-ray observations were potentially a more powerful tool than optical observations to study the physical conditions of a cluster. They were more sensitive in detecting clusters at large distances, because the x-ray emission of the cluster is much greater than the sum of the emissions from the individual galaxies. We could derive from spectroscopic observations the temperature, mass, and metallic content of the cluster. By measuring red-shifted iron lines, we could measure the cluster distance. Finally, we could study the morphology of the cluster with high statistical precision.

The Attempt to Build a Wide-Field X-Ray Telescope

At this point I began to wonder whether x-ray surveys of clusters of sufficient depth and size could be carried out with existing spacecraft, such as ROSAT, or those under construction, such as AXAF and the X-Ray Multi-Mirror, and realized the need for a different mission that would give us the capability we needed. What was necessary was to explore a small but contiguous region of the sky (10 × 10 degrees) with great sensitivity (3 × 10^{-15} erg/cm^2 · sec) and a region ten times larger with ten times less sensitivity. ROSAT was to carry out an all-sky survey at low sensitivity in the 1990s. The X-Ray Multi-Mirror did not have the required angular resolution. AXAF could certainly achieve the sensitivity, but its very small field of view, about 200 arcmin^2, would require more than a thousand exposures to cover an area of 10 × 10 degrees. Thus AXAF would require three years of dedicated time to complete the survey, an impossible request, considering the expected oversubscription of observing time.

I started thinking whether it might not be possible to dedicate a smaller mission to this one task, and realized that we could make up for the lesser sensitivity of a smaller mirror with wider field optics, which needed only to be invented. An x-ray telescope with half the diameter of AXAF (the same as that of Einstein) would have a quarter of the area and therefore a quarter of the sensitivity, but if it could be designed to have a much larger field of view (1 × 1 degrees) and an angular resolution of better than 5 arcsec, and if the entire time of the mission could be dedicated to this one program, it would do as well as or better than AXAF. It would also be much less expensive, because in space ventures cost increases with weight and weight with volume, the cube of the linear dimensions; therefore, halving the linear dimensions would reduce the cost of the mission by a factor of eight.

I discussed this possibility with STScI staff members Chris Burrows and Richard Burg, and we realized after much work that it could be done. Chris developed a computer program for generalized optics systems based on two-reflection grazing incidence, in which the reflecting surfaces could be high-order polynomials rather than conics. He searched for solutions of the equations that would maximize the angular resolution over the field, weighted by the solid angle on which it could be achieved. He was able to obtain solutions that would achieve angular resolutions of 2.5 arcsec over a field of 28 arcmin radius. We submitted these results in April 1990 in a letter to *Astrophysical Journal Letters,* which was not published until June 1992.[1]

Having convinced ourselves that a system using these optics could be built, Burg, Burrows, and I together with engineers from Ball Brothers designed a low-cost mission that was superior to AXAF for this single purpose. For me it was also an important demonstration of methodology in the design of scientific projects, and of the feasibility of doing first-rate science without always relying on multibillion-dollar missions.

We proposed the mission for the Wide Field X-Ray Telescope in January 1993, just after I had left STScI for the European Southern Observatory, with Richard Burg and myself as co–principal investigators and a stellar cast of astronomers as co-investigators. The mission was attractive to them because its goals were not only to discover at least 2,000 clusters but to find 20,000 quasars. The co-investigators included George Ricker and Mark Bautz of MIT; Maarten Schmidt of Caltech; Christine Jones, William Forman, and John Huchra of CFA; Robert Rosner of the University of Chicago; Chris Burrows, Rodger Doxsey, and Brian McLean of STScI; Bruce Margon of the University of Washington; Jeremiah Ostriker of Princeton University; and Guido Chincarini and Oberto Citterio of the Observatory of Brera. The participating institutions were the Johns Hopkins University, MIT, the Observatory of Brera, Alenia Spazio, and Ball Optics Cryogenics Division. The proposal was submitted in response to a NASA announcement of opportunity of August 31, 1992. It was rejected by a peer review committee convened by NASA, for reasons that were vague and unconvincing.

I felt sufficiently disturbed that I wrote to the NASA administrator, William Golden, pointing out that though I believed in the peer review system, it sometimes made bad mistakes. I stated that in my opinion this decision was as bad as the one in 1961, when NASA had turned down my proposal to search for x-ray stars on the grounds that they did not exist. Golden did not answer, but I received a letter from Wesley Huntress, then associate administrator for space science, explaining that the committee had discussed our proposal to observe clusters at large redshifts ($z > 2$; that is, in the distant past), but had decided that clusters did not exist at redshifts greater than 0.3–0.5 and therefore it was useless for us to search for them.

The proposal was resubmitted later to the European Space Agency and to the Italian Space Agency but without success. The broad survey that was proposed has not yet been performed, although Piero Rosati of ESO has done an incredible job of searching and finding distant clusters at increasingly large redshifts. I felt bad for all the people who had done such a marvelous job in designing the experiment, and I am still proud I was associated with it.

This tale can perhaps explain some of the bitterness I still feel for the shortsightedness of review committees that are dominated by the theoretical fads of the moment and do not hesitate to pontificate on what, in their opinion, can and cannot exist in nature. Nature is mercifully much more imaginative than they are, and it goes without saying that x-ray clusters have now been found at $z = 1.4$, equivalent to 90 percent of the age of the universe.

The worst aspect of such episodes is the devastating effect they can have on young people involved in a program.

My Italian Engagements

During my tenure at STScI, I carried out two additional activities outside the institute.

In 1987 I was invited by Mario Schimberni, president of Montedison, one of the largest chemical and energy companies in Italy, to help in turning their Research Institute Donegani into a center for excellence. I accepted and was making some progress in changing the culture of the institute by introducing free research by the staff, when in 1988 Montedison was bought by Raul Gardini for the Gruppo Ferruzzi. I was summarily discharged. This operation was typical in Italy during the corrupt regime of Bettino Craxi, whose main purpose was to put illegal funds into the coffers of the dominant political parties. A few years later Raul Gardini committed suicide just hours before he was to be arrested for fraud and corruption. The net result was that Montedison was dismembered.

A much more positive experience was the offer in 1991 of a professorship in physics and astronomy by the University of Milan "per chiara fama" (for clear fame). I accepted and gave a course of lectures in x-ray astronomy together with Luigi Stella, a distinguished young astronomer and a genuinely nice person. I greatly enjoyed this teaching experience and continued to hold the professorship until 1999, when I resigned to make room for a young scientist.

I had received permission from the AURA board for both activities on the basis of unpaid research leave for up to six weeks per year. NASA requested and received a formal audit of my activities, an episode that I found a little shabby and which reminded me of the Nixon era.

But now it is time to turn to the much more inspiring subject of the Hubble Telescope results and their impact on science and on the public's perception of space astronomy.

Hubble in Its Splendor

When the spherical aberration of Hubble was discovered and fully understood, we at STScI became very much involved in the program that would restore its capabilities through the servicing mission. We also felt that we should carry out whatever science could still be done.

Volume 369 of the *Astrophysical Journal Letters* of March 10, 1991, was dedicated to the first results of the Hubble Space Telescope. The lead article by Chris Burrows et al. described the spherical aberration and the pointing control degradation due to this effect, as well as the severe pointing disturbances during portions of the orbit stemming from thermal shock. Nevertheless, Hubble could still do unique science by high-resolution imaging of low-contrast objects through deconvolution techniques. These techniques relayed detailed information on the point-response function (which was achieved by modeling) and on the comparison between ground-based and in-flight data. In some cases, where photometric accuracy is not important, one can obtain an angular resolution approaching that of the design goal. At some wavelengths, the angular resolution obtained was much superior to what could be obtained from ground-based observations, even though only a fraction of the light (about one-fifth) was concentrated in the core of the image and the remainder was spread over an area a thousand times bigger. This high angular resolution was used by the authors of the other twelve articles in the same volume of the *Astrophysical Journal Letters* to study phenomena as varied as the central regions of the galaxies NGC 1068 and NGC 7457 and the core of globular cluster M 15; to obtain highly resolved pictures of Saturn; to obtain a visible light picture of the jet in the radio source PKS 0521; to resolve the gravitational lens system G223 G2237+0305 into four quasar images as well as the center of the galaxy; and to clearly reveal the expanding ring produced in the supernova SN1987A and the complexity of the Orion Nebula.

Apart from their scientific interest, these observations served to demonstrate that the instruments were working well, and that the ground operations systems at Goddard and STScI were fully capable of issuing commands, providing the necessary guide stars, receiving and processing the data, archiving them, and distributing them to the observers. It was clear, therefore, that Hubble could realize its potential for science if the servicing mission should prove successful. This required replacement of the WF/PC, replacement of the floppy solar cells, and completion and installation of COSTAR.

I left the Hubble Program on December 31, 1992, as I will later describe. By that time STScI was a mature organization, ready to fully discharge its responsibilities toward the community. The construction of COSTAR was proceeding smoothly, so that Jim Crocker felt he could go to the European Southern Observatory before me, as a sort of advance scout, before I joined him there for several years. The joint effort of the astronauts; the Goddard, Marshall, Johnson, and Kennedy space centers; and STScI to prepare for the service mission were sustained by the enthusiasm of the participants and by NASA's full commitment to succeed.

I hope I can be excused if instead of describing the slow grind that culminated in the successful launch of the Hubble service mission in December 1993, I describe the extraordinary impact of the new data on all of us in the research community.

From the first servicing mission in December 1993 through the four missions that followed (to 2006), Hubble has undergone notable increases in its scientific capabilities. New instruments have been installed: the high-resolution spectrograph, the advanced camera for surveys with wide field and high angular resolution, a new version of the WF/PC with improved performance, and a new infrared camera. A new on-board computer and digital recorder have permitted a sixtyfold increase in data retrieved from Hubble with respect to its capabilities at launch.

Hubble has had several key advantages over most other astronomical facilities: an unprecedented angular resolution, spectral coverage from the ultraviolet to the infrared, a dark sky, highly stable images unaffected by any weather (which permit precision photometry), wide fields of view, a high data rate, and pipeline processing (including calibration and archiving capability). Hubble had cost too much for anyone to be satisfied with mediocre science; thus we were allowed to strive for excellence, a situation that does not always prevail. The Hubble data are proprietary to the original guest observer for only a year; after this time, they are made available to the entire community through the archives, which by 2005 had grown to about 20 terabytes of data. These factors combined have made the Hubble Space Telescope one of the most powerful and best-used astronomical facilities in history.

By all standards Hubble has been extremely productive: as of January 2005 it had provided the data for 4,634 refereed papers. The rate of publication is continually increasing and is now six hundred papers per year, almost twice as great as for its nearest rivals, the European Southern Observatory's Very Large Telescope (VLT) and Chandra.[2] The number of citations per Hubble paper is also higher than for Chandra, the VLT, and Keck. Of the two hundred most-cited papers per year in all of astronomy, only Chandra appears more often than Hubble, followed by VLT and Keck. Though these numbers do not tell the whole story, they are still quite impressive. Moreover, the most important Hubble discoveries are among the most significant in astronomy in the past decade.

One of the objectives of Hubble has been the measurement of the Hubble constant, H_0, which is necessary to calibrate the distance scale and size of the universe. The uncertainty of a factor of two that existed in the past has been reduced to 10 percent—high precision in cosmology.

Even more important for cosmology were the very deep survey images obtained by summing exposures obtained from many orbits. The deepest to date is the Ultra Deep Field survey, obtained with the Advanced Camera for Surveys in 800 exposures, taken over 400 Hubble orbits from September 24, 2003, to January 16, 2004 (Plate 9). (The total exposure of one million seconds equals that of Chandra's Deep Field South.) In this observation, Hubble peers into the "hoary deep" to a depth of billions of light-years, observing the universe when it was only 800 million years old. The nearly ten thousand galaxies observed there are the first large structures to form, and they appear at a much earlier time than previously expected. These distant young galaxies are different in appearance from nearby galaxies, giving us direct proof of the evolution of the universe.

The discovery that structure formed so early in the history of the universe was one of the most unexpected and important contributions of Hubble, and it could not have been achieved with the angular resolution available from the ground. I remember having in my office at STScI the deepest picture of a region of the sky from ground-based observations made by Tony Tyson of Bell Labs, showing galaxies to magnitude 27. The galaxies seemed to occupy the entire field and to be a barrier to deeper observations. Hubble has not yet reached its confusion limit, even though it can detect objects with magnitudes that are six times fainter.

Hubble firmly established that quasars exist in galaxies, thanks to the work of John Bahcall and his colleagues. It also confirmed what had long been suspected, namely, that the energy source for active galactic nuclei and quasars is the gravitational infall of gas onto a supermassive central black hole (Plate 10). To be exact, I should note that the doughnut shown in this wonderful picture of NGC 4261, a picture I have often used in my own lectures, is not the actual accretion disk. The size of the dusty disk we can see is about 400 light-years across, whereas the actual accretion disk is less than one-hundredth of that size, so small in fact that it would be unresolvable from the central source. In this case, however, the pedagogical value of this picture is great enough to excuse the imprecision.

Spectroscopic observations from Hubble revealed that black holes can trigger massive waves of star formation; established the existence of invisible filaments of matter linking galaxies over millions of light-years and providing the bulk of the baryonic matter in the universe; and finally measured the distribution, chemical composition, and physical state of interstellar gas.

Hubble provided a new method to find planets around other stars by detecting with highly accurate photometry the partial eclipses of a planet passing in front of a distant star. The study with Hubble of distant supernovas extended to very large distances the findings by ground-based astronomers of the dimming of the observed emissions, thus strengthening the theoretical conclusion that there has to be a dark energy, whose negative gravity accelerates the expansion of the universe. If confirmed, this discovery would rank among the most significant in physics of the past hundred years. And Hubble observations confirmed the extragalactic origin of gamma ray bursts and may yield important clues to the physics of black holes.

I was particularly struck by the visual impact, the beauty, and the power of three Hubble images: the Eagle Nebula, the Herbig Haro 30 object, and the remnant of the Crab Nebula supernova. The Eagle Nebula picture (Plate 11) has been published in every science magazine in the world and has become an icon for Hubble. It shows gas and dust pillars in M 16 (named the Eagle Nebula), which have been called the pillars of creation. These pillars are dense clouds of molecular hydrogen and dust that have survived the vaporization caused by blasts of ultraviolet light from hot newborn stars, which are located just outside the upper edge of the field. These pillars are light-years in dimensions; as they are eroded away by the ultraviolet stellar light, they expose small globules of gas at higher density that have been called EGGs (evaporating gaseous globules). The acronym for once fits, because inside the EGGs are embryonic stars. The eerie three-dimensional effect that gives the photograph its visual impact is due to the fact that the surface of the pillars glows under the same ultraviolet light that erodes them. No astronomer had ever tried to visualize the complex process of star formation. This picture is a treatise in and of itself.

Another aspect of stellar and planetary formation that had never been resolved was accounting for the angular momentum and the magnetic field contained in the primordial nebula as a star is formed by gas spiraling toward the center of the protostar. Hubble's picture of several young stars surrounded by protoplanetary disks shows what is happening to the gas (Plate 12). As the gas collapses it indeed carries its magnetic field and angular momentum along. In a process not yet fully understood, the central object loses the excess energy by producing jets and blowing off matter from the accreting gas. I hope that the studies of these relatively tame objects will ultimately provide clues to understand the process, which is repeated again and again on all scales, from stars to the active nuclei of galaxies.

In Plate 13, the beautiful optical picture of the Crab Nebula by Hubble (red) is shown with that obtained in x-rays with Chandra (blue). What is seen here is a different type of emission from stars at the ends of their lives. As the nuclear fuel is exhausted, a star of one solar mass collapses into a neutron star, an object 10 km in radius, which is maintained in equilibrium against further collapse by the pressure of a gas of neutrons. The neutron star in the Crab Nebula rotates on its axis at 30 revolutions/sec, and its intense magnetic field accelerates electrons and protons to relativistic velocities. The interaction of the particles with the field of the nebula produces light by synchrotron radiation. The neutron star and its wind power the entire nebula, thus dissipating its kinetic energy of rotation. The picture from Chandra matches perfectly the features seen in visible light, except that the higher-energy electrons that produce the x-rays have much shorter lifetimes; thus they disappear closer in to the central source.

What I find remarkable is that these highly sophisticated pictures have been taken to heart by the public at large. The pictures convey information that is apprehended intuitively and can be confirmed by analysis to be basically correct. I am still startled by the incredible wealth of information that can be gathered by observations unimpeded by the absorbing and scattering effects of the atmosphere. The clocklike regularity of the Hubble observations at all times, the stability and reliability of the measurements, and the speed with which reduced and calibrated data are in the hands of the observers are stunning to most astronomers. The pioneering use of disk archives to place information at the disposal of the entire world makes it possible for many astronomers to participate in this venture.

It is fair to say that Hubble and STScI established the standards by which any future astronomical enterprise will be judged. The methodology developed for Hubble has been adopted by VLT and Chandra. It is not perhaps by chance that these two facilities are the only ones rivaling Hubble in scientific productivity.

I did not get to use Hubble for my own science while I was at STScI, but I made up for it when I used Hubble, VLT, and Chandra observations combined to study and resolve the problem of the XRB in 2001.

Getting Away from the Past:
From STScI to the European Southern Observatory

In September 1991 I was asked by the AURA board to serve another five-year term as director of STScI. With some reluctance, I accepted. The reluctance was because of a family tragedy: the loss of our son Marc in an automobile accident in August of that year. He was a loving, intelligent, and handsome young man, who had gone through more than his share of troubles. My wife and I were devastated. I decided to go on working at STScI for many reasons: I had no other choice, I loved my work and the people I was working with, I was proud of what we had accomplished, and I was determined to do my part to fix Hubble.

However, it was hard to continue living in Baltimore, where so many things reminded us of him. My wife and I routinely had to drive by the tree against which he had smashed the car and the hospital where he had been taken. We had built a beautiful house near the Johns Hopkins University campus, but it, too, was filled with memories and pain. Added to that was my remorse for not having done enough for him. Although I was still trying hard to give my best to STScI, the "fire in the belly," to use Vincent Severo's expression, had been reduced to cinders.

So when I was approached by the president of the European Southern Observatory Council in May 1992, asking me whether I would be interested in becoming director general of the observatory, I said yes, and when offered the position, I accepted.

I was familiar with this observatory because of my frequent visits there in connection with ROSAT research. The institution is located in Garching, near Munich, next to the Max Planck Institutes for Astrophysics and Space Research. I had always liked the southern Tirol (where I used to go skiing and climbing when I was young), and Bavaria reminded me very much of it. Munich is in some respects a large mountain village, with a great deal of culture, in addition to its folklore. After more than 36 years in the United States, it was a return home to old Europe with its soothing beauty. There was also the lure of the Atacama Desert in Chile, where the European Southern Observatory had its first observatory (La Silla) and where it was planning to build a second one on Paranal. Finally, I had a keen professional interest in the new venture because, by charter, I would have to build the observatory rather than just operate it, and there promised to be less of a continuous struggle to do so. I hoped that I could accomplish both tasks better than had been done on Hubble.

As it turned out, the six and a half years we spent at the European Southern Observatory was a sort of vacation from the stresses of our previous life for both my wife and myself and a place where pain was suspended.

SIXTEEN

Joining ESO—Fusing ESO into a Coherent Whole—
Personnel Issues: Overhauling the System—Shaping Up: The La Silla Observatory—
Catching Up: CCD Detector Development—Transforming Data Management at ESO—
War and Peace: Relations between Chile and ESO

The European Southern Observatory

Joining ESO

The European Southern Observatory is an intergovernmental European research organization for astronomy in the southern hemisphere. It was created by international treaty in 1962, is governed by a council composed of representatives of the governments of the member states, and is funded by these states, normally through their foreign ministries. Currently the member states are Belgium, the Czech Republic, Denmark, Finland, France, Germany, Italy, the Netherlands, Portugal, Spain, Sweden, Switzerland, and the United Kingdom. More countries are expected to join in the next few years.

The ESO headquarters is in Garching, near Munich, Germany; its observing stations, located in Chile, include the VLT at Cerro Paranal (the largest array in the world) and several medium-sized telescopes at La Silla. Currently ESO is developing, jointly with North America and Japan, the Atacama Large Millimeter Array (ALMA) at Llano de Chajnantor in the Atacama Desert, Chile.

Prior to my accepting the position of director general of the organization, I was given the opportunity in 1992 to visit ESO's offices and residence in

Santiago, the La Silla Observatory, and Cerro Paranal. Harry van der Laan, my predecessor, was my gracious host in Munich, and Daniel Hofstadt, the director of La Silla, squired me around Chile on the chartered ESO plane.

Harry was understandably saddened by the decision of the council not to renew his appointment, as he felt he had done his very best for ESO. He published a farewell article in *The Messenger,* a widely circulated ESO quarterly journal.[1] Dutch astronomers have had a long relationship with ESO; the very concept of a European astronomical organization arose through discussions at Leiden in 1953 between Jan Oort and Walter Baade. The long stewardship of ESO by Lodewijk Woltjer resulted in the construction of the New Technology Telescope (NTT) and the studies that led to the ESO Council's approval of the VLT project. Harry himself had initiated the VLT project in 1988.

In his farewell message, Harry mentioned the two basic ideas that inspired him during his tenure that are still valid today: ESO's function is to serve the European astronomical community, and the community must be given the opportunity to participate not only through use of the facilities and by providing scientific guidance, but also through instrument development and construction. These ideas I completely share.

ESO was confronted at the beginning of the VLT program by several daunting challenges. Although the organization had been successful in developing the NTT at La Silla in the 1980s, this instrument was very different in scale and technical challenges from the VLT; its cost was about $15 million, or about one-third the ESO budget at the time, and it was fully funded at the start as an add-on to the normal expenditures. VLT, on the other hand, represented at least six times the organization's yearly budget, and because the funding was approved on a yearly basis, management of the cash flow became an important issue. In addition, NTT was not as radical a technical departure from established practice as VLT, even though it provided a test of some of the concepts that were later fully developed in VLT. The construction and operation of the La Silla Observatory and the development of NTT had created an extremely competent scientific and technical staff, but they had not resulted in an organizational structure suitable for leading a project of the magnitude of the VLT.

The situation became quite clear to me as I visited ESO headquarters and the sites in Chile in the summer of 1992. In discussions with many of the scientists and managers, I found an organization that was not unified through communication or a shared vision. The operations in Chile seemed disconnected from headquarters and did not benefit from the scientific and technical competence available there. The lines of authority were clear only in that everybody reported to the director general, but there were no unambiguous statements of tasks or assignment of responsibility at the project level. There was little hands-on management at the senior levels of the organization, where an old-fashioned and rather paternalistic approach had developed early on and continued unchanged over the years. ESO had the atmosphere typical of a government bureaucracy.

There was no "faculty" (in the sense we had one at STScI) that "owned" the observatory. There was no clear process for the hiring or promotion of the scientific staff and little involvement of the staff itself in the process. There was no yearly review of the entire staff by means of written documentation or face-to-face discussions for evaluation of performance on goals established in the previous year.

In addition to these general issues, there was an urgent and serious problem concerning VLT. The program had been initiated in 1988 with in-house management under Massimo Tarenghi; then Harry van der Laan had hired a new manager with industrial experience to lead the project. This person was subsequently fired, and the project was reassigned to Massimo and Daniel Enard with no clear delineation of responsibilities. Massimo was an able scientist and manager who had led the successful NTT project, and Enard was a highly competent chief engineer who essentially controlled all of ESO's technical resources. In the administrative shuffle, Massimo had been named VLT project scientist.

Early on, during one of my visits to Garching in 1992, Raymond Wilson invited me to dinner and we had the opportunity for a long discussion about ESO's problems. Ray was a distinguished optics designer who had participated in the Hubble Strategy Panel Study three years earlier and had retired from ESO around the time of my arrival. As head of the Telescope Group at ESO, he had led the design of both NTT and VLT. He has written an article for *The Messenger* that is a fascinating description of the progress of active optics from the early 1970s to the realization of NTT and VLT.[2] Ray had the highest confidence in Massimo's technical and managerial competence and considered him the only person at ESO with the ability to carry VLT to completion. Soon after my arrival at ESO in 1993, I named Massimo as VLT program manager, and he did indeed carry it to success.

In evaluating the situation from a management point of view I was greatly helped by James Crocker, who had actually reached ESO some weeks before I did and who had, at my request, undertaken the same assessment tour in

Garching and Chile that I had. As I described in previous chapters, he had led the Operations Division at STScI, invented the COSTAR solution for Hubble, designed COSTAR in detail, and supervised its successful construction. I offered him a position at ESO for as long as he wanted. He told me that he accepted because he felt challenged and he wanted to acquire some international experience. He proved himself extremely valuable at ESO for several years before returning to Baltimore, where he contributed significantly to the successful completion of the Sloan Digital Sky Survey Program at the Johns Hopkins University. He is now a senior vice-president at Lockheed, where he is responsible for all civilian space programs of the corporation. Apart from his managerial and technical ability, Jim is blessed with a great facility to relate to people and to communicate with them. This particular virtue helped immensely to balance my rather stiff approach, which I am afraid hardened with time and events. Jim also strongly believed in the type of close cooperation between scientists, engineers, and managers that I had initiated at AS&E and had insisted upon since then.

Over the next few years, Massimo was given the full managerial responsibility for VLT, and Jim and I supported and guided his efforts, particularly in managerial and financial aspects, the presentation and defense of the program to the ESO Council and Finance Committee, and the re-engineering of the ESO organization so that the observatory as a whole could successfully complete the program.

Fusing ESO into a Coherent Whole

Although there was a wealth of engineering and scientific talent at ESO when I joined it, it was not, in my opinion, fully utilized. In large part this stemmed from a feeling of separation between upper and lower management, among the different teams, and between staff in Garching and Chile. I began the process of unification by holding a two-day meeting in Garching in early February 1993, a month after starting the job. At this review, all aspects of ESO's work were presented and discussed. The annual ESO-wide review became a yearly event to which members of the ESO Council were invited; the last I held was a four-day affair on February 1–4, 1999.[3] This was an expensive proposition in terms of travel costs, but I always felt that the gain in effectiveness and cohesion more than paid for it.

This first review revealed that the staffs knew very little of one another's work. Thus the review helped them, as well as me, to understand the different aspects of ESO's work and the purposes (or lack thereof) of the various activities. Little by little, over the years, the concept of a single observatory in which everybody contributed to achieve the same goals became accepted, and teams were formed to carry out the work utilizing talent wherever it could be found, whether in Garching, La Silla, or Paranal, or sometimes from outside institutions.

My basic approach was along the following lines: a drive for excellence in all aspects of our work; emphasis on quality, not quantity; involvement of the scientific community; placing priority on items that would yield high scientific returns; cooperation with other institutions worldwide; and finally, effective use of our resources. Over time we developed a clear methodology based on participative management; for instance, all division managers participated in a joint decision on the yearly budget assignment. We established clear procedures for advancement and promotion for all staff and also unambiguous procedures for peer review of scientific appointments by a "faculty," which was created during my tenure. No secrecy was allowed about technical issues, and open communication and debate were encouraged. Authority and responsibility were delegated, but with accountability based on agreed-upon goals. In other words, Jim Crocker and I proceeded to reengineer the organization as required to achieve success. Finally and most important, we tried to create a shared long-term vision, and to carry out the strategic and tactical planning required to bring it to reality. In what follows I illustrate how we accomplished some of the most important goals we had set for ourselves.

Personnel Issues: Overhauling the System

Above I mentioned that the process of hiring and promoting scientists at ESO was not clear. What I should have said is that ESO's personnel policies were modeled on those at the European Organization for Nuclear Research (CERN), and that both organizations resembled civil service agencies rather than research institutions.

The hiring procedures at STScI resembled those of academic institutions, with the fundamental difference that our job descriptions entailed service to the research community rather than teaching. This service required a coordinated effort by many scientists, and therefore emphasized particular qualities for the people on the STScI tenure track. Given this difference, we were as strict in the selection process at STScI as I ever experienced at Harvard.

For tenured appointments, we solicited ten letters from distinguished astronomers around the world. We also solicited the views of functional supervisors, and we made no offers unless a functional slot became available. We shared the outside letters and the views of the supervisor with the senior tenured people (the "faculty"), and they voted on the proposal for tenure. We required that the candidates satisfy three out of four of the following criteria: they had to be good scientists, they could lead others in research, they could do functional work themselves, and they could lead teams in functional work. The director was required to submit, along with his own recommendations, all letters received and the results of the faculty's vote to the Space Telescope Institute Council (chaired during my tenure by Lyman Spitzer). The council would then make a recommendation to the AURA board. The director could disagree with the recommendations of the council and appeal directly to the board, something I never did during my tenure at STScI. I am still proud that the Space Telescope Institute Council accepted all my recommendations for tenure. This process was made deliberately tight, so that the STScI scientific staff would be of a quality comparable to that at any major university in the world.

In contrast, the process at ESO was rather informal, and its success depended almost entirely on the scientific taste of the director general. Over the years no equivalent of a faculty had been identified, and because appointments were in a sense rather arbitrary, differences in rank, such as senior astronomer, associate astronomer, or assistant astronomer, made little sense.

I tried hard, with the help of Jacqueline Bergeron, who led this effort, to remedy this situation by creating an Office for Science, which was to carry out the functions of a university department and of the Academic Affairs Division at STScI. I appointed Jacqueline associate director for science in 1994, and she managed this activity. The task of the Office for Science was to advise the director on scientific matters; guide the expansion of the ESO scientific staff in the VLT era through a single Scientific Personnel Committee for all of ESO; and provide a forum for debate of major scientific, technical, or operational issues with a view to developing a unified ESO position. The Office for Science was to take initiatives in the pursuit of active scientific programs, including fellowships, visitors, workshops, and joint astronomy colloquia with all academic and research institutions in the Munich area. Finally, it was to promote collaborations between European and Chilean astronomers.

Many of the activities undertaken by this group were quite successful. I was only a little disappointed by the length of time it took to convince the forty-five members of the newly created faculty to take ownership of the work and of the future of the observatory. This clarification of the career path and of the meaning of being an astronomer at ESO probably helped create opportunities for the younger staff, whose contributions turned out to be essential to the results we achieved.

This entire re-engineering process was certainly helped by the introduction of yearly performance reviews for all employees, including a face-to-face discussion between staff member and supervisor and written documentation of the evaluation. Unfortunately, this revamping required the removal of the head of personnel, who did not believe in any of the above. Given ESO's regulations, his removal was an expensive proposition, costing the equivalent of five years' of salary, but I felt it had to be done. I was supported in this decision, as on all other administrative matters, by two wonderful associate directors for administration, first Gerhard Bachmann, who had served in that position with two of my predecessors and was an important stabilizing influence, and then, after he retired for health reasons, by the capable Norbert Koenig.

My philosophy was that unsatisfactory performance by a member of the staff did not cause just a null but a negative contribution by adversely affecting the work of others. This negative influence is well understood in well-run corporations where, during mergers or acquisitions, the negative performers are culled out at the outset of the reorganization process, as Machiavelli had suggested centuries ago in his much-maligned *The Prince*. Following this philosophy, some five senior scientists, who were not making either scientific or functional contributions, were encouraged to leave early in my tenure.

A completely different problem existed in personnel relations with the Chilean staff, something I discuss later in the chapter.

Shaping Up: The La Silla Observatory

When I first visited La Silla in the fall of 1992, the observatory had fifteen telescopes. It was a source of pride for ESO that La Silla was thus the largest observatory in the world (Plate 14). However, I quickly found out that many of the telescopes were obsolete and were not contributing significantly to advances in astronomy. The worst example was the Schmidt telescope, in which there appeared to be little scientific interest or participation. Plates were taken by dedicated technicians, but often they were much inferior in quality to the plates taken by scientists at the Anglo-Australian Schmidt at Mount Stromlo Observatory, Australia, and they were not used. Other telescopes were also underused, in need of urgent repairs to their control and drive mechanisms, or equipped with obsolete detectors. My immediate impulse was to close some of the telescopes and concentrate the available resources on upgrading the remaining ones. However, the many national interests of the member states made such an action tricky.

For this reason, the creation in October 1994 of a special working group of the Science and Technology Committee, chaired by one of this committee's members (Johannes Andersen of the University of Copenhagen), was very important. The group included Jacqueline Bergeron, Jim Crocker, and Jorge Melnick of ESO; Michel Denenfeld and Hans Shield of the Users Committee; and Sergio Ortolani of the Science and Technology Committee. The charter of the group was to study and make recommendations regarding the role appropriate for La Silla in the VLT era. The group solicited the opinions of all European astronomers and held a number of meetings, after which they issued a comprehensive report (in December 1996) with a clear statement of priorities that was submitted to the council for their approval. The council's decision, based on these recommendations, allowed me to close all obsolete facilities and concentrate all available resources on bringing those that would continue to operate up to state-of-the-art standards. If this upgrade could be done while also helping the VLT program, so much the better (Figure 16.1). The result of this procedure was to rationalize and streamline the existing facilities for scientific research while helping to instill a spirit of cooperation and strategic thinking in the ESO staff and gaining experience that proved valuable in developing VLT.

A splendid example of this kind of synergism was the complete upgrading of the command and control system of the 3.6-m NTT by using and testing the software developed for VLT. This upgrade was successfully done by a mixed Garching and La Silla team under the leadership of Jason Spiromillo, a relatively young astronomer who was given complete authority and responsibility for this work, to his great amazement. Traditionally, such responsibility was given to senior scientists. On the other hand, I have always believed in giving full rein to the enthusiasm of brilliant young people, before it peters out into complacency. The original suggestion for this enterprise came from Joseph Schwarz, a scientist and software engineer, who had worked with me on the Einstein program at Harvard and had then joined ESO. He pointed out, with old-style Uhuru sincerity, that we were crazy if we thought we could succeed with the new and sophisticated VLT software when we had so little opportunity to test it in the field before proceeding to install it.

The project led by Jason was so successful that NTT became convinced that it was Unit Telescope 5 of the VLT array, and we had to teach it that it

FIG. 16.1. *The timeline of planned (1999) telescope closures and instrumentation improvements for the La Silla Observatory.*

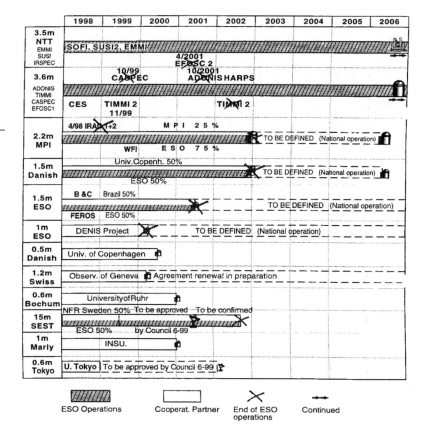

was at the latitude of La Silla rather than Paranal. New calibration, operation, and maintenance procedures were introduced. The new VLT autoguider system was installed on NTT so that we could make better use of the active telescope control. New CCD detectors for the optical and a new instrument for the infrared were installed. Finally, the reflectivity of the telescope was improved from 85 to 93 percent. Rumor has it that part of this improvement was due to the removal of a mummified condor that was found when the telescope was taken apart (Figure 16.2).

The NTT, thus revamped and improved, was used to install SOFI (son of ISAAC), a smaller version of the powerful new Infrared Spectrometer and Array Camera (ISAAC), which was being built for VLT. Both ISAAC and SOFI were built under the scientific leadership of Alan Moorwood in the Garching laboratories. SOFI was built with very modest funds in only two years, this speed being essential to test the advanced technology required for ISAAC. The dedication and enthusiasm that motivated Alan and his team to work so hard and achieve so much in such a short time was due in no small part to the fact that, while we were so seriously engaged in VLT, SOFI was almost a lark that promised immediate scientific returns.

A similar story—but one less directly related to VLT—can be told regarding the old 3.6-m La Silla Telescope. For years it had given poor angular resolution on the order of 1.0–1.5 arcsec, whereas the seeing limitations on La Silla are on the order of 0.5 arcsec. This effect had been ascribed to "dome seeing" stemming from the temperature gradients in the observatory dome. To compensate, the huge dome of this telescope was air conditioned to very cold temperatures, such that the observers had to wear heavy parkas, but without real improvement in the resolution. In addition, the instrument suffered from poor pointing performance.

FIG. 16.2. *The New Technology Telescope of the European Southern Observatory, showing the actuators for image improvement.*

Jorge Melnick, who succeeded Daniel Hofstadt as La Silla director in 1996, took a radical approach. He commissioned a mechanical engineering firm to develop a finite element model of the telescope and its support structure that revealed a severe design fault: the telescope was too heavy for the supporting structure, which could not provide sufficient rigidity at all pointing angles. He re-engineered the telescope and lightened it as much as possible. Light plastic replaced steel whenever possible, but in addition, he introduced a system of active forces to counter unwanted deformations, including an active lateral pad support. He also repaired the gears of the pointing control, and the pointing control system was upgraded to VLT standards, including the VLT auto-guider.

Most importantly, the telescope was realigned and refocused, because it appeared to suffer from coma, astigmatism, and possibly spherical aberrations. Subsequent monitoring showed 0.5-arcsec images (that is, at the limit of seeing) and a good, well-behaved pointing system with acceptable errors of 8 arcsec on average. Dome seeing had never been the problem at all, and suitably placed small fans were sufficient to control the mirror and instrument seeing. The mirror was cleaned and recoated, and the reflectivity improved from 85 to 91 percent. We essentially had a new state-of-the-art telescope—and the total upgrade had been achieved with only 35 man-years of effort and an expenditure of only $1.1 million!

Both the NTT and the 3.6-m telescope were equipped with state-of-the-art CCD detectors with the new ESO controller developed for VLT. By 1998 La Silla had become one of the most up-to-date, scientifically productive, and cost-effective observatories in the world. The number of papers from all La Silla telescopes in refereed journals rose from 330 to 419 in the four years from 1995 to 1998. The enthusiasm created by these achievements and the full employment of the remaining staff in highly productive work was made possible by the closure of seven of the fifteen telescopes during the same period. These closures also reduced the personnel costs of La Silla by about 25 percent (from 1997 to 1999) and freed resources for VLT and Paranal. Taking advantage of the VLT tide, all boats rose at ESO during those years.

The agreement with the user community (which allowed so much change in an old observatory) would not have been possible without the broad-based consultations and discussions instigated by Johannes Andersen's special working group of the Science and Technology Committee. This type of scientific evaluation was continued, and the chairperson of the committee for 2000 was Birgitta Nordstrom, Andersen's wife, who had been informally helping with the previous work.

Catching Up: CCD Detector Development

ESO and European astronomers in general had lagged behind the United States in the development of high-quantum-efficiency CCD detectors and low-noise fast controllers. These devices increased overall operational efficiency by factors of two to ten. It made little sense to develop the largest array of optical telescopes in the world and then be handicapped by low efficiencies. Sandro d'Odorico was in charge of optical instrumentation at ESO and was concerned about this problem. He succeeded in persuading Jim Beletic of MIT Lincoln Laboratories to join ESO in 1995.

Jim took over the full responsibility of improving the performance of ESO's systems to equal or surpass world standards. The aim was to meet all the scientific requirements imposed by the new generation of instruments for VLT: high speed, low-noise readout, multiple-port readout systems, and CCD detectors with high quantum efficiencies over a broad wavelength range, large format, and high cosmetic quality (that is, few blemishes). His goal was realized as early as 1999. By that time, ESO had succeeded, through close collaboration with U.S. industry and research institutions, in obtaining the CCDs it needed. The manufacturers included Loral Space Systems (Palo Alto, California), EEV/E2V (San Diego, California), and MIT's Lincoln Laboratories (Lexington, Massachusetts). The testing was often done in collaboration with the University of Arizona and Lincoln Laboratories. Figure 16.3 shows the high quantum efficiencies of the devices we obtained.

Jim's group also developed a new and advanced electronic controller called FIERA (Fast Imager Electronic Readout Assembly) that produced little noise and permitted us to operate at high speeds. These improvements, when combined with multiport readout, gave us a gain in observational efficiency of about ten. When this high performance became established, ESO started upgrading all of the La Silla instruments. Between 1998 and 1999, ESO produced eighteen systems, including one $8{,}000 \times 8{,}000$ 15-μm mosaic for the Wide Field Imager at the 2.2-m Max Planck–ESO survey telescope. Beletic's group also started development for the even larger $16{,}000 \times 16{,}000$ 15-μm array for the wide-field imager Omegacam, a new instrument to be used at the VLT survey telescope. In effect, ESO became the standard provider of CCD detector systems for most of the VLT instruments, whether built in-house or by other European institutions. While this intense work was proceeding, Jim's group held workshops attended by astronomers and technicians of all nations, with representatives from most leading manufacturers and major astronomical observatories. Two of the members of his group

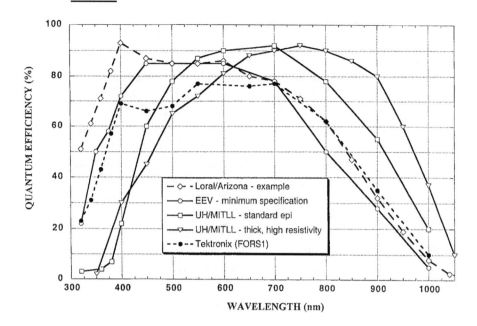

FIG.16.3. *Quantum efficiency of the new European Southern Observatory charge-coupled device detectors plotted as a function of wavelength.*

were pursuing PhD theses at the observatories of the universities of Munich and Heidelberg, and younger students were frequent visitors.

In 1999, ESO had the best CCD systems in the world. This success was achieved by a closely knit group of professionals from seven different countries working in Garching, La Silla, and Paranal. These individuals combined skills in detectors, solid state physics, electronics, computer software, cryogenics, optomechanics, and image analysis. To quote Jim: "The optical detector team's strongest asset is our group interaction, trust, and team work. The formation of this team has been our greatest achievement and is the foundation of our success."[4]

Transforming Data Management at ESO

In the processing and archiving of data, in 1991, ESO was just as far behind the times as optical astronomers in the United States. The exception at ESO was the European Coordinating Facility, a group of scientists and software experts led by Piero Benvenuti. This group had been created by the European Space Agency (ESA) to support European astronomers in the use of the Hubble Telescope and had been hosted at ESO since its inception. The group had participated with STScI and what became the Canadian Astronomy Data Center in the development of the Hubble data archive. It was therefore thoroughly versed in the state-of-the-art concepts that guided the development of the end-to-end data system for Hubble.

Apart from that, when I arrived at ESO in early 1992, I found a small group at headquarters that had been working on the development of the Munich Image Data Analysis System (MIDAS), a command language for data analysis and image processing. The main disadvantage of this language was that it was being developed almost entirely by the ESO group without outside contributions. The choice by STScI of the alternate language IRAF, the command language developed by the National Optical Astronomy Observatory, assured us at STScI that this situation would not occur. The later adoption of IRAF by the Chandra and Spitzer Space Telescope programs meant that several groups were working on it and thus IRAF could develop more rapidly with many application software programs, while MIDAS lagged somewhat.

A greatly stepped-up effort was required to handle the expected data flow from VLT and the revamped La Silla. It was anticipated that as the four 8-m telescopes came into operation on Paranal, the data rate would grow from 19 gigabytes/night in 1999 to 70 gigabytes/night in 2002, the main contributors being the Wide-Field Camera on the 2.2-m telescope at La Silla and Omegacam for the VLT survey telescope. These rates were comparable to or greater than those of the Hubble data flow, even after the servicing missions that upgraded the space-based instrument.

Similar problems to those discussed in Chapters 13 and 14 with regard to Hubble would occur in the science operations of VLT. These problems could only be solved by a high degree of automation, starting with the handling of proposals for telescope time, and going on to program selection, program scheduling with observational blocks for all instruments (twelve on the VLT and five at La Silla), on-line data calibrations, pipeline data processing, data quality control, archiving, and then data distribution.

I decided to create an Operation and Data Management Division which, under the leadership of Piero Benvenuti (the head of the European Coordinating Facility), was charged with the task of bringing the required developments about. When the full magnitude of the effort became clear, it became apparent that Benvenuti could not serve in these two demanding jobs. Peter Quinn, an Australian astronomer who had worked at STScI, took on the leadership of the division in 1996. Over time, the group grew to more than eighty people, with about twenty-four full-time equivalent positions required to develop VLT instrument-specific software (two people for each of twelve instruments), fifteen for data pipeline processing, six for archiving, eighteen for the user support group, and nine for quality control and operations. This ambitious program was sufficiently developed that by 1998 highly automated procedures supported all major instruments on the NTT and the 3.6-m telescope at La Silla and the three instruments on the first VLT telescope (Unit Telescope 1). I will not discuss this whole area in any further detail except to point out the massive transfer of operational and data management experience from STScI to ESO in the years from 1993 to 1999.

Such a methodology of transfer to ground-based observatories did not occur in the United States until late in the 1990s, when the Gemini program adopted some of the processes of its sister organization STScI. The transfer was facilitated in part by the fact that both Gemini and STScI were under AURA management. The Keck Observatory could not adopt similar procedures because of the cost of retrofitting the software to fit existing hardware, and its scientific productivity may have suffered from it. The ALMA project, which is currently under development, has adopted this methodology in its entirety.

The people who stand out in my mind in the software area are Gianni Raffi, whose group provided superb control software for all VLT telescopes; Anders Wallander, who made invaluable contributions in the commissioning of the new NTT, VLT, and VLT interferometer; Franz Koch, who developed the finite element model of the VLT; and Joseph Schwarz, for his suggestion to test the VLT concepts on NTT.

I believe that Michael Rosa made the most original intellectual contribution to the field of calibration with a potential for much wider applications. Michael called what he did "predictive calibration and forward analysis of spectroscopic data."[5] He introduced the concept that during calibration, instruments need not be considered as just black boxes that give a certain response for a certain input, as we had done at STScI. Much could be gained if one used computer models of astronomical instruments based on knowledge of how they are built and operated and of their pertinent parameters, such as reflection efficiency and diffraction constants. He likes to call this an approach based on first principles. This concept, which seems at first of purely theoretical interest, proved its practical value in the recalibration of the FOS on Hubble and the prediction of the two-dimensional spectral formats of echelle spectrographs, such as the Space Telescope Imaging Spectrograph on Hubble and the Ultraviolet High-Resolution Spectrograph (UVES) on VLT. ISAAC also benefited from this approach.

It is not too bold a step to think that such models can be constructed before any metal is cut. Furthermore, one can integrate the model into the data analysis software that will be used to analyze the data, and optimize instrument design and software to obtain the scientific performance one needs. In a modest way, this is what Burrows, Burg, and I had done to optimize the design of the proposed Wide Field X-Ray Telescope; perhaps it was my own experience that made me so sympathetic to Michael's efforts. Modeling is important for both telescope construction and instrument design, and I believe that it will be used more and more in the future.

War and Peace: Relations between Chile and ESO

The biggest problem I encountered at ESO was the poor relations ESO had established with the Chilean government after the departure of the Pinochet regime in 1990. ESO had received many concessions under the Pinochet regime, just as it had before 1973. Its international status had been recognized since 1963, and ESO had received the same immunities and customs and tax exemptions that were granted to the United Nations Office for Economic Development in Santiago. The Chilean government had donated a wonderful tract of land in Vitacura (near Santiago), where the ESO offices could be built. ESO had purchased a large tract of land in the mountainous and desert region around La Silla, which had no commercial value but would to some extent protect the observatory from light pollution.

The problem for ESO occurred when it decided to build VLT on Cerro Paranal. Years of studies had determined that this region had the best seeing in Chile and possibly in the world. However, the ownership of the land was not at all clear. This situation was common in Chile, where Crown grants from the time of the conquest, international border shifts, unclear titles, and poor recordkeeping created a situation of total confusion. The situation was not favorable for economic development, and in the early 1980s the Pinochet government passed a law whereby claimants to land titles had a period of 11 years in which to present their claims. If no claimant came forward, the ownership of the land would be transferred to the Ministry of National Assets. After the 11 years had expired, ESO received a donation from the ministry of ten thousand hectares of land around the Paranal summit. Under the same law, a narrow strip of land near the La Silla property was donated to be used by ESO chartered planes ferrying the staff back and forth from Santiago. The legal position of ESO with respect to ownership was unassailable, but the decision to build a new observatory in a different location became the casus belli for a rather virulent political debate. This debate took on many different aspects.

There was resentment on the part of the Chilean astronomical community because ESO, although granting observing time to Chilean astronomers, had never granted them privileged, guaranteed time, as the U.S. observatories of Cerro Tololo and Las Campanas had done for years. Our colleagues in Chile felt that, in view of the many donations and concessions made by their government, this time was something owed to them. Questioning the applicability of the 1963 ESO-Chile agreement for a new location gave them the opportunity for negotiations that would change this situation.

There were also long-festering issues in labor relations. As an international organization, ESO recognized only its own internal regulations, just as CERN and ESA do. This policy is essential to avoid having to adopt the labor laws of the different host countries. In addition, international organizations are immune from prosecution in the host country, so that all disputes are referred to an International Labor Court. This practice makes a great deal of sense in Europe, but it presented particular problems in Chile. The trade unions, which had been greatly weakened during the Pinochet regime, were becoming strongly and vociferously engaged in the political arena. With respect to ESO, they took the position that as Chilean citizens, the Chilean employees had a constitutional right to work under Chilean labor laws. The U.S. observatories had adopted Chilean labor legislation for their Chilean employees since the beginning of their operations.

Finally, after work had started on Paranal in 1992, a claimant appeared for the land. A firm of well-connected and powerful lawyers represented the interest of the Latorre family, which was descended from Admiral Juan José Latorre, a naval hero of the nineteenth-century Chilean war against Peru and Bolivia that gave Chile control over the rich nitrate zone of the Atacama Desert. Chilean and even U.S. astronomers were found who testified on the importance of good seeing for astronomy, underscoring the value of the Paranal site. The claim was made that the value of the land should be set at 10 million deutsche marks (at the time equivalent to about $6 million). It should be noted that the normal value of land in the desert is one-hundredth that number, as demonstrated by the purchase price of the lands for La Silla, Cerro Tololo, and Las Campanas.

Certainly ESO had made few friends in Chile. Accusations by the staff of violations of human rights by ESO, which they described in letters to the newspapers, and even of the use of corporal punishment, were given credence by some of the Chilean press. Such accusations were absurd, but the aloof attitude of ESO toward the host country (including its astronomers), and the rather privileged treatment received by ESO under Pinochet had created a vast reservoir of resentment. Politicians waded in, since this situation afforded ample opportunity for dramatic statements useful for re-election campaigns. A few thoughtful and influential politicians, particularly Senator Arturo Alessandri, tried to ease tensions, but with only limited success.

At this point I was badly advised by our legal council in Chile, who insisted we should fight these battles in the Chilean courts. What we should have done was to stand on our immunities and ask the Chilean government to intervene. Instead we ended up in the supreme court with three reversals of the decision on our case by one-vote margins within two weeks. Although this vacillation could have been expected from a weak judiciary in the wake of the transition to democratic rule, and the decision could be reversed by parliament, it was not in the interest of the Chilean government to further weaken the court.

In 1994 we had started negotiations with the Chilean government on all these issues. I led a delegation of the ESO Council, which met with representatives of the Chilean government headed by Ambassador Mario Artaza (later the Chilean ambassador in London and Washington). He took a very hard position at our first meeting, and I was equally adamant in defending ESO's rights. With the help of many people of good will, we made some progress in understanding each other's positions. But it was hopeless to try to make real progress, given the political situation.

On March 30, 1995, the situation came to a climax with the violation of our immunity on Paranal by the illegal forced entry of a Chilean magistrate accompanied by the police. The ambassadors of the ESO member states, who had been kept informed of what was happening in a series of meetings, made formal protests, and the German ambassador intimated that this action could affect relations between Germany and Chile. The Swiss ambassador was equally forthright in his statements. I gave an interview to one of the major newspapers in which I compared the action on Paranal to the seizure of the American embassy in Tehran, a behavior more in character with third-world countries. Although the result of unbearable frustration and a verbalization of what many other observers from the ESO countries felt, my words were too harsh, and for a moment I considered whether I should resign for the good of the organization.

At this point, the good offices of Claudio Teitelboim, a well-known physicist who was the science advisor to the president of Chile, proved providential. I was introduced to him by a common friend and colleague, Remo Ruffini of the University of Rome. Claudio arranged a private luncheon meeting between José Miguel Insulza, the Chilean foreign minister, and me in the offices of his institute. There followed a meeting at the house of Eduardo Frei Ruiz-Tagle, the president of Chile, also attended by Minister Insulza, Daniel Hofstadt, and myself. It helped the discussion that President Frei had gone to the Polytechnic Institute in Milan and Minister Insulza to the law school of the University of Rome during the Pinochet era, so we could occasionally switch to Italian, in a rather informal atmosphere. I was reassured by the president that the Chilean government intended to fulfill all of its obligations under the treaty, and that the Paranal issue would be taken over by the Chilean government as an internal matter. I was able to assure him of ESO's intent to establish a much closer relationship with Chilean astronomers and to explore ways to resolve all other outstanding issues. Minister Insulza took on the responsibility to come to a fair negotiated agreement with ESO and to present it to parliament for approval. This meeting was the turning point: there followed my apology to the Chilean public if my remarks had offended them, and regrets expressed by the Chilean government for the violation of our immunities on Paranal.

These positive steps culminated in the signing of an "Interpretative, Supplementary and Amending Agreement" to the 1963 convention between the government of Chile and ESO, by Chilean Plenipotentiary Ambassador Robert Cifuentes and myself on April 18, 1996. Our agreement was approved by the Chilean senate on September 5, 1996, and soon thereafter by the ESO Council. Minister Insulza and I exchanged instruments of ratification on December 2, 1996. With this event, peace broke out! The agreement dealt with all the major issues. ESO staff regulations for local staff would be retained but modified to incorporate the principles of Chilean legislation regarding collective bargaining and freedom of association. ESO's immunities were confirmed and extended to all properties and possessions of the organization located in Chile. The issue of Paranal was settled between the Chilean government and the claimant, and ESO's ownership was reaffirmed. ESO would increase its contributions to Chilean astronomy and local communities. Under the agreement, Chilean astronomers were granted privileged access to observing time, up to 10 percent, on all present and future ESO telescopes. They would be represented on all technical and scientific committees of ESO. Although it was not easy to convince the ESO Council to accept some of these concessions, I felt they were essentially fair.

As a public expression of good will, we hosted a foundation ceremony with the deposition of a time capsule on Paranal, attended by His Excellency the president of Chile; the foreign minister; the ambassadors of all member states; the president-elect of the ESO Council, Henrik Grage; the retiring president of the council, Peter Creola; religious, military, and political authorities of the Second Region and of the nearby towns of Antofagasta, Paposo, and Taltal; members of the ESO Council; and, somewhat unexpectedly, Their Majesties King Carl XVI Gustaf and Queen Silvia of Sweden. All the speakers, who included President Frei, Minister Insulza, President Grage, and me, emphasized the significance of the event for astronomy, for Europe, and for Chile (Figure 16.4).

I was in agony until the arrival of the Chilean president was announced. He was to arrive by military plane, and the ESO Paranal staff had prepared a landing strip in the desert especially for him. The Chilean Air Force had rehearsed landings and takeoffs with some of our local staff as test passengers (to their great amusement), so that everything would be in order. But until the last moment, I feared a cancellation. I was therefore particularly touched when President Frei referred to me in his speech as his "estimado amigo."

FIG. 16.4. *Foundation ceremony at Paranal, 1996.* Left to right: *Foreign Minister José Miguel Insulza, President Eduardo Frei, Mirella Giacconi, and the author.*

At the conclusion of the ceremony, President Frei, Minister Insulza, and most of the Chilean dignitaries left, but my wife, Mirella, and I hosted, on behalf of ESO, a sit-down luncheon for 150 guests, including Their Majesties the king and queen of Sweden. It was somewhat eerie to sit comfortably under a canopy and eat a catered lunch in the middle of the Atacama Desert, but all went off very well.

President Frei was gracious enough to accept our invitation to the inauguration of the Paranal Observatory on March 5, 1999. He was accompanied by his wife, Martita, and Minister José Pablo Arellano Marin. Also present were the Italian minister for research, Luciano Guerzoni; the ambassadors of the member states; the members of the ESO Council; and the religious, military, and civilian authorities. The participants heard speeches by President Frei, ESO Council President Grage, and me. Carlo Rubbia, Nobel Prize winner in physics for 1984, gave one of his most technical lectures, which, though not understood, was warmly applauded by the assembled and rather stunned dignitaries.

As part of the celebrations, I awarded on behalf of ESO a prize (a 6-inch telescope) to 17-year-old Jorssy Almanez Castilla from Chuquicamata near the town of Calama, who had won the competition among schoolchildren for the naming of the four VLT unit telescopes. The names in Mapuce (a native Chilean language) are: Antu (the sun), Kueyen (the moon), Melipal (the Southern Cross), and Yepun (Sirius). Jorssy wrote: "Todas esta palabras estan relationadas con la luz y para mi la luz significa paz y vida. Seguida de la idea de que todo tiene relation con el universo" [All these words are related to light, and for me light means peace and life. This follows the idea that everything is related to the universe].

Since these events, relations with Chile have rapidly improved, which has been important not only for La Silla and Paranal, but also in establishing the new world observatory for millimeter-wave astronomy (ALMA on the Llano de Chajnantor). On several other occasions I met with Minister Insulza and later with the new president of Chile, Ricardo Lagos, in an atmosphere of great cordiality.

In the next chapter I describe VLT, a glorious achievement for European astronomy. Much of what I have described here is reported in the collected presentations by myself and thirty members of the staff at the ESO-wide review on February 1–4, 1999,[6] and in a series of ESO *Messengers* from December 1992 to December 1999. Significantly, while all the events described in this chapter were taking place, the schedule for VLT was maintained to within six months over five years, whereas the previous experience had been delays of a year in the predicted first light date for each year of work (see Figure 18.1).

Building the Very Large Telescope

The Very Large Telescope was designed to be an array of four identical 8-m telescopes. They could work independently, be combined, or act as an interferometric array. Interferometry could be obtained by combining the beams of two or more unit telescopes and/or using unit telescopes in combination with smaller auxiliary 1.8-m telescopes in an array with a maximum dimension of 120 m.

The primary mirror of each unit was to be a thin meniscus of Zerodur, a ceramic glass of low expansion coefficient produced by the German company Schott, 8 m in diameter but only 17.5 cm thick. The correct shape of the mirror was to be achieved by active optics. A total of 150 axial and 70 radial support actuators would be constantly controlled to maintain the mirror in the proper shape. This design concept was a radical departure from that used for the NTT primary mirror, which was thick enough to retain its shape to good accuracy independent of the controls. The VLT mirror is fifty times more flexible than the NTT mirror and relies entirely on its autoguider and the appropriate software, in closed loop, to function.

The VLT is supported by an alt-azimuth mount. The relatively light mirror (23,000 kg [50,700 lb]) is supported by a structure weighing only 21,000 kg (46,300 lb), making it possible to pivot the telescope about an altitude axis

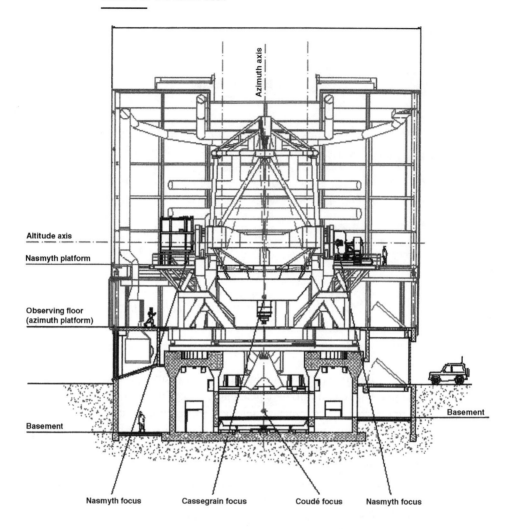

Azimuth axis

Altitude axis

Nasmyth platform

Observing floor
(azimuth platform)

Basement

Basement

Nasmyth focus   Cassegrain focus   Coudé focus   Nasmyth focus

FIG. 17.1. *A section view of one of the Very Large Telescope unit telescopes and enclosure. The placement of instruments at the Nasmyth and Cassegrain foci is shown.*

coinciding with the Nasmyth focus. This in turn permits the mounting of large instruments weighing several tons on the Nasmyth platforms (Figure 17.1). The VLT optical layout is of the Ritchey-Chrétien type, and the instruments can utilize Cassegrain, Nasmyth, or Coudé foci (Figure 17.2).

From the beginning VLT was designed to be used for interferometry, and the positioning of the telescopes on the mountain was the result of a careful study to maximize the imaging quality of the interferometric array. An underground tunnel houses the optical delay systems, and an underground laboratory contains the instruments used to detect the interferometric fringes.

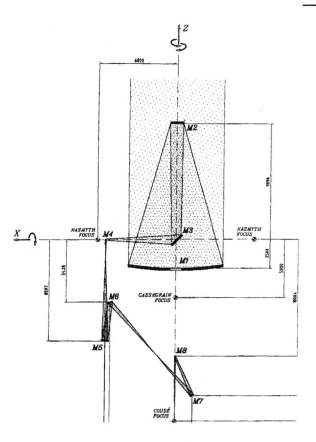

FIG. 17.2. *The optical path to the Coudé focus, which is used to combine the light from the four unit telescopes for interferometry.*

The auxiliary 1.8-m telescopes that extend the array are placed at pre-selected stations, and their beams are combined in the underground laboratory.

All power, electrical, thermal, and access connections to the four 8-m unit telescopes and the control building are underground as a precaution against the high winds on Paranal. Complete descriptions of VLT and all its instruments can be found on the ESO web site and in ESO *Messenger* publications. Here I only discuss a few topics in which I was personally involved either technically or managerially.

Paranal: Seeing . . . and Believing

VLT is an observatory that was designed to be located at Paranal in the Atacama Desert, a place well suited for astronomy but not at all for humans. Its location (latitude 24°40′ S, longitude 70°25′ W) is one of the driest places on Earth, with a yearly precipitation of less than 10 mm. Cerro Paranal is a mountain 2,635 m above sea level, 81 miles south of the town of Antofagasta and 7.5 miles from the sea. The only human activity nearby consists of strip mining for nitrates. The only animals I ever saw there were foxes attracted from miles away by the scraps of food from our camp. There is no water at all and no electrical power lines. The site can be reached from the Pan-American Highway over 40 miles of hard-packed road in the desert.

The first time I visited Paranal in 1992, Peter de Jonge drove me in an unforgettable cross-desert drive with plumes of dust raised by the car following our trail. I think I fell in love with the Atacama Desert right there and then. The colors, the lights and shadows, and the fields of walking stones produced an eerie but beautiful, alien landscape. My daughter Anna, upon returning to Santiago after a visit to Paranal, commented in her usual pithy way: "It is good to be back on Earth."

The Paranal peak had already been blasted away by 1992, removing the top hundred feet of the mountain to create a plateau large enough to contain the four unit telescopes and the control building (Plate 15). However, no final decision had yet been made on whether Paranal or La Silla should be the VLT site.

The great advantage of La Silla was that it offered an established base of logistic support. There was a source of water at the bottom of the La Silla mountain, and a pumping station had been built to bring water to the observatory; a diesel power plant furnished the electricity required. Housing for the construction crew would be easily available. If we wanted to proceed with Paranal, I had to convince myself and the ESO Council that the opening of an entirely new observatory in the middle of the desert was worth the cost.

At first I had the feeling that one of the motivating factors for choosing Paranal was the desire to leave La Silla, with its aging telescopes and its difficulties with the local staff. Gradually, however, I became convinced that Paranal was the right choice: the data on seeing that had been collected since 1987 appeared to show consistently better conditions (Figure 17.3); the site was 400 miles north of La Silla, and thus it would be less affected by the secular shrinking of the desert if the trend persisted; there was less light pollution to worry about at present and in the foreseeable future; and construction on La Silla would completely disrupt the scientific operations of the existing telescopes. Thus I made my first decision on VLT. I have never regretted it.

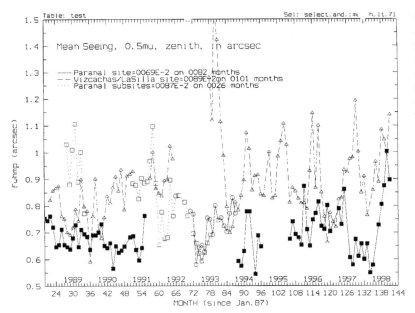

FIG. 17.3. *Comparison of seeing measurements at Paranal (filled squares) and at La Silla (open squares) from 1987 to 1998.*

M1 and Its Support Structure: A Collaboration of Industry and Research

The contract to Schott for the development of the primary mirror (M1) blanks was initiated in 1988. The casting and slow annealing of the 8-m-diameter blank was a formidable technical challenge. In 1987 Schott had developed a spin casting technique that was used to minimize the machining of the mirrors by casting them approximately to the right shape. The process has been described by Philippe Dierickx, who supervised the contract for ESO (Plate 16).[1]

A short aside is in order here. For the VLT program, ESO had chosen to rely heavily on industrial subcontractors while retaining the prime responsibility. I completely agreed with this approach, which is not only much more effective but also better and cheaper. The contributions of European industry to the VLT were essential to the success of the program. I consider many of the efforts made by astronomers to produce in house those things better done by industry as a waste of precious research funds. That research institutions still persist in this approach is due, in my opinion, to poor accounting, which blinds both them and their sponsors to the real costs, and to poor systems engineering, which puts emphasis on the wrong things. For instance, in the case of mirrors, it is my experience that their cost never amounts to more than 7–10 percent of the total cost of an observatory, whether on the ground or in space; thus, any savings on mirrors alone (if it can be achieved at all) produces hardly any change in the overall costs.

Schott was ready to deliver the first blank to the French firm REOSC on June 25, 1993. The blank had to be transported from the Schott plant in Mainz to the specially constructed REOSC polishing facility in Évry, near Paris (Plate 17). The blank traveled by truck to the Mainz Harbor, then on the barge *Eldor* down the Rhine to Rotterdam, south to Calais, and then up the Seine to Paris and then Évry.

REOSC did a magnificent job of polishing the mirrors to 1/60 of a wave, and in the active mode we achieved an average accuracy for the wave front of 40 nm. The results were so satisfying that the Gemini program chose to have their mirrors polished in the same plant, and ESO was gracious enough to allow them to schedule one of their mirrors in among fabrication of the VLT units.

An interesting problem arose with regard to the M1 support structure, which had to be relatively light but still hold rigidly a mirror of weight equal to the support itself. There was some question on how to proceed with its design and construction. For its 10-m mirror Keck had used a whiffle-tree structure made of tubes soldered at their joints. This is a relatively difficult but proven technique, which requires expert welding of steel tubing at complex intersections. As it turned out, we learned that the expert technician who had done the Keck work would soon be retiring, and we did not know how well his successors would fare in this process, which is more art than technique. There was also an entirely new approach, proposed by a consortium of French industries, GIAT (Branche GITECH) and SFIM Industries, which seemed promising but had never been tried before; as things stood at that point it was difficult to make a decision.

I suggested to Massimo Tarenghi that we should award two simultaneous contracts, one for each of the concepts, in competition. Under these contracts, each contractor would provide a complete design, proof of their fabrication techniques testable in the laboratory, and a full-scale mock-up of a section of the structure to be erected behind the ESO building in Garching under an Oktoberfest-style tent.

ESO had never used this approach of parallel competitive contracts, but it is common at NASA and the U.S. Defense Department, and in the aerospace industry in particular. This apparent extravagance was made possible because the cost of the contracts would be amortized over the four units. An additional advantage of the full-scale prototype was that it focused attention on operations early on.

We awarded the contracts in April 1993, and the prototypes were erected some months thereafter. We were thus able to compare the two approaches for ease of accessibility to and maintenance of the 150 axial actuators, whose mean life was then unknown, and in general for convenience in cabling and operations. We also received samples of the beams that would be produced by GIAT using a fully automated laser cutting and welding process. This new approach not only used about 25 percent less steel than the former technique (which made the support structure cheaper) but also gave the structure greater rigidity and was much better for operations.

I met with the team of young engineers from GIAT who had led the proposal and design efforts. I was glad we could give them an opportunity to be rewarded for their innovative thinking. In February 1995, ESO awarded the contract to GIAT, which did a wonderful job (Figure 17.4).

FIG. 17.4. *Two views of the 8-m primary mirror light-construction support cell at GIAT. The cell is created with hollow square tubing that is cut and welded by laser.*

The Coating Plant for the 8-m Mirrors

I had nothing at all to do with the mirror coating plant that was being designed for ESO by Erich Ettlinger of the German firm Linde under the direction of Michael Schneermann of ESO, except to worry about whether it would meet the stringent requirements we had imposed on this facility.

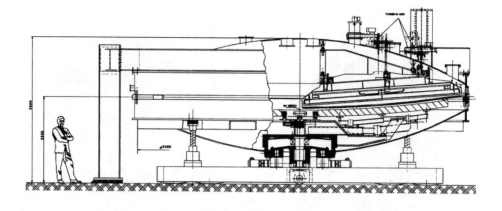

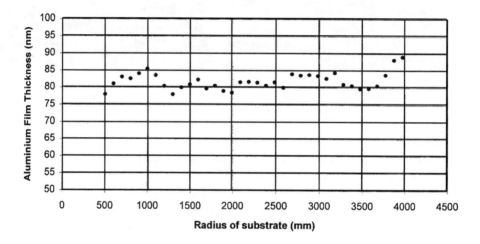

FIG. 17.5. (Top) *The mirror-coating vacuum vapor deposition tank.* (Bottom) *The precision of coating obtained with the evaporation facility.*

The contract was signed in August 1995, for delivery at the end of October 1997.

New technology permitted the use of a vacuum chamber 9 m in diameter but only 4 m high, which could be transported relatively easily to Paranal. The 8-m mirror was coated with a single rotation of the sputtering arm that houses the DC planar magnetron source. This source uses magnetic focusing of electrons onto the sputtering target, which is made of aluminum (Figure 17.5). The measured reflectivity of the mirror was 92 percent at 400 nm and 97.5 percent at 2,500 nm. The film thickness was 79 ± 2.9 nm. The region where the coating overlapped was less than 20 mm, corresponding to less than 1/1,000 of the mirror area. The entire operation was carried out with only five and a half hours of downtime![2]

Stubborn Problems: M2

The secondary mirror unit was designed so that a single mirror could be used for all three foci, Nasmyth, Cassegrain, and Coudé. The switch between different foci was achieved by motion of a tertiary mirror. M2 is a 1.2-m-diameter mirror with f/15 aperture at the Nasmyth focus. As part of the overall active optics design, the secondary can be positioned in three coordinates to correct for focusing and centering. In addition, it can be controlled in tip-tilt to provide field stabilization. The design was quite demanding, because all motions had to be counterbalanced, and temperature control was required. The engineering for the M2 unit support resembled closely that used in spacecraft design. It is therefore quite natural that after initial studies, a contract, which included the integration of a mirror, was awarded in November 1994 to Dornier, a German aerospace corporation.

The mirror itself was also very demanding in its design. It had to be quite stiff so that it could be moved at relatively high frequencies and made of a low-density material with a low thermal expansion coefficient. The choice for the mirror material, made in about 1991 by Daniel Enard, the ESO chief engineer, had been silicon carbide, which would have had ideal properties in terms of rigidity and weight. Unfortunately, the material had not been fully tested except for military applications. With the end of the Cold War, military funding had dried up and, in view of the lack of civilian applications, those industries in the West that had initiated its development abandoned the research. The only mirror of 1-m diameter that been constructed under contract with the U.S. Air Force had explosively shattered, presumably because of internal stresses.

Beryllium, which was an excellent alternative, had not been seriously considered because of its cost, potential for creep, and possible health effects if inhaled during fabrication or not properly handled. However, both Jim Crocker and I were familiar with beryllium, the material that I had used for the Apollo Telescope Mount Skylab x-ray mirrors in 1973, and that Jim had used in other aerospace applications. We were convinced that the concerns were unfounded.

In September 1995 we were already late in placing the M2 contract, and Sandro d'Odorico, then head of the Instrumentation Division, shared with me his serious concern that it could become a show stopper. Sandro and I were able to convince Massimo of the need to stop dithering and proceed immediately with placing a contract for a beryllium mirror. Jim Crocker found manufacturers in the United States willing to take on the job of building the largest beryllium mirror ever made on a tight schedule. The U.S. companies involved were Brush Wellman, which produced the blank by the sintering technique called hot isostatic pressing; Loral American Beryllium, which handled the machining of the blank; and Tinsley, which fine-turned the surface to the desired aspheric shape. Loral was also responsible for the nickel coating of the mirror prior to the final grinding and polishing at REOSC. Dornier engineers were responsible for the integration of the mirror into the electromechanical unit they had built. After thorough testing, Dornier delivered the unit in December 1997.

The performance of the M2 units met or exceeded all specified requirements, resulting in less than 0.01 arcsec residual image blur, a centering accuracy of 0.45 arcsec, spurious tilt of less than 0.02 arcsec on average, and field stabilization of 0.05 arcsec on average at 5 Hz. The mirror mass is only 52.5 kg, the inertia is 4.17 kg · m^2, and the characteristic frequency is greater than 365 Hz. The microroughness of the surface was 1.5 nm!

The preliminary delivery of the electromagnetic unit and M2 in December 1997 (final delivery took place on December 12, 1998) was essential to achieve first light on May 16, 1998.

The Main Structure of VLT: A Rough Start and Smooth Moves

The main structure of VLT is shown in its full beauty in Figure 17.6. The contract had been awarded to a consortium of Italian companies in the early 1990s, including Ansaldo Energia in Genoa, European Industrial Engineering in Venice, and Soimi in Milan. Ansaldo was a state-owned electromechanical and energy engineering company in the IRI (Institute for Industrial Reconstruction) Finmeccanica Group. IRI was intended as a social effort to avoid layoffs of workers in bankrupt industries. Massimo Tarenghi and I went to visit Ansaldo in 1993, and I realized that although they had able engineers, technicians, and machinists, their management was very poor. No substantial progress had been made, and there were no plans for recovery of lost time. The Ansaldo Energia program manager in Genoa had no concept of time schedules or of the management tools normally used in industrial projects; he did not even know how many drawings he needed to complete

FIG. 17.6. *European Southern Observatory Council members in front of the support structure of one of the unit telescopes at the Ansaldo plant in Milan.*

before manufacturing of all the various components could be started. I was invited to lunch in a beautiful villa overlooking the sea, and that was supposed to be enough to dispel my worries.

I came away truly concerned about what this state of affairs meant for the VLT project. I reasoned that, because Ansaldo was state owned, the Italian government would care enough about the repercussions of a contract cancellation to remedy the situation. Italian Minister of Universities and Research Umberto Colombo was kind enough to see me promptly. He listened with great attention to what I had to say and assured me of Italy's interest in seeing ESO and VLT prosper, and particularly in the success of the portion of VLT entrusted to Italian industry. Somewhat to my amazement and delight, within a week the leadership of the effort was transferred to Ansaldo in Milan, where it was entrusted to their most competent management team. This team had been responsible for the NTT, which it brought to successful conclusion in December 1987.

The main VLT structure weighed 430 metric tons (948,000 lb). It supported, in addition to the M1 mirror in its cradle, the M2 and M3 electromechanical assemblies and provided platforms for the instruments at the Nasmyth focus. It rotated in azimuth on a circular track isolated from the structure by hydrostatic oil pressure. The oil thickness stability was 0.1 μm, and the position of the mount was measured with optical encoders to 0.01 arcsec. The driving torque was provided by electromagnetic drivers on the perimeter of the structure. The same system was used for the altitude mount. This technological innovation was quite important, because it provided essentially frictionless motions, much smoother than those that could be achieved by traditional cog-and-gear mechanisms. We believed this smoothness would prove important for the Very Large Telescope Interferometer (VLTI) applications. The structure also had to survive the earthquakes prevalent in the volcanically active region of the site; in its early life, VLT came through undamaged by an earthquake of intensity 8.4 on the Richter scale.

Ansaldo's completion of this impressive structure was a good thing for Italy as well as for ESO. The technical performance was superb, even though I am convinced that Ansaldo lost money on the VLT project. The company, however, acquired an outstanding reputation in the astronomical community, which led to additional contracts, including the structure for Roger Angel's Large Binocular Telescope in Arizona.

The Telescope Enclosure and Systems Engineering: Modeling Winds and Earthquakes

As is the case for NTT, the telescope enclosure for VLT rotates around the azimuth axis of the telescope. It is built as a cylindrical structure 29 m in diameter. It consists of a fixed part and a rotating part, which rises to 36 m, the height of a twelve-story building (a small enclosure for an 8-m telescope). The rotating part of the dome has a 10-m-wide slit that can be closed by two doors. Ventilation is maintained by regulating the air flow by means of a wind screen and louvers installed in the walls around the circumference. Given the high wind velocities experienced at Paranal, wind tunnel testing and computational fluid dynamics simulations were performed, which established the optimal flow-control strategy to ensure thermal stability and low turbulence under a variety of wind speeds and directions.

Although some systems engineering existed at ESO as part of the engineering group, I felt that more attention should be given to this discipline and new skills brought to bear on the systems problems. I set up the Systems Engineering Group as a separate entity in 1995 under the leadership of Torben Andersen, a gifted Swedish engineer already working in the VLT group. Torben's group, which rapidly grew to five members in 1996, was composed of specialists in modeling, systems analysis, and synthesis. Thermal and wind models computed by the newly formed group contributed significantly to the management system for optimum seeing.

The group started their work with the model of telescope optics computed by Ball Brothers Research Corporation. Jim Crocker knew of these models (which had been commissioned by the U.S. Air Force) through his work on COSTAR for Hubble. I thought that such models would be essential to the operations of VLT and VLTI and crucial in making decisions on critical and noncritical performance specifications.

The Ball models soon proved to be insufficient for our requirements, but they were quickly followed by a much more sophisticated finite element ESO model for the structure with 100,000 degrees of freedom, which was reduced to 40,000 before solving the equations. There followed a simulation model that included optics, structural dynamics, mirror support systems, atmospheric effects, detectors, and active optics. The model was used to study a number of effects: performance of the active optics system under wind loading, the effect of different time constants in the control loop, the response to temperature changes, and seeing management optimization. Figure 17.7 shows the finite element model of the unit telescope structure and Plate 18 the results of modeling of the autoguider and of the active system performance. The optimum control strategy for louvers, doors, and wind screens for different wind conditions is shown in Plate 19.

Some important results were obtained by Torben and his group regarding VLTI. The four M1 mirrors had been polished with slightly different curvatures, and we were concerned about the potential problems this variance might cause for VLTI. To rectify this difference by actually repolishing the mirrors would have been very expensive, and in any case, we needed to know how critical it would be. The systems group was able to show that there would be only negligible impact in accepting the mirrors as they were.

Torben's systems group also took on the onerous responsibility of maintaining the configuration control system for all of VLT and VLTI with computerized tools developed at Garching and Paranal. I had anticipated that the Systems Engineering Group would play a role in the early planning of new facilities, such as the Overwhelmingly Large Telescope, but the departure in 1996 of Torben and some of his best staff, and the heavy commitment of his

FIG. 17.7. *Finite element model of a Very Large Telescope unit telescope, including optics and support. The model has more than 100,000 nodes, which allow the structure to be simulated with considerable detail.*

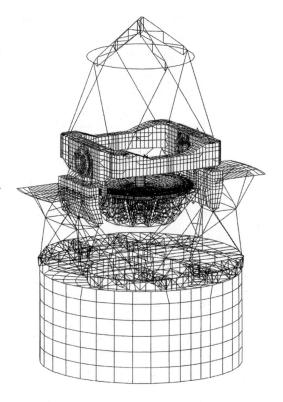

successor Richard Kurz to the ALMA project, made the systems engineering effort weaker than one would have wished.

Instrumentation: Rationality and Rationalizations

Most of the instruments for VLT were built by research groups not at ESO, but with ESO's financial support for the hardware. Although some control could be exercised over the in-house ESO instruments (the test camera, ISAAC, and UVES), it was much more difficult to do so with those built at other institutions. It did not help that Guy Monnet (head of the Instrumentation Division) and I did not always agree in our philosophies on instrument building.

In a forum held in Garching in April 2003 and chaired by Monnet and Roland Bacon of the Science and Technology Committee, a number of the principal investigators of the different instruments met to discuss how to improve instrument development and procurement. Monnet and Bacon stated: "As a result of this whole effort and for the first time in a century, the observational capabilities of European ground-based astronomy have overtaken the other side of the Atlantic."[3] This statement was wonderful and probably true, but I am concerned that the misunderstandings that surrounded the instrument construction effort in the past still linger on, and that although the authors applauded the result, they still do not appreciate the disciplined effort that was required to make VLT/VLTI so successful. In their article, Monnet and Bacon refer to "VLT standards" as something of a dirty word. I suppose that by "VLT standards" they mean the requirements that ESO tried to impose (with some success) on quality control, configuration management, software and hardware compatibility, simplicity of design and maintenance, timeliness of delivery, pipeline data processing, reduction of instrument modes, and all other such bugaboos that were so odious to some scientists both within and outside of ESO.

I remember attending one of the conceptual design reviews for a particular instrument. The principal investigator was proud of the dozen ways in which the instrument could be configured and also of the fact that it had to be manually fine tuned to achieve the best performance. He had no clear idea about cost or schedule for the project. This case was rather extreme, and later ESO canceled the instrument because of lack of progress. I was concerned not so much by the principal investigator's point of view as by its acceptance by the very ESO people who were supposed to understand the overall vision of what we were trying to do.

In a sense the situation with VLT was similar to that I had found with Hubble in 1982. The Hubble principal investigators contributed their unique ideas and skills to the building of instruments, but their purpose was different from that of STScI. STScI and ESO saw their missions as providing the entire community with easy access to the first-class observational capabilities being built. They also considered it their responsibility to guarantee at some level the quality of the scientific data produced.

Furthermore, the importance of a data archive suitable for research was clearly understood at STScI by 1995, where more than one-third of the publications came from archived data. At ESO, where no usable archive of the data obtained from the ground-based facilities had ever been established, the idea of an archive was a foreign concept whose utility had not been grasped. Compounding the problem, the instrument builders found it rather hard to accept that their data should be utilized by the whole community rather than by themselves alone. Although VLT itself was proceeding on time and within cost, and was meeting or exceeding all specifications, the instrument builders felt that the "fast track" approach (as they defined doing the job within the established schedule, performance, and cost parameters) did not apply to them.

I believe that this feeling by some groups that unreasonable constraints are being imposed on them is only the result of poor management at many of the research institutions, as well as at ESO. I consider it dangerous that rather than improving their ability to produce on schedule powerful, simple, and cost-effective instruments, such groups still appear to be trying to rationalize why they fail to do so.

The First Four Instruments to See Light

It was fortunate that the first four VLT instruments to come into use (the test camera, Focal Reducer Low-Dispersion Spectrograph [FORS I], ISAAC, and UVES) were built by dedicated groups that delivered, on time, technologically challenging, superb instruments, which have proved to be highly productive scientifically.

The test camera was never intended as a full-fledged scientific instrument. It was built by the ESO VLT group itself as an engineering tool to be used for the commissioning of the unit telescopes, and it actually produced the first light images in May 1998. Although it had been built with the engineering-grade CCDs available at the time, which had severe cosmetic defects and were not intended for scientific research, it was available when needed and was used during the commissioning and science verification observations in August 1998. With this instrument, Jason Spiromillo and Anders Wallander succeeded, under optimal conditions, in obtaining images of 0.26-arcsec resolution—an amazing accomplishment (Figure 17.8).

The first scientific instrument to see light on VLT was FORS I, built under the leadership of Imo Appenzeller of the Landessternwarte Heidelberg by groups at Heidelberg, the Universitäts-Sternwarte Göttingen, and the Universitäts-Sternwarte München. Imo was justifiably proud of the accomplishment of his group. The instrument was built to ESO specifications regarding electromechanical functions, image motion stemming from flexure, optical performance, calibration units, and instrument-related software. Extensive tests were conducted at the Deutsche Forschungsanstalt für Luft- und Raumfahrt facility near Munich with telescope and star simulators. FORS's detectors were based on a 2,048 × 2,048 CCD array with 28-μm pixels, which were provided by Jim Beletic's group at ESO. The instrument saw first light on September 15, 1998, at the Cassegrain focus of Unit Telescope 1 (the tele-

FIG. 17.8. *Image obtained with the test camera on Antu (Unit Telescope 1) on August 19, 1998, with a resolution of 0.26 arcsec during the science verification program.*

scope named Antu), and it has proven very successful. It and its twin have become workhorses for extremely successful extragalactic research programs at Unit Telescopes 1 and 2 (Plate 20).

The second instrument was ISAAC, the cryogenic infrared spectrometer and array camera, which was developed by an ESO group under the leadership of Alan Moorwood. It saw first light at the Nasmyth focus of Antu on November 16–17, 1998. As soon as it started operations, ISAAC proved to be the most powerful infrared camera in the world. Its data were processed by pipeline for several modes (for imaging: jitter, twilight flat, creation of dark and bad pixel maps, illumination frames and zero-point computations; for spectroscopy: nod on slit, flat field, spectroscopic response function, star trace calibration, and slit position; Figure 17.9).

The third scientific instrument to see first light on VLT (on September 27, 1999) was UVES, the high-resolution spectrograph, built at ESO under the leadership of Sandro d'Odorico with the collaboration of the Trieste Observatory, Italy, for the development of the control software. UVES is a complex, two-channel spectrograph weighing eight tons, which operates at the Nasmyth

FIG. 17.9. *An early Infrared Spectrometer and Array Camera image produced by the fully automated imaging pipeline.*

focus of Unit Telescope 2 (Kueyen). It can achieve a spectral resolution of 100,000, as shown in Plate 21.

The Virtues of Sticking to Schedules: Competition and Costs

These instruments were built within constrained schedule and cost and without sacrifice in performance. My strong push for timeliness and operability of the VLT instruments was not due to some psychological compulsion but rather what I considered very good reasons. First, delays are expensive. This concept is not understood in institutions where the staff salaries are assured and not added to the cost of the project, but it is clear that delays consume resources available to do science. Second, a leisurely schedule for instrumentation construction does not appear to be focused on specific scientific problems to be solved and ignores the competitive nature of research.

For instance, the 10-m Keck I telescope had been in operation since 1992, and Keck II had joined it in 1996. Given the great competence of the astronomers using this facility, it was obvious that the cream of the new observations had been skimmed by them. I believed that VLT could catch up and be competitive by providing a more effective observational system and making better use of the data generated.

The first condition was met by VLT's speed in acquiring targets (90 sec) and its ability to stay pointed indefinitely, thanks to the autoguider. This fact alone makes a difference of a factor of two or more in observing efficiency. In 5 years, VLT would accumulate more observing time on the sky than Keck could in 10. The second competitive edge is that VLT operates like Hubble, and Keck like Palomar. The VLT automated data systems can make available to the scientist reduced and calibrated data ready for analysis, whereas at Keck data reduction and analysis are very much the responsibility of the observer.

A lack of understanding of these basic facts was fairly prevalent in the European astronomical community, where extremely talented instrumentalists are delighted to build elegant and complex systems that produce little science. As a result, I am concerned that the desire to build ever more complicated second-generation instruments for VLT may detract from full utilization of the existing ones in the study of the problems of greatest current astrophysical interest. The drive for excellence in science rather than in instrumentation per se was my constant goal at ESO.

Phasing in VLTI

When in 1994 the decision was made to delay the effort on VLTI, to free up resources for the completion of VLT itself, the ESO Council granted me a great deal of freedom on how to use some of the money left in the program and any VLT contingency we did not need. Massimo Tarenghi and I agreed, however, that we could not defer the construction of the infrastructure that was necessary to implement VLTI. To delay the infrastructure would have meant having a construction crew on site at Paranal for years while we were commissioning the VLT telescopes—an unattractive prospect. Even Peter Gray, an Australian systems engineer who brilliantly led the commissioning effort and was ready to take on any challenges, might have balked at this one. Anything we could do in advance of a full resumption of the VLTI program would greatly shorten its development.

Tarenghi and I therefore decided to proceed with all required civil engineering work, which included the interferometric tunnel that would house the delay lines (168 m long by 8 m wide), the tunnels from the Coudé foci of the unit telescopes to the Coudé laboratory and to the interferometric tunnel and the interferometric laboratories. In addition, we proceeded to build the entire platform for the auxiliary telescope stations, the rails to move them, and the optical ducts to transfer light to the tunnel that housed the delay lines. We were greatly aided in making this decision by the presence on Paranal of Jorg Eschway, an experienced construction engineer and one of the unsung heroes of the VLT/VLTI project. He managed all the civil engineering activities by the subcontractors and our own crews on Paranal, and he succeeded in having the work done so effectively that we actually could prepare the entire necessary infrastructure for VLTI within budget.

We had to provide the mechanical structures supporting the Coudé foci optical trains, which would be used only for VLTI but would be very difficult to install after the main telescope structures were set on their base. This much preparation for the interferometer we could afford to do without too serious an impact on VLT's costs or schedule (Plate 22).

But we had more ambitious aims. A group of European scientists interested in high-resolution astrophysics and interferometry was established by ESO's Scientific and Technical Committee in 1995: the Interferometric Science Advisory Committee, whose task it was to review the development of VLTI, define a key science program, and recommend changes to the VLTI's early concept in view of the financial difficulties. This group convened in May and October 1995 and April 1996. The recommendations of Oscar von der Luhe and his colleagues on the interferomtry committee[4] were approved by the Scientific and Technical Committee, and the ESO Council agreed that we could proceed, provided the work could be done without additional funds.

As early as December 1995, ESO had decided that in view of the expected competition from Keck II, then under construction, and the satisfactory rate of progress on VLT, we could and should reinstate the VLTI program as quickly as feasible. The process was much helped by additional funds of 10 million deutsche marks made available for VLTI under a 1996 agreement between ESO, the Centre National de la Recherche Scientifique (CNRS), and the Max Planck Society; this accord was a follow-up to a general agreement of cooperation on adaptive optics and interferometry signed in 1993.

Soon thereafter ESO and the Netherlands Organization for Applied Scientific Research–Institute of Applied Physics agreed to collaborate on the VLTI program for building the delay lines with a contract to Fokker Space in Leiden. This agreement was followed by a cooperative effort between ESO and the Netherlands Research School for Astronomy (known as NOVA), a consortium of astronomical institutes from the universities of Amsterdam, Groningen,

Leiden, Utrecht, and Nijmegen (where the Dutch graduate schools in astronomy are concentrated). This cooperation eventually reached a new level, in May 2000, with the creation of the NOVA-ESO VLTI Expertise Center in Leiden.

The new plan for VLTI emphasized that the program would proceed in steps, with early focus on near-infrared and thermal infrared wavelengths. Interferometry at infrared wavelengths is much simpler than in the visible, because both auxiliary and unit telescopes are diffraction limited at these wavelengths, with the tip-tilt compensation provided by M2 and without the need for adaptive optics. Simplifications were also introduced in the requirements regarding fields of view, the use of dual feeds, and the design of the Coudé optical trains.

In the implementation of this new program, we proceeded first with long-lead items. A contract was awarded to Advanced Mechanical Optical Systems (AMOS) in Liège for two auxiliary 1.8-m telescopes in June 1998, followed by an additional unit in September 1999 and a fourth one in September 2001. This contract included all necessary support hardware for the thirty stations to which the auxiliary telescopes could be moved by rail. The AMOS auxiliary telescopes produced fringes soon after delivery. In March 1998, we signed a contract with Fokker Space for two of the delay lines for VLTI. The delay lines were of a completely new design that owes a great deal to Fokker's aerospace experience, and they have performed extremely well. Two sidereostats of 40-cm diameter were commissioned from Halfmann Telescoptechnic in Nassau near Augsburg, to be used as test mirrors. Design and manufacturing of these units took just a year.

Meanwhile, the instruments that would be used in testing and commissioning the VLTI were being developed:

The VLT Interferometer Commissioning Instrument (VINCI) is an ESO-built conceptual copy of the near-infrared interferometric instrument by the Grenoble Group installed on the interferometer at Mount Hopkins. It was intended to be used for commissioning.

The Mid-Infrared Interferometric Instrument (MIDI) was built by a consortium led by the Max Planck institute for Astronomy at Heidelberg.

The Astronomical Multiple-Beam Recombiner (AMBER) is a near-infrared instrument that operates at wavelengths between 1.0 and 2.5 μm and requires near-infrared adaptive optics for unit telescope observations. It was built by a consortium of institutions led by the universities of Nice and Grenoble.

Multi-Application Curvature Adaptive Optics (MACAO) is an ESO-built adaptive optics system with a bimorph mirror controlled by sixty actuators and a wavefront sensor that detects curvature. Its deformable mirror replaces one of the mirrors of the Coudé optical train, thus requiring no additional optical elements.

Phase-Referenced Imaging and Micro-arcsecond Astrometry (PRIMA) is a dual-feed system that adds faint-object imaging and an astrometric mode to the VLTI capabilities. Two objects separated in the sky by as much as 1 arcmin can be observed simultaneously; one can be used as a reference star while the other is the science target. This instrument should be able to reach microsecond astrometric accuracies. A development effort was initiated at Dornier in Friedrichshafen and at Onera in Paris, which resulted in a report in July 1999 and plans to proceed in 2000.

In addition, a great deal of care was expended in ensuring that we would provide a solid and steady platform for interferometric measurements. Studies were carried out as early as 1994 to analyze the potential sources of disturbances that could affect VLTI operations.[5] Interferometric devices are extremely sensitive to vibrations. Displacements at the submicron level may blur the fringe pattern and decrease the fringe contrast observable in interferometric studies. The sources of vibrations include natural microseismicity, ground motions caused by earthquakes, wind buffeting on the enclosures, and human activity (such as rotation of the enclosures, operation of heavy hydraulic pumps, and vehicular traffic).

Knowledge of the ground-vibration transfer function is essential for predicting how these various disturbances can affect the VLTI elements. In situ measurements to determine this function were started in September 1993 with the collaboration of the Instituto de Ingegneria y Diseño de Edificaciones Modulares, a center for the investigation of materials and the mechanics of soils of the Faculty of Physical and Mathematical Sciences at the University of Chile.

In 1995 the studies of microseismicity were completed. Bertrand Koehler and his colleagues described the statistical characterization of microseismic activity on Paranal.[6] They utilized the measured values of the transfer function and the 40,000-node finite element model of VLT developed by Franz Koch of the Systems Engineering Group to predict the effects of the disturbances on interferometric measurements. Their preliminary conclusion was that such disturbances would affect VLTI measurements 2.5 percent of the time in the worst case. They were able to pinpoint the cause: most of the noise was transmitted not through the overall VLT structure, but rather through the attachments of the mirrors M5, M6, and M8 in the Coudé optical train, something that could easily be remedied. The results also showed the advisability of adding to the sight-seeing monitor installed on Paranal a seismic measuring device whose measurements would be continuously piped to the control room, allowing for real-time interventions by the operators.

These efforts and the collaboration of so many scientists and institutions in Europe allowed ESO to recover in large measure from the early delays brought on by financial woes.

In a similar fashion, the Keck interferometer became a NASA-funded joint project carried out by the Jet Propulsion Laboratory of Caltech and the W. M. Keck Observatory. JPL and the California Association for Research in Astronomy carried out the development of instrument subsystems and associated infrastructure. First interferometric fringes between the two 40-cm sidereostats were achieved in February 2001, and the first fringes for the two Keck telescopes in March 2001. For VLTI, those two landmark dates were March 2001 for the sidereostats and October 2001 for the first fringes between Antu and Melipal, two of the unit telescopes (Figure 17.10). Considering that Keck II had seen first light in 1996 and Melipal in January 2000, four years later, it is a testament to the close cooperation and dedication of the European research community that the two interferometers achieved their first results only seven months apart.

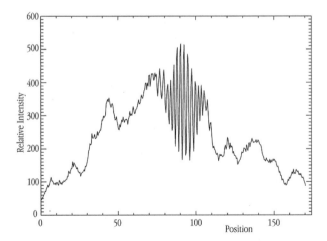

FIG. 17.10. *First interferometric fringes obtained in the K-band with Antu and Melipal on October 30, 2001.*

At Ease: The Residence at Paranal

My last act as director general of ESO in 1999 was to sign the contract for the construction of the ESO residence and offices at the base of the Paranal mountain. Tarenghi and I had delayed the construction of permanent quarters because we had considered it of lower priority than getting VLT completed, and both contractors and ESO staff had lived in trailers for more than five years. But now that VLT/VLTI was reaching completion, we had to think about housing visiting astronomers. We wanted to construct a facility worthy of the technical excellence of the observational facility. The Paranal Observatory is different from most other observatories in the world in that its design resembles more that of particle accelerators than that of traditional observatories. All of Paranal was designed at once as a machine to do astronomy.

We wanted to build an office building and a residence that would be useful over a period of at least 20 years, the time when the scientific research would prosper. We also wanted to build a place that would not destroy the natural beauty surrounding the observatory but would instead meld with the desert. Finally, we wanted to build a place where people could rest from work and from the harshness of the environment. I decided to hold an international competition among interested architects in the member states and Chile, out of which ESO would select the best design.

In the mid-1990s I had been visiting several countries to discuss the possibility of their joining ESO. They included Australia, Austria, Ireland, Portugal, Spain, and, informally, the United Kingdom. Portugal in particular had signed an agreement of scientific cooperation with ESO in the late 1980s and was seriously considering joining ESO in the near future.

It was during one of these visits that I first saw the Astrophysics Institute of Oporto, directed by Teresa Lago. I was impressed by the design of the facility they had constructed: it included research laboratories, a library, classrooms for young students, and a planetarium for the general public. The building was constructed on the grounds of the University of Oporto with contributions from the city and the national government. It was a simple but pleasant and functional modern building that had been built on a tight budget, a fact that immediately endeared it to me. The low-budget building succeeded in giving a feeling of serenity and welcome. Its architect, José Manuel Soares, was from the University School of Architecture, which is very well known, and he consented to act as our consultant. He visited Paranal and helped Tarenghi and me to study and select the most suitable project. His help was essential in persuading me and the ESO Council, in June 1999, to adopt the innovative proposal by Auer + Weber Freie Arkitekten of Munich, with Dominik Schenkirz as the principal designer.

Taking advantage of an existing depression in the ground, the building is entirely underground except for a façade facing the Pacific Ocean (Plate 23). Particular care was taken to create a refuge from the blinding sun, the dryness, and the absence of any living creatures (aside from the staff and the desert foxes) by providing an oasis in the desert where the air is moist and warm, and there is greenery and even a swimming pool.

I must say that, although there were many rational arguments in favor of a swimming pool as a provider of humidity and for visual relief, I was afraid that it would be considered extravagant by the ESO Council and visiting astronomers. A pool also presented technical problems in that Paranal experiences frequent earthquakes. Pools in Chile are always kept full to prevent cracks, but the result is that during an earthquake, a miniature tsunami is formed that overflows the pool. Thus an outflow channel to the desert had to be planned in the design of the building.

It took a concerted effort by the staff to persuade my wife, Mirella, to lend her very influential support to the idea of a pool, and at that point I could not but agree. So the Residencia has a pool, and it has proven to be a blessing (Plate 24). I was particularly proud to sign the contract for the Residencia in July 1999, which completed for me the construction of the perfect observatory:

> ESO's concrete star-catcher in the Atacama Desert is much less grand than the Pantheon. It is a place to stay as pure, as economical and as beautiful in form as a Bedouin tent or a Coptic monastery in a rocky east African desert. It is one of those rare structures, such as the Roman aqueduct at Nîmes or a bridge by Eiffel, that enhances rather than damages remote landscapes. It has, or so I see from Roland Halbe's photographs, a sense of rightness about it: it appears to belong to the Atacama Desert.[7]

The Role of ESO in Major European Astronomy Programs

EIGHTEEN

ESO Management: Optimizing Resources and Improving Communication

I have left management issues to the end of the VLT/VLTI story, even though in my opinion they are very important. When I arrived in 1992, most astronomers at ESO considered management a useless and expensive waste of resources that could better be employed in pursuing research. This attitude was not unusual, because the idea that management tools can be used to make more research possible within a limited budget seems not to have occurred to many scientists anywhere in the world. But the fact is that much of the success at ESO during my tenure was due to better management, particularly in a situation of limited resources.

We were also forced by the ESO Council to improve the management system at the very beginning of my term of office. The council and its finance committee had lost confidence in the ability of ESO's management to carry out VLT successfully. As a result, in 1993 we were asked to undergo an external audit by an ad hoc group composed of outside managers, engineers, and accountants chosen by the council to review our plans for the VLT/VLTI project. In retrospect, it was a most useful exercise, because it set a new baseline in which both the council and I had confidence. To prepare for this audit, Jim Crocker and I used the experience we had acquired over the years during many similar audits by NASA; building on the work done by Massimo Tarenghi and his team, we established a Work Breakdown Structure (WBS) for VLT/VLTI. This system is widely used in space projects and had been installed at STScI with good results.

WBS consists of the following procedures. The program, project, or activity of an organization is divided into work packages whose sum represents the totality of the task. The packages can be quite large (such as building the VLT telescope support structure) or quite small (such as the small-team effort required to develop a particular piece of software). Each item has a designated person who is responsible for the task. He or she is given an agreed-upon level of staff and resources to accomplish the task in a given time. Tasks are often dependent on the completion of other tasks, and such dependencies are noted and reported through a management information system available to all managers. Tasks that are on the critical path for successful completion of the mission can easily be identified, and if they are late, remedial actions can be taken.

There can be tasks at different levels of detail in which the entire program can be viewed, level 1 being the entire program, level 2 its major components, and so forth. Typically each level component is made up of seven smaller components (seven is not magic, but it is the average number of people easily managed by a typical person). If N is the number of levels, then $7^{(N-1)}$ is the total number of packages required to describe the program in its entirety. Only four such levels are required to describe programs costing up to a billion dollars, with the smallest unit consisting typically of some 10 man-years of effort plus materials and overhead. The chain of command follows the work structure and results in a flat pyramidal organization with easy communications up and down the rather short management chain.

The most important aspects of this approach are its consequences for human relations. It requires all managers and team leaders to plan their work and to explain to the next level what needs to be done, with a commitment to provide the necessary resources. The person responsible commits to do the work within a given time. Although this structure seems at first glance to be stifling, it is in fact a liberating one, because at each level, individuals are free to approach the work as they see fit, provided they deliver. It is also a structure that encourages the formation of teams to carry out a specified project, with voluntary team participation of its members, who can be drawn from anywhere in the organization.

WBS is a system suitable for providing information about progress or problems, but to do so requires a software system that keeps track of manpower and expenditures, and compares actual numbers to the planned ones—a system that ESO did not have in place. The ESO system was essentially designed to ensure recovery of costs rather than as a management tool. It literally took years for a complete ESO-wide management information system to be put in place, but even the fledgling structure we set up in 1993 was sufficiently useful for us to successfully pass the audit.

The main finding of the external review was that ESO had somewhat underestimated the total cost of the VLT/VLTI program. With ESO Council approval, Jim Crocker, Massimo Tarenghi, and I decided to reduce the program by some 5 percent by 1994, thus creating a contingency for VLT that was under my control. In a program that ultimately cost 900 million deutsche marks (equivalent to about $530 million), this fraction provided a contingency of some 45 million marks. The decrease in program costs was achieved by postponing some aspects of the implementation of the interferometric use of VLT telescopes. The decision was in part the result of the opinion of VLTI Project Scientist Jacques Beckers, who had contributed to the design of the interferometric array, but who believed that interferometry with VLT could not be achieved in less than 25 years, which meant 2018. Thus expenditures on VLTI could easily be postponed.

As I mentioned earlier, neither Tarenghi nor Crocker nor I believed him, and we agreed with the council that we would proceed to build on Paranal the civil engineering infrastructure required for interferometry. Moreover, any contingency money not used for VLT could be used to advance the VLTI project. This tactic gave the entire staff a great incentive to proceed efficiently with the VLT work and resulted in ESO interferometric fringes between two of the VLT unit telescopes in 2001—some 17 years earlier than Jacques had predicted. We were very much helped in this effort by voluntary contributions in research and resources from ESO member states.

Due in part to the new management structure but mostly to the hard work of Massimo and his team, we were able to almost maintain a schedule and improve on cost through the ups and downs of the Chilean negotiations (discussed at the end of Chapter 16), which had cost us at least a six-month delay, as shown in Figure 18.1 from Tarenghi's presentation in the 1999 ESO-wide review.[1] This timely completion of the work permitted us to build VLT/VLTI at less than the 1994 planned cost.

Many U.S. astronomers have chosen to believe that the success of VLT/VLTI was due to the large funding ESO dedicated to this program, rather than to the performance of the ESO staff and European industry. To be quite precise on this matter, I presented a report to the ESO Council on September 15, 1998.[2] Two summary graphs (Figure 18.2) compare the expenditures pre-

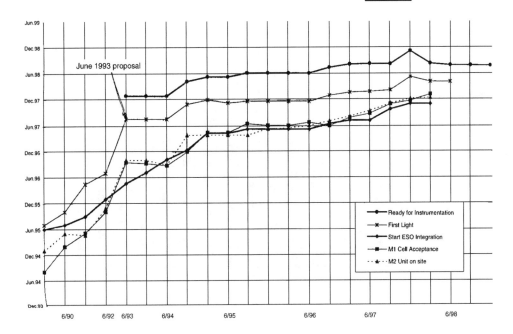

dicted by the 1994 audit and the actual or estimated expenditures up to 2003 for the VLT/VLTI itself and for site construction. For site construction, the expenditure planned in the "Blue Book" of 1987 is also shown. The change between 1987 and 1994 is mainly due to changes of scope.

The total estimated cost from 1994 to 2003, when VLT and VLTI would be fully functional, shows a decrease from 1,060 to 990 million deutsche marks. At an exchange rate of 1.7 marks per dollar, this amount turns out to be equivalent to $582 million, or $145 million per telescope. Considering that this cost includes the opening of a new observatory, the refurbishment of La Silla, and the building of twelve first-generation instruments, it compares favorably with the costs accrued on U.S. research projects.

The technical and managerial success of the VLT/VLTI program was vital in establishing ESO's capability to carry out major programs on behalf of the European astronomical community. It is this new confidence that has permitted the start of the ALMA and Large European Telescope programs.

FIG. 18.1. *The delay expected in achieving important milestones in the completion of Unit 1 of the Very Large Telescope, from specific reports. Prior to 1993, The European Southern Observatory was losing nine months for each year of work; after 1993, we lost nine months in five years.*

Tackling Problems and Changing Mindsets

To implement the management changes described above, I created a Program Office reporting directly to me as part of the Director General's Office

FIG. 18.2. (Top) *Very Large Telescope/ Very Large Telescope Interferometer expenditures: actual (1989–98) and predicted costs (1998–2003) compared to the audit predictions. (Bottom) Paranal site expenditures: actual (1989–98) and predicted (1998–2003) compared to audit predictions and "Blue Book" plans.*

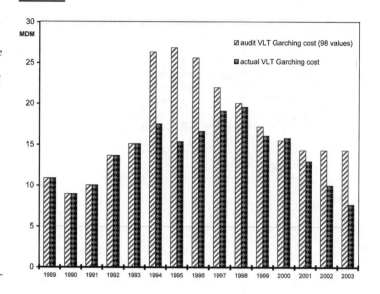

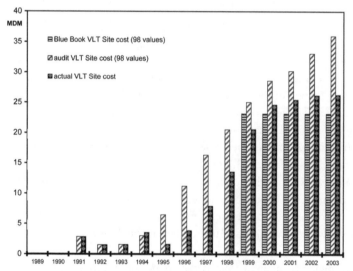

on January 1, 1993. It had only three employees: Jim Crocker; a deputy, Armelle Cabillic (who succeeded Jim after he left); and a secretary. Jim's experience I have described previously; Cabillic had been ESO's administrative representative at La Silla, where she was well known for keeping a rifle in her locker. Possibly this was to scare off the wild donkeys that used to chew on our cables at the observatory, or perhaps it was a memento of her solo bicycle trip from east to west across Africa.

The Program Office worked in close cooperation with the administration to introduce the ESO-wide WBS system and also the management information system that supported it; but it had much broader responsibilities. It helped all program managers or team leaders who had no other administrative support in setting up their own WBSs for inclusion in the ESO-wide system; it helped me and the various managers set up the periodic reviews of their programs (typically once a quarter); and it helped me think through and prepare the material that I would use for presentations to the ESO Council or the Finance Committee (Jim himself gave very effective presentations to the same groups).

Jim was often able to suggest engineering or managerial solutions to technical problems. Furthermore, both he and Armelle were experts in the operational side of running an observatory and were indefatigable in looking for means to reduce operational costs both at La Silla and at Paranal, where they could affect the planning process. They prepared the draft of the yearly budget, which was circulated to all heads of divisions prior to a plenary meeting at which the budget was agreed on.

By keeping the office small, Jim's view was that the burden of planning and documentation would be placed on the technical and administrative managers themselves, so that they would learn how to do it. The major benefit of Jim's work came from his ability to sit down with people and teach them how to see their tasks in a different light and how to maximize results through planning, value analysis, and new thinking. Jim was so obviously unafraid to tackle any problem that his attitude convinced people that they could also risk being innovative and that it was in fact fun to do so.

Setting Goals and Evaluating Performance

The new WBS and management information system, when coupled with yearly performance evaluations, permitted a much more objective evaluation of the staff's efforts. Having defined goals and objectives for the previous year in a face-to-face discussion between staff member and supervisor, followed by written documentation signed by both parties, the evaluation was based on the performance of work that had been agreed upon. The main virtues of this approach are that it forced supervisors to sit down with their staff at least once a year, and it eliminated as much as possible personal likes

327

and dislikes from the evaluation. It also made it easier to differentiate exceptional from mediocre performers and reward them accordingly.

The idea of rewarding people for performance took a while to be accepted. At STScI, after the AURA board had approved a given total sum, based on many factors (including market conditions, inflation, and performance), I was given the authority to approve the individual pay raises. I was, however, required to show the statistical distribution of the raises, which were expected to show a double-peak distribution. AURA wanted to be assured that we could identify good and bad performers and rewarded them accordingly. At ESO, on the other hand, the normal practice had been to give a more or less flat increase to everybody, an approach typical of a civil service mentality, and probably in line with Eurosocialism.

Another way of decreasing ESO's costs was to reduce the salary increases, which had become excessive. Over the years, median salaries in Germany had grown not only to take into account inflation, but also to share in the general growth of the nation's wealth (that is, they were indexed to the growth in the country's gross national product). ESO's salaries had grown at an even faster rate; thus it was possible to slow the increases a little without too much pain and convince the ESO Council of our earnest desire to save costs everywhere we could.

Even more important than the savings themselves were Jim's, Massimo's and my efforts to re-establish the council's confidence in ESO. We did this successfully by providing them with prompt, clear, and unvarnished communications on all the various initiatives we undertook. We invited council, Finance Committee, and Science and Technology Committee members to attend the yearly ESO-wide reviews. Many of them took advantage of this opportunity to become fully informed about our doings. I was given the authority to borrow funds if it became necessary for cash-flow reasons, the highest statement of trust I could imagine. Fortunately, I never had to use that authority, but I was grateful for the vote of confidence. I always received steady support from the ESO Council under the successive presidencies of Franco Pacini, Peter Creola, and Henrik Grage.

Norbert Koenig, who succeeded Gerhard Bachmann as head of administration in 1996, was successful in reducing overall ESO costs, thus freeing money for research. Surely the most unpopular saving measures were introduced at La Silla, where we stopped shuttling the employees every two weeks back and forth from Santiago, where their families typically lived, by private plane and paid instead for transportation by commercial airlines, a much cheaper if less convenient arrangement. But what really created a stir was that, in our attempt to reduce the cost of restaurant services—although we decided to continue to provide tea and cake in the afternoon—they would no longer be served at the tables, forcing the staff to endure the indignity of self-service.

What I believe was more important than all of the above was the enthusiastic response of the young scientists to the new opportunities opened to them; by joining or leading one of the various teams, they had the chance to work hard and succeed in something they really wanted to do. I believe that the management reforms we introduced were instrumental in encouraging this process and in making it possible.

ESO and Europe: Struggling with Budgets and Building Consensus for New Methods

The great change that the successful completion of the VLT/VLTI would bring about in European astronomy was already clear in February 1998, when I presented to the ESO Council at its extraordinary meeting a document titled "The Role of ESO in European Astronomy," which was unanimously approved.[3] At that time, however, the transformations taking place at ESO were not understood at Europe's funding agencies (or, possibly, by European astronomers), and I had a hard time maintaining ESO's budget at an appropriate level.

In 1996 the German government decided that it would no longer contribute more than 25 percent of the budget of any international research organization for the years 1997 to 2003. Over the years, Germany's gross national product had grown faster than that of other European states and therefore Germany's contributions (which were indexed to this measure of economic size) had increased to 27–28 percent of the budget, which Germany considered too high. This decision affected CERN, ESA, and ESO, among others. The response of the ESO Finance Committee and Council to this decision was to decrease all contributions proportionately, leaving ESO with a 7 percent shortfall in 1997.

In 1997 Claude Allègre, the French minister for Education, Research, and Technology during 1997–2000, when Lionel Jospin was prime minister, decided abruptly to cut France's contribution to ESO by 20 percent. Mercifully, he did not have the legal power to do so, given the nature of ESO's treaty instruments (and in any case, the budget was under the control of the Foreign Ministry).

Still, this issue festered, with a request by the French delegation, in December 1997, to postpone the vote on the 1998 ESO budget until a working group of the ESO Council could discuss further savings at ESO (beyond the

7 percent reduction) that might be suggested by the French delegation itself. Vincent Courtillot, a special advisor to Minister Allègre, suggested certain items for consideration in a letter to the president of the ESO Council. ESO had to make available a pile of documents, which included an executive summary, organization and work assignments for 1997–2003, a new VLTI plan, a plan for restructuring La Silla around self-management teams, VLT schedule documentation, and a summary long-range WBS plan for 1996 to 2003.[4]

The management reforms that ESO had carried out since 1993 allowed us to answer all the questions with a clarity scarcely paralleled in the European research environment, and the ESO budget was passed by the ESO Council at its extraordinary meeting of February 18, 1998, without modifications. But I considered this exercise, probably done to allow Minister Allègre to save face, to be extremely unfortunate at a crucial time in the development of VLT.

Faced with the budget reduction by the council on one side and problems in Chile on the other, I believed that VLT was in a critical situation. Jim, Massimo, and I were continuing to push the program full steam ahead, because we thought that delay or stoppage would be suicide. However, we did not want to have a unit telescope ready but no place to put it. And it would have been nice to have financial help through the accession of a new partner.

In 1996, desperate to find an alternative site for at least one of the VLT telescopes, I had initiated an ESO seeing survey on the Gamsberg Plateau in Namibia, where the Hess Observatory is now located. Although this action was more of a tactical move in the negotiations with Chile than a real possibility, the seeing on the plateau is excellent in the dry season.

A much more attractive and serious possibility was that of Spain joining ESO in 1997 or 1998. There was a Spanish plan to build a 10-m copy of Keck, to be placed on the Canary Islands as the major facility of what they hoped could become the European Northern Observatory. I discussed with Francisco Sánchez Martínez, director of the Instituto de Astrophysica de Canarias, the possibility of placing one of the VLT units on La Palma. The cost of a fifth VLT was estimated by ESO to be 110 million deutsche marks (approximately $65 million), which could be approximately covered by Spain's entrance fee to ESO. The telescope could have been in operation by 1999.

However, the funding arrangement for the Gran Telescopio Canarias, which included contributions by the European Union to build a mirror factory in the Canaries as an aid to an underdeveloped region, could not be changed and diverted to ESO membership; thus Spain became the only European country to lead an effort to build such a large telescope on its own. The seeing in the Canaries does not compare with conditions in Chile or Hawaii, but it is sufficiently good to support some arguments for the creation of a European observatory in the northern hemisphere. In any case, the construction of the Gran Telescopio Canarias has proceeded to a ceremonial first light in July 2007, with instruments to come. Spain has recently become a member of ESO.

To focus the attention and efforts of European astronomers on the completion of VLT, I thought that it might prove useful to stress the vital role that ESO's VLT/VLTI would play in optical ground-based astronomy in Europe as the only facility that could compete with Keck I and II. I pointed out in my presentation to the ESO Council on February 15, 1998, that with the construction of VLT/VLTI, Europe would, for the first time in a century, become competitive with the United States in optical ground-based astronomy. I saw ESO's role as a continuing one, to provide members with technology and ground-based facilities that would otherwise be beyond the reach of national groups. I tried to stress once again the profound change in methodology for data reduction, distribution, and archiving that was being adopted by ESO and which I believed was fundamental to progress in astronomy in the twenty-first century.

I am not sure whether my article[5] helped to clarify the situation, but it is a fact that the VLT/VLTI was completed, and the future I was advocating is actually happening now, with ESO's participation in ALMA and in the European Large Telescope studies; I hope what appeared controversial in 1998 may be generally accepted now.

I am comforted in this view by the statements made by Gerry Gilmore, a New Zealand astronomer who is now a professor at the Institute of Astronomy in Cambridge, on the occasion of the United Kingdom's joining ESO. The caption to his illustration of the development of astronomical telescopes reads: "The telescopes of Galileo, Newton and Birr Castle, reflecting European technological innovation and dominance in astronomy until the twentieth century. Mt. Wilson and the two Keck telescopes are typical of the dominance of telescope technology by the private U.S. observatories through the twentieth century. Finally, the VLT sets the standard of excellence at the start of the 21st century."[6]

Radio Waves: From the Large Southern Array to ALMA

My first interest in a large millimeter and submillimeter array arose from humble beginnings. A 15-m-diameter millimeter (that is, in the radio wave band) antenna had been installed at La Silla as a cooperative effort between ESO and Sweden's Onsala Space Observatory. While examining the overall ESO budget in 1995, it came to my attention that some improvements had been suggested for the instrumentation by our Swedish colleagues. Peter Shaver seemed to be the only scientist at ESO who had some knowledge of this program, and I asked for a briefing on what had been done and what was planned for the future. I found what he told me to be rather modest, and I asked for a vision of the potential of this branch of astronomy. If there was no clear role for ESO, there was no point in continuing the program; but if there was a possibility that ESO's developing capabilities could be of use to the radio community, then we should by all means study what needed to be done. I soon found that there was a large project under discussion, namely, the Large Southern Array (LSA), an array of about sixty antennas of 15-m diameter, with an area totaling up to 10,000 m^2.

My rather dynamic approach and the fear that ESO would play an overbearing part in LSA required some discussion with members of the radio astronomy community. In particular I had several encounters in 1995 and 1996 with Roy Booth of the Onsala Space Observatory in Sweden and Michael Grewing, then director of the Institut de Radioastronomie Millimetrique (IRAM). IRAM is a collaborative effort between the Max Planck Society and the French CNRS, and it operates two observatories: an interferometer on the Plateau de Bure near Grenoble, consisting at the time of four 15-m antennas, and a 30-m antenna at Pico Veleta in Spain, in collaboration with Spain. The Plateau de Bure interferometer was at the time the largest in the world. Michael would have preferred that the future project proceed under the leadership of IRAM, which had experience in this field. However, this scheme did not fit well with the plans of the management at Max Planck to decrease rather than increase its commitment to astronomy, which already took up more than 10 percent of its budget. On the other hand, the possibility of creating a European intergovernmental entity like ESO for radio astronomy seemed remote.

I went to visit Plateau de Bure in 1995, because I wanted to acquaint myself with the technology and the methodology current in this field. What I saw there convinced me that a project of the scope of LSA would be well beyond the managerial and technical capabilities of IRAM. What they had achieved was remarkable but quite expensive: a hundred technicians working for 10 years had produced four antennas, with two more under construction. To produce 15 times more with the same techniques (if at all possible) would require 15,000 man-years or more than a billion and a half dollars.

What I found even more distressing was that they had given very little thought to how they could themselves produce the six hundred receivers for each polarization of ten different wavelength bands required for the array. Some of the receivers they were using required manual tuning, which implied prohibitive personnel costs for operations and maintenance. As I later found out, their approach was similar to that of most millimeter-wave astronomers worldwide.

While learning about these problems, I had also become convinced of the great scientific potential that a millimeter or submillimeter array would have in studying a variety of astrophysical phenomena, ranging from organic molecules in interstellar space to the red-shifted emission from the most distant galaxies in the universe.

The situation came to a head in 1996 with the discussion initiated by Paul Vanden Bout, then director of the National Radio Astronomical Observatory (NRAO), with scientists in Leiden about the possibility of U.S.-Dutch cooperation on the Millimeter Array, which NRAO had studied for some time and which had been proposed to the National Science Foundation. The Millimeter Array was envisaged at the time as a thirty-two-element interferometer of 12-m-diameter antennas, totaling about 3,600 m^2 of area. The general reaction by French, German, and Swedish millimeter-wave astronomers to this idea was that they did not want to become minor partners in an American project; furthermore, the U.S. project seemed too modest to satisfy the needs of the European community.

NRAO seemed interested in the shortest possible detectable wavelengths, down to 0.3 mm, which required an observatory at very high altitude above sea level to avoid absorption by water in the atmosphere above the site. In addition, they were interested in relatively large fields of view for stellar astronomy, which required a compact central array, and were less concerned about total area. On the other hand, the Europeans, who already had a 1,000-m^2 array in the works at Plateau de Bure, felt that a tenfold increase in area and long interferometric baselines would be required to achieve many of their scientific objectives in extragalactic astronomy.

Given the clear expression of interest by the National Science Foundation in having an international partner and the large cost (in both money and time) for Europe to proceed alone, it seemed natural to study a compromise that would satisfy the scientific requirements of both sides. The result was a June 1997 statement of intent by ESO (representing all ESO member states) and NRAO to collaborate on a single program on the basis of equal shares. The program consisted of an array of sixty-four 12-m antennas (a total area of 7,238 m^2) located on the Llano de Chajnantor, a 5,000-m-high plateau near the village of San Pedro de Atacama in Chile. The document was signed by Paul Vanden Bout as director of NRAO and by me as director general of ESO.

Japanese astronomers had long been interested in millimeter-wave astronomy and had cooperated with U.S. colleagues in the exploration of the Chajnantor site since the early 1990s. There was uncertainty in Japan, however, on whether the scientific interests of Japan could best be served through international collaboration or competition. In 1996, when I first met Norio Kaifu, the future director of the National Astronomical Observatory of Japan (NAOJ), who was then directing the construction of the Subaru Telescope project in Hawaii, he seemed completely uninterested in a collaborative effort. Yet he was the Japanese expert on radio astronomy and was widely regarded as the future leader of any major Japanese project. However, Masato Ishiguro, then director of NAOJ, and several other senior Japanese astronomers (including Minoru Oda, my colleague in x-ray astronomy and a friend) were quite keen on an international collaboration on the basis of parity. Thus, soon after the ESO-U.S. memorandum of understanding was signed, discussions on a three-way partnership culminated in 1999 in a document titled "Resolution Concerning the Expansion of the ALMA Partnership to Merge the Atacama Large Millimeter Array and [the Japanese] Large Millimeter and Submillimeter Projects." This document was signed by Martha Haynes, president of AUI, for the United States; Masato Ishiguro, director of NAOJ, for Japan; and me as director general of ESO. Because funding for the Japanese contribution was delayed, a two-way partnership for the construction of ALMA started in December 2002, with the signing of an agreement between the United States and Canada on the one hand and the ESO (representing all the member states plus Spain) on the other. The funding having materialized in Japan in 2003 or 2004, a three-way collaboration is now in place. The ALMA program thus enlarged would satisfy all the scientific requirements and interests of the different communities. It would have a minimum of sixty-six and up to eighty-two antennas (depending on available funding) distributed both in a compact array and in a large array, placed on Chajnantor: it would cover all atmospheric windows between 350 μm and 10 mm. An angular resolution of 10 milliarcsec would be achieved by the largest separation between antennas, to be 10 km. When completed (first light for some units by 2009; completion by 2010–12), ALMA will be the premier array in the world in millimeter and submillimeter astronomy and one of the most powerful instruments in the entire astronomical arsenal.

Looking Ahead: The Overwhelmingly Large Telescope

No sooner had we reached success with the first light on VLT than Roberto Gilmozzi and his colleagues presented a proposal for studying the feasibility of a 100-m telescope at a SPIE (formerly known as the Society of Photographic Instrumentation Engineers) symposium held in Hawaii in 1998. The idea that the next step in ground-based astronomy would require a five- to tenfold increase in size was not novel. What made the case so interesting was that it seemed scientifically desirable because of the development of active and adaptive optics, which would permit angular resolutions on the ground comparable to or finer than those achieved in space, coupled with the much greater sensitivity resulting from an area larger by a factor of twenty-five to one hundred. The continuing development of increasingly capable computers for control of the optics and the industrial manufacturing capabilities demonstrated in VLT made the project appear difficult but feasible. In fact, there was a convergence of the mirror technology needed for projects like the Overwhelmingly Large Telescope (OWL) and that needed for laser-fusion experiments then being undertaken in Europe. Thus both Schott and RE-OSC declared themselves able and willing to take orders for the new project. A more up-to-date and I assume politically and technically correct description of the project is contained in the article by Roberto Gilmozzi and Philippe Dierickx (Figure 18.3).[7] The reason I mention political correctness is that I understand, thirdhand, that several European groups have shown an interest in becoming involved in a large telescope project. The name has changed from OWL to Extremely Large Telescope with prospects for European Union involvement. Current debates seem directed to making the size of the telescope as small as possible rather than as large as feasible. ESO decided in December 2006 to proceed with a 42-m telescope.

As I understood it, the OWL project made a great deal of strategic sense. Clearly the United States may still regain the lead in ground-based optical astronomy by constructing in the near future a 30-m diffraction-limited telescope with adaptive optics (candidate projects include the Thirty-Meter Telescope by a consortium of Caltech and Canadian universities and the Giant Magellan Telescope by the National Science Foundation). Because the present sociology in the United States is adverse to a truly national cooperative effort, it may still come about as a private effort and will therefore be limited in size and underfunded for instrumentation and data analysis. Trying to compete in the short term with telescopes of similar size means that

FIG. 18.3. *Artist's conception of the Overwhelmingly Large Telescope (100-m diameter).*

Europe would still be playing a catch-up game. I believed then and I believe now that the correct approach for Europe is to work together, as was done for VLT/VLTI, and build a telescope that will dominate world astronomy in the next generation of telescopes. To take an intermediate step now could cost leadership in the future.

Apart from these strategic considerations, I was most interested in how we should proceed at ESO on this project. There was no question that a new major European program in optical astronomy would be needed to succeed VLT/VLTI in 15–20 years. It was also obvious that, although ESO was and still is the only organization in Europe that could carry such a program to a successful conclusion, this fact would not be recognized until VLT/VLTI proved entirely successful and produced significant scientific results.

We were already committed to a very large project in radio astronomy, ALMA, which is intrinsically of great scientific importance and which hopefully will also allow ESO to retain engineering and management capabilities for the next 10–15 years. (The ALMA project is relatively easy technologically, the greatest challenges being the truly global aspects of the collaboration as well as the problem for ESO of leading such a diverse group of European institutions.)

On the other hand, the Extremely Large Telescope project also presents serious technical challenges in adaptive optics and computation capabilities, besides requiring funding at least three times greater than that for ALMA. Perhaps Europe should look to international collaboration on a global scale for its execution; ESO would retain the lead role, as CERN does for the Large Hadron Collider, its new particle accelerator.

In sum the prudent course of action was for me to encourage the Extremely Large Telescope studies to proceed at a reasonable level, which could be increased beginning in 2002 with the expected drastic reduction of La Silla costs. But in any case I recognized that the responsibility for the next program would fall on my successor.

Doing Science at ESO

While at ESO I tried to continue my personal research and also teach at the University of Milan. But I found it more difficult to take time off from my duties at ESO than I had at STScI, and I had to give up most of my teaching commitments.

For my research, I continued the collaboration on ROSAT x-ray surveys with Joachim Trümper, Günther Hasinger, and Maarten Schmidt, and actively participated in the Chandra Working Group meetings as an interdisciplinary scientist. I described in Chapter 15 how I had competitively earned guaranteed observing time on that facility.

As the time of Chandra's flight approached (1999), I thought more carefully about how I could best utilize my time and decided that, as I had proposed in 1987, I would devote my entire 500,000 sec to a very deep observation of one field to determine once and for all the nature of the x-ray background. There remained to choose a field in the sky to carry out this research. I had originally thought that it would be one of the Hubble deep fields, because the optical counterparts of the faintest sources detectable with Chandra were expected to be extremely faint in the optical domain (R [red] magnitudes greater than 26). But the more I thought about it, the more I became persuaded that this approach was not optimal.

The Hubble fields had been chosen with different scientific and operational questions in mind. The field in the northern celestial hemisphere, for instance, contained a bright quasar and had been chosen in part to allow for long exposures with Hubble, but no particular consideration had been given to hydrogen absorption (which occurs in the visible, infrared, and radio wavelengths) in that line of sight through the galaxy.

I thought that we should choose the best region for our purposes based entirely on scientific reasons and leave the problem of follow-up to a later date. I started discussing with Piero Rosati the possibility of finding a region in the southern hemisphere which could be observed using VLT, had low line-of-sight hydrogen density, was similar to the Lockman hole in the north (a break in the galactic hydrogen clouds that surround us), and had few bright stars or nearby bright active galactic nuclei (AGNs). He and Roberto Gilmozzi consulted the literature on radio astronomy and found a region with the lowest hydrogen density in the south right on top of Paranal ($N_h = 8 \times 10^{-19}/cm^2$; at right ascension 3 h 32 min 28.0 sec, declination −27 deg 48 min 30 sec; J 2000).

Optical plates of the region were scanned to find a specific area where there appeared no stars brighter than magnitude 14. Rosati and Gilmozzi carried out an imaging survey by using the FORS-1 camera on the VLT Unit 1 telescope Antu (program ID 64.0-0621) in the three wavelength bands V, R, and I. They obtained exposures of 4,000–7,000 sec in four adjacent fields of 6.8×6.8 arcmin, which covered a large fraction of the 16×16 arcmin of a Chandra field. Four additional R-band fields of about 1,200 sec each were obtained with the FORS imager.

Rosati was able to obtain spectra for a few objects using FORS-1 multislit capabilities in March 2000, just before the Chandra deep-field south became inaccessible. He also obtained data from the ESO Imaging Survey and from NTT's SOFI in J and K bands and in the ultraviolet bands with NTT's Superb Seeing Imager 2 (known as SUSI-2). A 30×30-arcmin image in the B band was obtained during commissioning of the Wide-Field Imager camera at the ESO's Max Planck Gesellschaft 2.2-m telescope, the wonderful instrument built at ESO by Dietrich Baade and his group.

This observing program was the first request for observing time I had ever made for ESO facilities. ESO's Time Allocation Committee refused to grant time for this and two other extragalactic programs of observations, even though all three programs had received very high scientific ratings. My program in particular was refused on the ground that the Chandra observations had not yet taken place. For once I used my prerogative as director general and in 1998 approved all three programs on discretionary time. I have no regrets. The Chandra deep-field south has become one of the most studied regions at all wavelengths, and the paper that resulted from the observations of this deep field was for a while the most quoted paper on VLT.

Coming Home

At the end of my first five-year term at ESO, in January 1998, I was re-appointed for only two more years (to January 1, 2000), because of my advanced age (67 at the time). I would have liked a full second term to see the completion of the VLTI, the start of the ALMA project, and the completion of the studies on OWL. The ESO Council, however, felt bound by the mandatory retirement age of 65, which they had already stretched by four years, I was told. In early 1999 I received an offer from AUI in the United States to become president of that corporation. AUI is the managing organization that had built and operated Brookhaven National Laboratory for 50 years as well as the National Science Foundation's NRAO. AUI was the lead center for ALMA in the United States. I accepted partly because, having seen ALMA's birth, I wished to continue nursing the project along and partly because it was my hope that as a manager I could help NRAO deal with the growing pains that I could foresee in the execution of ALMA.

Chandra was due to fly shortly, and I was looking forward to receiving the data on the x-ray background. Because I had maintained research collaborations with Colin Norman and other colleagues at the Johns Hopkins University and STScI, I would be able to complete my Chandra research project there. So between ALMA and Chandra, the United States was where I needed to be to continue to work in astronomy.

Radio Astronomy on the Radar

AUI

Associated Universities, Inc. (AUI) is a nonprofit science management corporation that was established in 1946 by nine universities: Columbia, Cor-
nell, Harvard, Johns Hopkins, MIT, the University of Pennsylvania, Princeton, the University of Rochester, and Yale. The charter of the corporation is
to "acquire, plan, construct and operate laboratories and other facilities" that
would unite the resources of universities, other research organizations, and
the federal government to create facilities so large, complex, and costly as to
be outside the scope of a single university. The facilities were to be made
available on a competitive basis to scientists without regard to affiliation as
well as to the resident scientific staff. The corporation is responsible for oversight and audit of the facilities it manages; it appoints their directors, approves the appointments of tenured staff and senior managers, reviews programs and budgets, and oversees long-range planning and proposals for new
major facilities.

The main impetus for research came from nuclear and particle physics,
and from 1947 to 1998 AUI was responsible for the construction and operation of Brookhaven National Laboratory, a research center located in Upton,

New York. Since 1956 AUI has also managed NRAO, which it proposed to and built for the National Science Foundation.

AUI is governed by a board of trustees, which elects corporate officers to carry out the business of the corporation. Because AUI is an independent corporation rather than a consortium, individual trustees serve as fiduciaries and do not represent their home institutions on the board. Over the years, AUI has taken on a broad national character, with a diversified board of trustees from universities and other research institutions across the country and abroad. The activities of the corporation are funded by management fees. The president of AUI is the chief executive officer of the corporation.

Two years before I joined AUI, the corporation had gone through a major upheaval: on May 1, 1997, the Department of Energy (DOE) had unilaterally terminated the AUI contract to manage Brookhaven for reasons that, according to Secretary Federico Peña of DOE, were in the best interests of the government. Thus came to an abrupt end the major management responsibility that AUI had exercised since the inception of the laboratory.

This action was the ultimate result of the discovery by Brookhaven in October 1996–January 1997 of a tritium leak into the groundwater at the center of the laboratory site (possibly emanating from the high-flux beam reactor), with concentrations exceeding those set by the State of New York for drinking water. There was a DOE investigation, much concern, agitation in the Long Island community (whether justified or not), and corresponding political grief for DOE. Notwithstanding the best efforts of AUI to remedy the situation, there followed on April 30, 1997, the retirement of Brookhaven Director Nicholas Samios, the assumption of the interim directorship by Lyle H. Schwartz, president of AUI, and finally the cancellation of the contract. Thus DOE resolved its political problems by blaming AUI for the event as well as for its poor handling of the public relations.

The immediate consequences for AUI were that AUI lost 90 percent of its research budget and therefore 90 percent of its revenues based on management fees. The contract of NRAO with the National Science Foundation was not affected by these events.

NRAO, the Premier Radio Astronomy Facility

Since 1956 AUI has operated NRAO, the premier facility for radio astronomy in the world. This preeminence is due to the early decision by radio astronomers in the United States to cooperate in establishing a national facility that could compete in the field worldwide. Thus NRAO has been able to lead the advancement of radio astronomy and radio technology in the United States and in the world. The observatory currently has a staff of about 450 and a budget of approximately $40 million, with additional funds for the construction of the Expanded Very Large Array (EVLA) and ALMA.

In 1999 NRAO was carrying out five different activities. The 100-m Green Bank Telescope was under construction in Green Bank, West Virginia, site of the first operations of NRAO in 1957. The Very Large Array (VLA), an interferometric array of twenty-seven antennas each 25 m in diameter located on the plains of San Agustin in New Mexico, had been in operation since its completion in 1981. The Very Long Baseline Array (VLBA), consisting of ten 25-m antennas spread across the country from St. Croix to Hawaii, was also continuing operation. Research in millimeter-wave astronomy was being carried out with a 12-m antenna located on Kitt Peak in southern Arizona. Finally, there was the start of development of the Millimeter and Submillimeter Array (MMA) that was ultimately to merge into ALMA.

While the scientific productivity of these NRAO facilities remained high, there were grave problems with their management and their funding from the National Science Foundation. Furthermore, the construction of the Green Bank Telescope had met with considerable difficulties.

The first problem was the scarce attention that the National Science Foundation has given over the years to its ground-based observatories, including NRAO. Although construction money has been made available to build the facilities, very little has been provided for their full scientific utilization, development of data analysis and archives, development of new instrumentation, modernization of obsolete equipment, or even facility maintenance. During the 1990s the NRAO budget decreased in real purchasing power by about 6 percent per year. Rather than decreasing services at its facilities, NRAO under the directorship of Paul Vanden Bout (its director from 1985 to 2002) continued the policies of his predecessors of avoiding staff reductions as far as possible. This goal was achieved by not developing the end-to-end software necessary to make radio astronomy data accessible to the community of nonradio astronomers, neglecting or postponing upgrades to the equipment, postponing maintenance of the facilities, and allowing staff salaries to decrease by 20 percent with respect to competitive levels. No effort was made by the National Science Foundation, AUI, or NRAO to set priorities in the execution of its tasks, so that everything was done as best one could. Considering that VLA was the largest and most productive interferometric array in the world, this approach was unfortunate.

The Green Bank Telescope

In 1988 the largest antenna on Green Bank, a 300-ft dish, collapsed due to structural failure. There was uncertainty in the scientific community about whether it should be replaced at all, but through the intervention of Senator Robert C. Byrd of West Virginia, who cared a great deal that the Green Bank Observatory should continue to play a scientific role in his state, a budget of $75 million was added to the Emergency Supplemental Appropriation Bill for the replacement of this telescope. NRAO was certainly anxious to proceed but wanted to ensure that the new telescope could do good science. The NRAO scientists decided that the best strategy was to adopt a new design: an off-axis feed from a 110×100-m section of a 208-m parent parabola. Although this design minimizes potential background sources and reduces side lobes, it presents significant construction difficulties because of the asymmetry of the reflecting surfaces (Plate 25).

The entire surface is composed of 2,004 panels, whose position can be adjusted by 2,209 actuators, which are used to control the overall shape of the mirror to an accuracy of better than 400 μm. The total weight of the structure is 7,300 metric tons. It rotates in azimuth on a track 64 m in diameter. The original receivers were designed to cover the frequency range from 100 MHz to 115 GHz, but higher-frequency receivers are under construction.

The construction of this telescope, the largest fully steerable telescope in the world, proceeded much more slowly than anticipated, with a total delay of six years from the originally anticipated completion date of December 1994 to the actual first light in August 2000. The delay had various causes, but was primarily due to changes in the ownership and management of the major construction contractor and initial underestimates of the complexity of the job. From the contracting point of view, the overrun was only about 5 percent, but for NRAO, the costs of supporting the staff of one hundred essentially doubled, adding another estimated $80 million to the cost of the facility.

When I arrived at AUI in July 1999, these financial problems left us with no choice: notwithstanding the severe technical problems still to be solved (such as designing and building the tracks to support the giant structure), the only option left to us was to complete the job as quickly as we could and declare victory. Thanks to the magnificent efforts of Philip Jewell, the director of Green Bank, and his staff, the Green Bank Telescope began early scientific operations in the spring of 2001 while commissioning was still under way. Routine science operations were achieved by the fall 2003. In the years since, the Green Bank Telescope has shown its scientific potential by producing a number of important results, including the discovery of new molecules in interstellar space, the study of large numbers of newly detected pulsars, the detection of water masers in AGNs, and the measurement of constraints on the cosmic evolution of the fine structure constant (a fundamental physical quantity).

VLA

VLA is an array of twenty-seven 25-m antennas, arranged along the arms of an equiangular "Y" configuration 21 km long (Plate 26). Its total area is equivalent to 13,000 m^2 and its angular resolution is as fine as 0.13 arcsec. It can be operated in four or more wavelength bands from 1.3 to 20 cm. Construction was completed in 1981.

Since that time VLA has provided a combination of greater resolution and sensitivity than any other existing radio telescope, and its scientific productivity has been comparable to that of any major observational facility in astronomy. This productivity could have been even greater if adequate computer facilities and data analysis systems had been available. David Heeschen, the NRAO director at the time of VLA construction, has described in detail the constraints under which VLA was built.[1]

Apart from financial difficulties, there had never been a strong commitment in the radio astronomy community to make access to their facilities "user friendly" for astronomers in other disciplines. The point of view still expressed today by some senior radio astronomers is that access should be limited to those willing to bring their own technical expertise to the same level as that of the staff. In my view, this approach is suicidal, because it limits the use of the facilities, now costing hundreds of millions of dollars, to a few hundred people and thus limits community support for their construction. In any case, the development of suitable data-handling software had not progressed to modern standards even 20 years after commissioning of the array. Generally speaking, this deficiency had to be corrected to allow proper scientific use of all major new projects, including ALMA.

When I joined AUI, I became aware of problems with the physical plant as well. The 60 km of railroad tracks, on which the antennas are moved to different configurations, rest on railroad ties that, for the sake of economy, were obtained from abandoned tracks in military bases. These ties degrade with time, and they have to be replaced every few years. Because there are about 60,000 of them, even replacing them at the rate of 6,000 per year it takes 10 years to do them all, whereupon one has to start all over again. The same goes for painting the antennas, which has to be done to prevent rust. Shortly after I took office at AUI, NRAO was able to obtain (with the help of Senator Pete Domenici of New Mexico) a special one-shot funding addition to remedy the neglect of the physical plant over the years.

Also with the senator's support, in May 2001 NRAO was able to initiate a program enthusiastically endorsed by the astronomy community, namely, the improvement of VLA as a whole. This program, which became known as the EVLA, consisted of two phases: the first was the modernization of several aspects of the existing VLA; the second was the addition of some eight additional 25-m antennas spread around New Mexico, thus increasing the baseline of the array and its angular resolution by a factor of 10. But this second phase was not funded.

In previous years every effort had been made to improve VLA's performance, utilizing the most disparate sources of funding, including funding from NASA to support the telemetry return from *Voyager 2* at Neptune (which provided for the addition of the 8.4-GHz receiving system as well as some infrastructure repairs), the Universidad Nacional Autonoma de México and the Max Planck Institüt für Radioastronomie (which funded the bulk of the 43-GHz receivers), Max Planck and AUI (which funded the fiber optics link to the VLBA Pie Town, New Mexico, antenna), and the Naval Research Laboratory (which funded most of the 74-MHz observing systems). While the NRAO staff was rightly proud of having received this support and of what they had been able to accomplish with it, I think that it is a sad statement that a successful and productive national observatory, unique in the world, should be forced to go begging at home and abroad to maintain its equipment.

Despite these efforts, much remained to be done to carry out improvements that should have taken place naturally over the years. The only way was to designate this work as a new program: EVLA. Among the goals of EVLA Phase I are to increase the sensitivity of VLA by a factor of 10, provide complete frequency coverage between 1 and 50 GHz, and enhance the spectral resolution of VLA. To achieve these objectives required the following:

replacement of the wave guide data transmission system with 60 km of fiber optics;
replacement of all electronic modules in the antennas with new wide-band modules;
installation of new wide-band feeds for eight frequency bands;

installation of wide-band receivers at the bases of the feeds;

implementation of a new wireless detection and ranging correlator (a fast, dedicated computer) from Canada; and

creation of a new back-end and data archive system.

The total expenditure for Phase I of VLA will be provided by the National Science Foundation ($51 million), Canada ($14 million), and Mexico ($2 million), and by cuts from NRAO operations of $14 million, for a total of $81 million.

Clearly this work was necessary and beneficial to a large number of astronomers. In 2003, 847 astronomers from 205 institutions used VLA, about half of them from the United States.

Phase II, which foresaw the addition of eight new antennas located throughout New Mexico, would have improved the angular resolution capability by a factor of ten, but its cost was foreseen at $116 million, and its start has not been approved by the National Science Foundation.

VLBA

VLBA is a system of ten radio telescopes controlled remotely from an Array Operation Center in Socorro, New Mexico. The large separation of the antennas, from St. Croix to Hawaii (5,000 miles), provides the highest angular resolution (milliarcseconds) currently obtainable from Earth (Plate 27). Each telescope has a 25-m diameter and weighs 240 tons. Each station records the observation of a celestial object on fast tape recorders with precise times given by atomic clocks. These data are sent to Socorro, where a correlator derives the image of the field. The construction of this system started in 1986 and was completed in 1993, for a total estimated cost of $85 million. This observational tool is even more difficult to use than VLA, and its user community is much smaller. Much work has been done to improve the recording system, but much more will be required for the data analysis software, and to make it more user friendly and a more productive tool for astronomy.

The Millimeter-Wave Telescope

The 12-m telescope was built in 1967 and placed on Kitt Peak, Arizona, where it operated for more than 30 years. It opened up the field of millimeter-wavelength molecular astronomy with the detection of dozens of molecular species from interstellar space, including carbon monoxide, whose lines are indicators of stellar formation in galaxies. In 1984 the telescope was given a new reflecting surface and supporting structure, providing greater sensitivity at shorter wavelengths. The result was the extension of the observations to interstellar gas clouds and star formation regions, to evolved stars, external galaxies, and even comets.

However, when even the first elements of ALMA become operational, each of its sixty-four 12-m antennas will provide better observational capabilities than the 12-m telescope, and their location at 5,000-m altitude in the Atacama Desert is much superior for submillimeter observations to Kitt Peak. For these reasons and because of financial exigencies, AUI, the National Science Foundation, and NRAO decided in February 2000 to close the observatory in July of that year. The staff at the observatory was already deeply involved in the ALMA program, and many members of the staff were transferred to Charlottesville, Virginia, to continue their work on the new project.

Central Development Laboratory

The Central Development Laboratory has been one of the gems of the NRAO program for many years. Even before accepting the appointment at AUI, I went to visit the laboratory in Charlottesville, where John Webber and a small team had, over the years, developed amplifiers and mixers of unique design for cryogenic operations that outperformed their commercial counterparts for observational radio astronomy applications. During my first visit in 1997, I asked John whether he also intended to tune his amplifiers by hand, as I had found at IRAM, and he pointed out that this would prove difficult to do, because his amplifiers were also used in space missions. I was vastly relieved to find, even on this first visit, that Central Development Lab technology was quite mature and suitable for transfer to production of large numbers of devices.

The activities of the lab included research and the creation of millimeter-wave integrated circuit components, heterostructure field-effect transistors, superconductor-insulator-superconductor mixers, local-oscillator multipliers, feeds and polarizers, and digital spectrometers.

New amplifiers centered at 22, 30, 40, 60, and 90 GHz had been designed and tested for the Wilkinson Microwave Anisotropy Probe (a NASA satellite launched in 2001). Their improved performance contributed significantly to the sensitivity of the survey and to the spectacular scientific results obtained by the probe. It is sad that the NRAO contribution to this program has received little recognition. However, the exercise was extremely useful in providing the Central Development Lab with valuable experience in building

FIG. 19.1. *The W-band Microwave Anisotropy Probe amplifier with cover removed.*

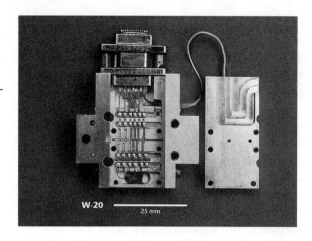

FIG. 19.1. *The W-band Microwave Anisotropy Probe amplifier with cover removed.*

rugged, stable, and high-quality devices. This experience is directly applicable to the ALMA program (Figure 19.1).

The technology described above also formed the basis for the refurbishment of VLA, and for EVLA and VLBA. Research in support of future programs, such as the Square Kilometer Array, is ongoing. It has been the policy of the lab and of NRAO to make advanced technology available to other institutions doing research in radio astronomy on a cost-reimbursement basis, thus advancing technology in the field.

Managing NRAO: Splinters of Excellence

Before discussing the largest program currently under way at NRAO, it may be useful to describe the situation as I found it upon my arrival in 1999. The main problem was that NRAO operated, rather than as a single observatory, as a collection of individual observing stations or programs. The great performance of these entities was due to the high caliber of the scientists that directed the efforts. Miller Goss, who had run VLA since 1988, is an outstanding scientist as well as a practical manager. Jim Ulvestad, who succeeded him in 2002, had directed the VLBA effort for years and pioneered VLB interferometry using satellites. Peter Napier, who was in charge of the EVLA effort, is also a highly talented scientist and engineer. Philip Jewell was able to carry through the commissioning and start of operations of the Green Bank Telescope by creating a dedicated group that he led to success. Robert Brown led the initial effort on MMA (later merged with ALMA) essentially singlehandedly.

It was quite clear to me—coming from ESO, which had just successfully completed VLT, the largest ground-based astronomy project yet undertaken—that ALMA would be even larger in scope (by a factor of two) and managerially more difficult to carry out than VLT had been. Considerable reorganization of NRAO would be required to utilize effectively the technical competence existing in the observatory as a whole to achieve success.

It was fortunate that the AUI–National Science Foundation cooperative agreement to operate NRAO was expiring; to renew the agreement, we had to pass an audit and a review, by a National Science Foundation ad hoc committee, of AUI's management of NRAO. This event gave me the opportunity to initiate a series of internal audits in all relevant management and administration areas. These audits led to reforms, which were implemented over the course of years, because it took time to overcome the reluctance of the staff (and often of the director) to adopt these newfangled management policies.

The progress we had achieved in two years was sufficient to pass the scrutiny of the National Science Foundation management review of AUI with flying colors. Here I give only some highlights of what had to be done.

The first and foremost problem concerned personnel policies, always the most delicate of subjects. The staff, hired at the time of VLA construction and retained through thick and thin, was aging. Because of the constant erosion of funds, there had been no opportunity to hire young scientists with new skills and interests. The result was particularly serious for the scientific staff. To give an example, there were hardly any millimeter-wave astronomers at NRAO at a time when the observatory was undertaking the largest project in the world in this subdiscipline.

The concept that NRAO's primary raison d'être was to serve the community and thus that research was a derived requirement had not been made clear to the staff. Seventy-nine members of the technical or scientific staff held continuing appointments or tenured positions, and they reported to the NRAO director, not their technical supervisors. This approach got to the point that site directors would be hesitant to propose meritorious people for advancement, for fear of losing them as contributors to operational tasks.

There were no yearly NRAO-wide personnel evaluations on the basis of agreed-upon goals and objectives, with written documentation. This system had been instituted at Socorro by Miller Goss, but it was considered too revolutionary to be introduced to the whole of the observatory without a period of testing. Because in addition there was no yearly observatory-wide review, the system was totally inadequate to convey to the staff a vision of the observatory's mission and their role in fulfilling it. In addition there was no affirmative action program or educational program for the staff. Increases in salary were essentially based on inflation and were rather uniform across the board. Salaries were below competitive levels by about 20 percent.

The personnel officer, who did not believe in the usefulness of yearly performance reviews, was let go, and appropriate policies and procedures were instituted to clarify to the staff what was expected of them and what their career paths might be. This process was particularly important for the tenure-track staff. In agreement with the director of NRAO, AUI tried to make early retirement more appealing to some of the senior members of the staff to leave room for new appointments. We were successful in half a dozen cases.

The Administration Division had to be completely revamped. Traditionally, NRAO had been concerned with maintaining financial data for the purpose of cost recovery rather than for management. A new cost accounting system had to be put in place before we could implement a WBS capable of reporting labor and material costs as well as schedule progress. A fortunate exception was that Bob Brown, on his own initiative, had introduced WBS in the management of the MMA project. (Currently NRAO has a full WBS and management information system, which apart from internal use, feeds directly into and supports the ALMA project-wide information system, with built-in schedules and dependencies. A new administrator, Ted Miller, and new staff were hired to carry out the transition.)

Material purchases and major contracts costing up to $200 million were planned for ALMA, yet there was no NRAO-wide purchasing department, staffed with contract and procurement specialists. Thus, each site and program had to have people on the staff doing this job as best as they could. This weakness was easily remedied, but it took the staff some time to believe that the new organization would actually help them.

Education and public outreach had been a major and continued concern of NRAO, but again this effort was carried out mainly at the sites, with no central coordination by headquarters. Given the importance of public outreach on a national basis, this deficiency is also being slowly corrected. It is interesting to note the reluctance of national observatories to follow well-established and successful approaches adopted by other organizations in the conduct of their affairs.

New impetus was given to the reorganization process in 2002 with the arrival of the new director of NRAO, Fred Lo, and the creation of a deputy director position, currently filled by Philip Jewell. I was also heartened by the appointment in June 2006 of Chris Carilli, a brilliant young scientist, to head the North American ALMA Science Center of NRAO, which I hope will play an important role in the scientific utilization of ALMA.

I have complete confidence in Ethan Schreier, my longtime collaborator and friend, who succeeded me as AUI president in 2004. I hope that under his leadership, the reforms I initiated at NRAO will make it a stronger and better organization, able not only to successfully carry out ALMA, but also to provide support for North American participation in the future Square Kilometer Array.

ALMA

NRAO started its design studies for the MMA in 1984, and the first science workshop took place at Green Bank in the fall of 1985. A second science workshop took place in November 1989 to update scientific goals and array designs to prepare for the MMA construction proposal, which was submitted to the National Science Foundation in September 1990. MMA received a strong recommendation from the National Academy of Sciences 1991 report titled "The Decade of Discovery in Astronomy and Astrophysics." In October 1991 the foundation's Advisory Committee for the Astronomical Sciences endorsed a two-stage approach: a development phase, with detailed designs and prototypes, and a construction phase. In November 1994 the National Science Board approved the development plan for MMA, and in 1995 NRAO began site testing at Chajnantor, Chile, in cooperation with the National Observatory of Japan.

I first became acquainted with the details of MMA during a review of the project by a National Science Foundation ad hoc group. Its purpose was to investigate the causes of a substantial increase in cost that had been communicated to the foundation by Bob Brown, without knowledge or approval by the director of NRAO, Paul Vanden Bout, or the president of AUI, Martha Haynes. There was no stronger indication that MMA had been run as a project separate from NRAO as a whole.

On the one hand, this independence may have been for the best, given the astonishing amount of technical and managerial progress that Bob Brown had been able to achieve; on the other, it showed a complete lack of the administrative control and direction one would expect for a program of that size. It was clear, moreover, that NRAO was not fully committed to a serious effort in millimeter-wave astronomy. Without such a commitment, it was not clear how NRAO planned to conduct long-term operations, or how it planned to ensure the involvement of the U.S. astronomical community so that it could reap the scientific returns from this project.

The MMA project consisted of thirty-two 12-m antennas with eight receivers for each antenna, covering the band from 30 to 600 GHz. The technical cost estimate had been done with great care from the bottom up by Bob Brown and his group (which at the time included Peter Napier), utilizing a WBS but assuming that the work would mostly be done in house.

The weaknesses in the administrative side of NRAO perhaps account for some of the initial errors in the cost estimate. The most glaring deficiencies were in the estimates of overhead costs (underestimated by about $15 million over 10 years), the cost of management and the management information system (about $17.5 million), and the cost of the necessary software development and the first year's operation costs ($27.5 million).

Although these potential increases were not fully known at the time, the National Science Foundation (with strong prodding by Congress) urged NRAO to find international partners to share in the cost of the project. Paul Vanden Bout succeeded in engineering a cooperative agreement between the National Science Foundation and Canada's National Research Council, which included financial contributions to MMA (by then renamed ALMA) of $20 million and in-kind participation in EVLA (the new wireless detection and ranger correlator). This contribution was certainly not enough to cover the growing costs, and Vanden Bout did not succeed in securing additional international partners in Europe on the basis of bilateral agreements, with NRAO retaining the lead role.

Instead, Europe responded by proposing that the MMA project be modified to more closely satisfy the LSA scientific requirements and offered to proceed on the basis of an equal partnership between Canada and the United States and Europe, as I discussed in the previous chapter. By joining efforts, the partners could carry out a more ambitious program that would satisfy the scientific communities on both sides of the Atlantic. The number of antennas was increased to sixty-four, yielding a collecting area of more than 7,000 m^2, which would permit the extragalactic observations of primary interest to the Europeans. The Chajnantor location, at 5,000 m above sea level, will permit the extension of the observations to submillimeter wavelengths, as advocated by the North American (and Japanese) groups.

Having secured the funding for the National Astronomical Observatory of Japan, Japan joined the ALMA project in 2001 as a third major partner and will contribute a compact central array of sixteen more antennas (Plate 28).

Additional cost increases occurred for ALMA because of the international nature of the project. In particular, there were delays mandated by the ALMA Governing Board (of a little more than a year, equivalent to about $20 million) and delays in the antenna procurement decisions (resulting in increased costs of materials). Thus the cost of the North American portion of ALMA rose from $276 million (in 2000 dollars) in January 2001 to about $420 million today; and of course the European part of the project suffered increases as well, so that the total project cost is now more than $800 million.

However the project is technically very advanced. All three prototype antennas (NRAO, ESO, and the National Astronomical Observatory of Japan) that have been built and tested at VLA have met or exceeded technical specifications; the construction of receivers and amplifiers is coming along quite well; and so is the first-generation correlator, while a second generation is on the horizon.

The software part of the project is proceeding as well, in a cooperative ESO-NRAO approach in which ESO is taking the lead, utilizing its VLT experience. I am quite confident that ALMA will be successful, and I hope to see the initial results of the first elements of the array in the not-too-distant future (first light is scheduled for 2008 or 2009).

ALMA as an Experiment in International Cooperation

ALMA is a scientific program that will bring millimeter and submillimeter observations to bear on some of the most interesting questions in cosmology, such as the study of the formation of structure in the universe and the origin of life. The program is truly an intercontinental cooperative effort, joining Europe, North America, and Japan, and it promises to be a wonderful experiment in cooperation on large-scale programs.

The fundamental management issue is whether nations are willing to put their resources into a truly supranational organization (such as CERN, the European Space Agency, and ESO in Europe), but now on an intercontinental basis. Should this level of cooperation be achieved, the way to proceed would be clear, hampered only by the expectations of industrial returns in proportion to contributions (as is the case for the European institutions mentioned above), which is a solvable problem. However, we have not yet arrived, at least in astronomy and particularly in ALMA, at this level of full internationalization.

If the partners on an international project are not ready to take this route, they can proceed on the basis of memorandums of understanding, as is done in collaborations between NASA and other space agencies. For such collaborations, there is normally an agreement to provide specified portions of the project for integration by the majority partner. This arrangement has worked well in most European Space Agency–NASA agreements, but there is always the possibility of unilateral actions by one of the partners at some point in the program, which may damage the other partner. The great advantage of the approach is that it allows the parties to solve their own problems and thus strengthen their technical and managerial capabilities.

Yet another approach was tried in the Gemini program, namely, to have the National Science Foundation act as the executive for a consortium of nations. A Gemini Board and Gemini project were created in 1994, neither of which had any legal authority to make project decisions. AURA, the managing entity for STScI and the National Optical Astronomy Observatory, took on the management of the project, creating in effect a virtual observatory in competition with the optical observatory. Although the project succeeded in building two 8-m telescopes (one on Mauna Kea, Hawaii, and one on Cerro Pachon, Chile), it did not bring to the U.S. community the technical and scientific benefits it had hoped for, and it pushed the national observatory to the sidelines. The negative effect of this approach can be perceived in the difficulties encountered with the start of the next major ground-based project in the United States. Having marginalized the National Optical Astronomy Observatory during the Gemini project, the National Science Foundation did not provide the observatory's staff with the opportunity to gain experience required for the next project.

ESO was certainly not planning to emerge from the ALMA project weaker rather than stronger, and neither was NRAO. Both organizations realized that they had to face new challenges, but these challenges were welcome, because they resulted in technical and managerial advancement. The NASA–European Space Agency type of collaboration is a model of how this technical growth can be achieved. Each partner is responsible for making tangible contributions to the program at its own expense and utilizing its own management capabilities. Project integration is provided by the larger partner in consultation with the other. Each partner is responsible for both its financial and technical aspects, and each enters into contracts with designated scientific and industrial groups to carry out the program, in accordance with the legal framework in which they operate.

The ALMA project was conceived by ESO to be carried out according to this model, and this approach was accepted by the North American and Japanese partners. Each partner would provide a specified hardware contribution with a value estimated in advance by agreement. The partners would receive observing time in proportion to the agreed-upon value of their contributions, removing potential problems having to do with differences in actual costs to execute the program in different countries and giving an incentive for each partner to be efficient.

The approach outlined above also permits the different scientific communities in the partner countries to retain their own styles in making use of their observing time, which are often quite different. The funding of data analysis is also different in Europe, Japan, and the United States. In Europe and Japan, data analysis is typically based on institutional support, whereas in the United States, grants for data analysis are given to individuals or groups on the basis of the observing time won competitively through the peer review process. These contrasts would make it quite difficult for a single time selection committee to take into account the different styles of conducting research in the different countries.

Although this programmatic approach for ALMA was not novel and had been used successfully in international collaborations in space, it was apparently somewhat of a novelty for a ground-based project. There was in fact little understanding in the scientific communities or in the funding agencies of the fundamental requirements necessary to make the project work. This lack of understanding manifested itself in many ways.

Much of the document that defines the agreement between the National Science Foundation and ESO follows the lines of the Gemini program structure. It creates an ALMA Board with authority to approve budgets for ALMA, but the board has no legal status or budget. The representatives of the funding agencies sit on the board and in fact have been sharing the chairmanship and vice chairmanship, but the funding agencies do not recognize the decisions of this board as having binding financial consequences.

To succeed, the ALMA project needed a central program management group that would coordinate activities during construction and operations. However, it was not intended that such a group would assume management or technical responsibility for each partner's share of the program. Nor was it intended that ALMA would become a new observatory, competing with ESO, NRAO, or the National Astronomical Observatory of Japan, but rather an observing station run jointly by the existing institutions.

This role was not clearly understood (or was deliberately ignored) by the ALMA Board and its search committees. As a result, the head of the Joint ALMA Office was called the ALMA observatory director, although this individual had no scientific responsibility for the program or observing time to distribute. This misleading choice of names has had many negative repercussions.

Another area of problems for ALMA has come about because the technical and scientific staff of the different partners prevailed in their desire to become involved in all aspects of the development work, including development and manufacturing of components. This failure to delegate creates such intricate interdependencies that the responsibility for timely delivery and cost can no longer be clearly established. The approach was followed to please the various constituencies but without any consideration of the management and cost burdens it generated.

Finally it was assumed, mainly by the United States, that ALMA would operate in a free market economy, and that the choice of subcontractors would be based entirely on technical performance and cost, to the benefit of the program. This assumption proved naïve, and the traditional European approach, in which nations expect to receive contracts in proportion to their financial contributions to the organization, prevailed when major decisions were made with regard to hardware.

When, as an ex-officio member of the board, I tried to point out the inconsistencies in what we were doing, I was considered uncooperative. But I still believe that the responsibilities of the ALMA Board and its creature, the Joint ALMA Office, must be clarified. At present, this system has proven to be a means for government bureaucrats to micromanage the program without assuming any responsibility. I also think that the management officers of the National Science Foundation were woefully unprepared to act in defense of U.S. interests in the international arena. But I still hope that the process of learning and gaining mutual understanding for all actors in this great saga will continue, and that ALMA will soon start producing science, even though more slowly and more expensively than I would have liked.

The Role of the National Science Foundation in U.S. Astronomy

When I started working at AUI in July 1999, I was confronted with many of the problems that ground-based U.S. astronomy has faced for decades in both radio and optical observatories. I have already commented in this chapter on the state of neglect in which I found NRAO.

In ground-based optical astronomy, the National Optical Astronomy Observatory has been less ambitious in constructing new facilities than could be expected from an observatory with a national charter. In fact, the National Optical Astronomy Observatory telescopes have been kept (perhaps deliberately) smaller than those already existing at the observatories of some individual universities, such as Caltech. The National Kitt Peak 4-m telescope, built in 1973, was smaller than the 5-m Palomar built in 1949. The response by the National Science Foundation to the Keck 10-m telescope built in 1993 was a 50 percent share in the two Gemini 8-m telescopes built as part of an international collaboration in 2000.

The National Science Foundation has chosen in practice not to provide optical observatories that could compete on the international scene, but instead to build second-tier observatories that are made available to the large number of observers who do not have access to private facilities. Is this policy the result of a conscious, high-level, political choice, or the result of a never-challenged propensity of the National Science Foundation bureaucrats to fund a large number of projects but not to seek excellence?

The problem for U.S. astronomy is that in Europe there has been a concerted and united effort to achieve supremacy in ground-based optical astronomy, which ESO has achieved with VLT/VLTI and apparently intends to maintain with the start of the 42-m European Large Telescope. The European effort has put an end to the supremacy that the United States had enjoyed in this field since the construction of the Mt. Wilson Observatory by George Ellery Hale a century ago.

As I have already discussed, the situation is much better in radio astronomy. NRAO has provided the entire community with VLA, VLBA, the Green Bank Telescope, and now ALMA (in collaboration with Europe and Japan)—facilities that are the best in the world. This success is due to the unity and cooperation of the radio astronomers themselves and not to the National Science Foundation's leadership following a clearly stated policy. It is not obvious to me, therefore, what will happen: Will there be a unified national effort to build the Square Kilometer Array, which is being proposed by the community? Will the National Science Foundation support the operations and data analysis effort necessary to fully utilize ALMA for scientific purposes?

There will be a splendid opportunity, in the forthcoming National Academy of Sciences decadal survey of astronomy for the second decade of this millennium, to fully discuss the foundation's plans and community expectations for major national endeavors in the context of international competition and/or cooperation. Such a clarification is essential if we hope to recover the ground lost in optical astronomy in the recent past.

TWENTY

First Loves and Last Words

Chandra: The Dream Comes True

After 36 years of proposals, work, cancellations, and restarts, on July 19, 1999, the Chandra X-Ray Observatory was finally launched. The 1.2-m x-ray telescope was placed into a highly elliptical orbit with an apogee of 121,279 km and a perigee of 27,539 km—the kind of orbit I had advocated in my paper in *Science* in 1987.[1] Its distance from Earth, one-third of the way to the moon, means that Chandra is outside the earth's Van Allen Belts most of the time and can spend 55 hours out of its 64-hour orbit obtaining useful, continuous observations. It is the largest and heaviest payload ever carried by the shuttle. To put it in its orbit, an inertial upper stage was required as well as an engine carried on the spacecraft itself (Plate 29).

The observatory has operated continuously for the past nine years without any serious degradation since its first light in 1999. Chandra's director Harvey Tananbaum and its project scientist Martin Weiskopf did a magnificent job, together with the principal investigator teams and the working group, of providing scientific guidance. Engineers and managers at the Marshall Space Flight Center and at TRW have given us a great platform from which to do science.[2] Chandra's sensitivity and resolution met or exceeded all expectations (Plates 30–32).[3]

For me, the use of Chandra time, 500,000 sec of my own and, thanks to Harvey Tananbaum's kindness, 500,000 sec of director's discretionary time, was a dream come true. I should note that attached to the discretionary time was the condition that the entire 10^6 sec of data would be made available to the entire community in less than a month from the last observation. Piero Rosati and I studied carefully how best to point our successive exposures, and we decided to always point to a single position in the sky. The different orientations of the spacecraft rotated the pitch with which we observed our field, but the point response function (which is a function of the distance from the telescope axis) remained the same for every source. Thus, we could distinguish between point and extended sources. We chose the position in the sky (right ascension 3 h 32 min 28 sec, declination –27 deg 48 min 30 sec; J 2000) as the one with the lowest intervening absorption. The sensitivity of the observatory in 10^6 sec was 5.5×10^{-17} erg/cm^2 · sec in the soft x-ray band (from 0.5 to 2 keV) and 4.5×10^{-16} erg/cm^2 · sec in the hard band (from 2 to 10 keV). Thus, in 40 years we had improved x-ray astronomy observations between one and ten billion times with respect to the first observation of Sco X-1, the same factor that measures the improvement from the naked eye to Hubble in the visible. I wonder sometimes how Tycho Brahe would have felt if he could have contributed to the development of Hubble and could have used it himself.

After many years of helping others do their science, it was great to be able to stare at my own data and let them flow through my fingers, as if panning for gold. I have always believed that contemplating data and keeping one's mind open to what nature is trying to tell us is the best way to discover the underlying phenomena, and just as it had occurred with Uhuru in the early 1970s and Einstein in the late 1970s, here was a chance to find something new by exploring a new region of discovery space. I must confess that I had at first approached the new data with a sort of arrogant conceit.

After all, I knew perfectly well that I would find the XRB to be made of individual point sources. Although I did not exactly know what objects they would be, I thought that by extrapolation from the work done with Einstein and ROSAT, we could confidently say that most of the objects would be galaxies with active nuclei or quasars. This in fact turned out to be the case, but nature, as usual, was richer and more inventive than I could imagine. Sometimes I think that nature is out there working busily to prepare samples of the most bizarre objects, just to reward whoever is curious enough to be looking.

Plate 33 shows the picture of our study region obtained by summing all of our exposures, for a total of 10^6 sec. All that is seen is a large number of discrete sources; their images are smaller in the center of the field of view, where the angular resolution of the telescope is best, and larger at the edges. As I discuss later in the chapter, most of the sources are supermassive black holes at cosmological distances from the earth. We detected in this field 346 sources altogether, corresponding to a sky density of about one source per arcmin^2. At this sensitivity, there are some 144 million sources in the sky, which is just what is needed to explain the uniformity of the XRB radiation observed with Uhuru and HEAO-1. The field of view of those detectors was large enough (on the order of square degrees) that the emission detected was the sum of tens of thousands of sources, yielding a pretty constant average (to the 1 percent level). The x-ray flux integrated over all of these sources, as well as brighter ones in other surveys, accounts for most (if not all) of the background, the ambiguity being in the precision of the measurements of the integrated background itself.

Rosati, Paolo Tozzi, and I still had to explain the famous spectrum discrepancy between AGNs and the background. This discrepancy had kept many astronomers busy explaining why point sources could not make up the background, and to reject the explanation for its origin that we had offered since 1979. We found the explanation by following two lines of inquiry.[4] We could not determine the spectrum for each of the sources individually, because the sources are very weak, and the faintest ones are detected with only eleven photons. However, summing up all the individual sources, we found the spectrum of the sum of the sources to be the same as that of the background.

The second line of inquiry was as follows: for each source, we could determine the ratio between high-energy and low-energy counts. Even with large errors, we could determine statistically that the spectrum of the sources changed with their intensity and became harder as the source intensity weakened. The explanation of the discrepancy became obvious: the sources that made up the hard part of the background could be detected by Chandra but not in previous surveys, such as the NRL (HEAO-1) survey, which were one hundred to one thousand times less sensitive than Chandra. Interpretations of these earlier surveys had generated the controversy.

Therefore, the big question of whether the background was made up of individual sources or due to diffuse emission was solved as I expected. Some remarks before continuing: There is no intellectual property in astronomy, and I had discussed what I intended to do with Chandra freely and openly. So it happened that, although I had worked on the problem of the XRB for years, the first Chandra results on the background were published by Richard Mushotzky of Goddard in early 2000, on the basis of thirty sources that he had observed for 10^4 sec. Niel Brandt and his colleagues of the University of Pennsylvania, the group that had built the CCD detector that we all used for these observations, published their results on the Chandra Deep Field North at about the same time as we published ours on the Chandra Deep Field South. The work by Niel Brandt and his collaborators and that of our group were mutually supportive in a long and serious research effort, which culminated in a collaboration in 1995.

Here I describe how our own (Rosati's, Tozzi's, and my) research progressed beyond this point. Once we had put to rest the controversy over the source of the XRB, the research became interesting for me in new ways: What were these discrete sources? How were they distributed in space? How luminous were they? And why were their spectra becoming harder as their intensities became weaker?

At this time (2002) we started using two sets of optical data: those we had obtained at ESO and new observations from Hubble. As I described in Chapter 17, we had carried out surveys of the Chandra Deep Field South with the FORS-1 camera on Antu (Unit Telescope 1 of VLT) at Paranal and also with SOFI on NTT and with the wide-field imager at the ESO / Max Planck Institute 2.2-m telescope on La Silla. The information from these surveys was used to identify the optical counterparts of the x-ray sources. Here the high angular resolution of Chandra (0.5 arcsec) comes into play; it permits us to securely identify the optical candidates with great accuracy both in the FORS-1 R images (Figure 20.1)[5] and then in the Hubble follow-up fields taken by Ethan Schreier and Anton Koekemoer (Figure 20.2).[6]

Once we made unambiguous identifications of x-ray sources with their optical counterparts, we plotted the ratio between the x-ray and the optical flux, which yields some information on the nature of the sources without knowing their distances (Figure 20.3). The dashed line on this graph represents the region where these values had happened to fall for the sources observed in ROSAT, which are typically identified with AGNs and quasars. The fact that the values measured by Chandra lie in approximately the same interval is a strong indication that most of our sources are of the same type.

However, Chandra's higher sensitivity pushes the x-ray fluxes detected to much lower values, and consequently the optical fluxes of the objects also become more faint, pushing the limits of sensitivity of both VLT and Hubble,

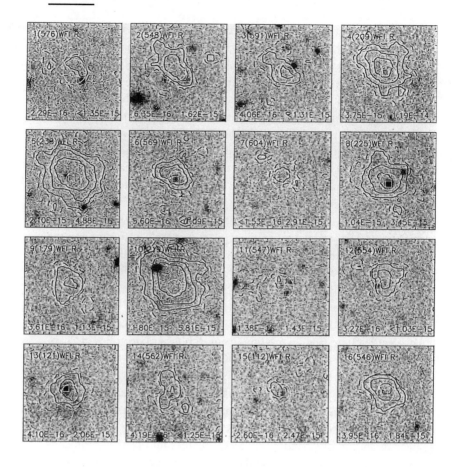

FIG. 20.1. *The identification of the x-ray sources obtained with the European Southern Observatory's Wide Field Imager. The x-ray isointensity contours are shown superimposed on the R band optical images.*

at R magnitudes greater than 27. Chandra is seeing x-ray sources so faint that it requires the best optical telescopes to see them in the visible. The graph also shows that we need to push to even fainter magnitudes to identify all the x-ray sources in the optical domain. At present, more than 98 percent have been identified in optical surveys, and Anton Koekemoer has used the Spitzer Observatory to identify the remaining seven sources in the infrared.

The ratio between x-ray and optical emissions is characteristic of the type of object that is being observed, and the objects to the left in the diagram have a lower ratio of x-ray to optical emissions. Such lower ratios indicate that they are either normal or star-forming galaxies. Thus, these findings open up a whole new field of observations in the study of star formation in galaxies.

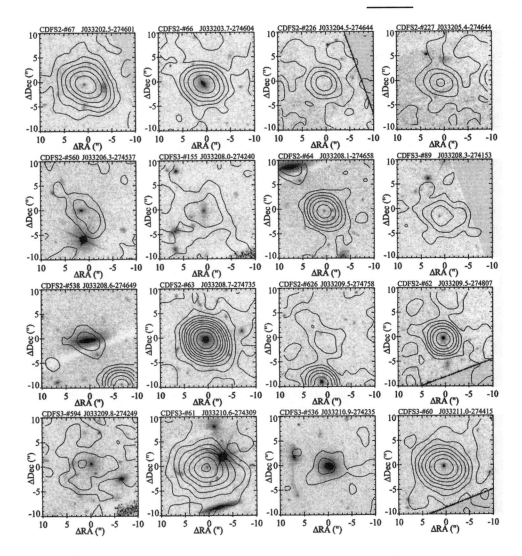

The next step after identification of the x-ray source with its optical counterpart is the measurement of the optical spectrum of the identified sources, a very difficult enterprise, given the faintness of the sources. Here the greater efficiency of the VLT compared to that of Keck came into play. Initially we were able to obtain spectra for only 88 sources, but these spectra were soon followed up by much more complete surveys of 288 sources with the FORS-1 and FORS-2 spectrographs on the Antu and Kueyen telescopes of VLT. This spectrographic information was complemented by photometric

FIG. 20.2. *More precise identification of the x-ray sources with Hubble than that shown in Figure 20.1. The higher resolution permits study of the source morphology.*

FIG. 20.3. *A plot of the x-ray versus optical flux. The various symbols correspond to different types of galaxies.*

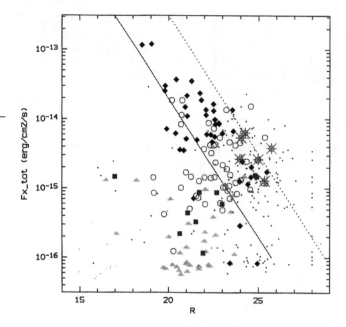

FIG. 20.3. *A plot of the x-ray versus optical flux. The various symbols correspond to different types of galaxies.*

distance determinations.[7] The combined results permitted us to plot the distribution in redshifts (which give an indication of distance) of the different types of sources, based on their optical characteristics. Figure 20.4 shows the distributions of Type I objects (AGNs plus quasars), Type II objects (absorbed AGNs plus quasars), and galaxies, as well as the total distribution. Knowing the distance, the intrinsic luminosity of these objects can be derived, which spans the range from 10^{42} erg/sec to 10^{45} erg/sec, thus placing them among the most luminous galaxies in the universe. Such high luminosities are explained by accretion onto supermassive black holes in the center of galaxies. Thus, Plate 33 is a picture of supermassive black holes.

A surprising aspect of the results is that these distributions do not at all fit the theoretical predictions of Roberto Gilli and others, which had been elaborated prior to Chandra's launch. These predictions required that most of the x-ray emissions would come from very distant sources, at redshifts of 2 or greater. This finding has caused a rethinking of the previously believed evolutionary path for active galaxies and quasars.

An additional finding with high-resolution spectra is that x-ray sources cluster in sheets at particular redshifts, as is the case for optical active galaxies. We were also able to study the question of whether the x-ray sources were pointlike or extended. The measurement of extent is a strong indication that

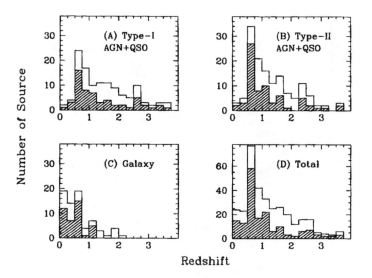

FIG. 20.4. *The redshift distribution of different types of galaxies observed in the Chandra Deep Field South.*

they could be either galaxies or clusters of galaxies. This observation was the basis for my proposal to search for clusters of galaxies with the Wide Field X-Ray Telescope, whose sad fate I described in Chapter 15. With Chandra, we could only explore 0.1 deg^2 (rather than the 10 deg^2 possible with the proposed wide-field telescope), but despite this difference by a factor of a hundred, we were able to determine whether the technique worked as anticipated. In our 0.1-deg^2 field, we found eighteen extended sources. Some of these appear to be groups or isolated galaxies, but at least three are identified as clusters of galaxies. An example is given in Figure 20.5, where the x-ray source is identified with a poor cluster (one with few member galaxies) at a redshift of 0.72. This work emphasizes not only the power of the technique, but also the fact that searching for clusters in the x-ray spectrum may reveal objects that cannot be found in optical surveys.

Finally, Maurizio Paolillo and I studied whether the x-ray sources vary with time. Using the fact that our 10^6-sec observations (which correspond to 10 days) of the Chandra Deep Field South were not all taken consecutively, but spread over about a year, and the fact that we always pointed in the same spot in the sky, we were able to show that at least 45 percent of the sources with good statistics (more than a hundred counts) exhibit significant variability on time scales ranging from a day to a year. The fraction of the sources found to be variable increases with flux, suggesting that those not observed to vary may just be too weak to give sufficient statistical accuracy. This logic suggests the possibility that more than 90 percent of all AGNs may vary on these time scales.[8]

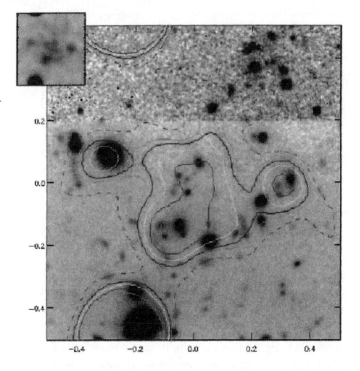

FIG. 20.5. *A poor cluster of galaxies discovered in x-rays at redshift 0.7 in the Chandra Deep Field South.*

This result is important, because it implies a source size of less than 1 light-year—a strong confirmation that the central engines of AGNs and quasars are supermassive black holes. Thus, what we are seeing in the Chandra Deep Field South is a field of black holes.

Most of the analysis on the x-ray data was carried out by Piero Rosati, Paolo Tozzi, and Andrew Zirm. The estimates of sky coverage at different sensitivities, which were critical in deriving the number densities of sources as a function of x-ray fluxes, were entirely their work. At one point, I found myself making an embarrassingly wrongheaded extrapolation for the number of expected sources, based on a misunderstanding of this quantity. They were patient enough to explain to me where I had gone wrong and to receive with grace my apologies for having been so hardheaded.

Having personally contributed to their realizations, it gave me great satisfaction to use three of the world's greatest telescopes—Chandra, Hubble, and VLT—to solve, after 40 years, a particular scientific problem. It was also a great joy to see Chandra finally in orbit and to be able to see my 30-year-old dream of a state-of-the-art imaging x-ray telescope become reality. I felt its success showed the validity of many of the points I had been making repeatedly in my research work: the nature of the XRB, the utility of high-resolution x-ray optics, the advantage of unmanned observatories in high-Earth orbits, and the use of x-ray observations to discover distant clusters.

There was also a sense of closure. As a young man of 28, I had invented the x-ray telescope; at 31, I had discovered the first x-ray star, Sco X-1, and the XRB. The nature of the x-ray binaries had become clear with Uhuru, the satellite that also discovered intergalactic plasmas in clusters in the early 1970s. Einstein, which flew in 1978, made the field of x-ray astronomy relevant to all astronomers. Now thanks to Chandra, the nature of the XRB was a problem that had almost been solved by 2002. It seemed as if my scientific life has come full circle (Plate 34).

I am grateful to live in this heroic era of astronomy and to have been able to participate and contribute to its evolution. Everything we know about the universe—its size and its age, the types of matter and energy that fill it, the formation of structures from the primeval chaos, the birth and evolution of stars and galaxies, and the great explosive events that end their lives—was learned in the past century. Not since the sixteenth and seventeenth centuries has there been such a profound revolution in astronomy. Astronomy is once again leading physics in posing the most profound questions on the nature of the universe we live in, on the unknown matter that makes up most of it, on the energies that drive its expansion. New physical laws and theories will have to provide answers to these riddles. We need another Newton or Einstein to bring about the new synthesis, which will fundamentally change our vision of the universe and ultimately produce as profound a change in our understanding of our place in it as occurred 400 years ago.

Some Comments on the NASA Space Program

I was quite fortunate to be involved in the space program at its beginning, when NASA was full of vitality, and I had great opportunities to carry out my most ambitious dreams as part of its rich program of scientific exploration, unparalleled in the world. However, over the past decades, significant problems have crept into both the space research and manned space exploration programs of NASA. Because I have worked in space research for more than 40 years, I have a powerful interest in the subject and feel a particular obligation to comment on it in the hope of contributing to a stronger program in the future.

Only here in the United States could I have been so critical of NASA in the past and yet be allowed to participate in and contribute to some of its greatest programs; I have also been recognized by NASA with its highest honors for my contributions. I hope I have been and will continue to be one of its most loyal critics, and that what I write here will be taken in the same spirit as in the past.

Space Science

I see several problems in scientific space research. One is that the space activities are not clearly separated from the manned space program activities. Management of the two requires, in my opinion, two very different styles. In space science, NASA should enable research rather than conduct it. There are strong scientific, technological, and managerial capabilities in the space science community at research and academic institutions. As in other scientific disciplines, these institutions should be allowed to take responsibility for scientific programs. NASA's insistence on hands-on, direct management of scientific space programs has had negative consequences.

The first such consequence is what I would call the "fascistization" of the science program, for lack of a better word. By this term I mean that the people charged with the administration of the science program have come to consider it their own program and to make scientific decisions without serious consultation with the scientific community at large to ensure that priorities are dictated by scientific interest. An example is the greatly increased program of research on extrasolar planets, stemming mainly from the personal interest of Daniel S. Goldin (a manager and engineer), NASA administrator from 1992 to 2005. This increased effort was welcomed by a minority of the astronomical community but happened without prioritization with respect to other fields in astronomy of much greater astrophysical interest.

In 2000 I was invited to attend a meeting of a National Research Council committee chaired by Norman Augustine, to review the advantages and disadvantages of transferring the National Science Foundation's astronomy responsibilities to NASA. During that meeting, I was rather surprised by the committee's ignorance of the progress made by European astronomers in ground-based optical astronomy and by the lack of strategic planning by the National Science Foundation to respond. I was even more surprised, however, by the comment made by the Goddard project scientist for the James Webb Space Telescope during a discussion of its instrument complement, that some instruments should not be considered because Ed Weiler, then head of the NASA astronomy program, did not like them; even more surprising to me was that nobody in the room considered this extraordinary.

Intermediate-level managers with modest scientific credentials seem to forget that their job is to provide opportunities for the scientific community and not to dispose of the science programs as if they owned them. Even more disheartening is the vindictive attitude toward NASA staff members who dare disagree; sometimes dissenting scientists outside NASA can expect a decrease in funding for the programs in which they are involved. What is appalling is the acceptance by the scientific community of this situation.

A serious problem is the ballooning cost of the scientific missions brought about by NASA's actions. For decades NASA has undertaken more missions than it can afford and has on the whole managed them quite poorly. Many overruns have occurred, and because of the oversubscribed agenda, every delay or overrun of one project has caused a domino effect of delays and overruns in others. I described a case in point in Chapter 9, where an overrun in the Viking program caused a 20-year delay and a huge increase in costs in LOXT. This program, which morphed into Chandra (much improved in technology over the original plan, although somewhat smaller in total telescope area, particularly at high energy), saw its launch date slip by more than two decades and its cost increase by an order of magnitude.

The other cause for ballooning costs is the insistence of NASA on managing scientific programs in house, with greater regard for the interests of the managing center than for efficient management. A clear example is the James Webb Space Telescope, whose cost has escalated from the initial NASA headquarters dream of $300 million to a reality of $3 billion, as executed by Goddard. The execution of this type of program has become a never-ending "workfare" program to support NASA centers and their subcontractors at exorbitant costs.

I have thought and still think that scientific research in space would be better served by a National Space Institute, modeled along the lines of the National Institutes of Health, that does only a small fraction of research in house. If this change could be brought about, there would be an opportunity to reduce costs by factors of two or three by better prioritization and management, thus enabling much more research to be done and simultaneously strengthening the technological preparation of a new generation of space scientists.

Manned Space Exploration

For the manned program of exploration, NASA must be given a rational, well-defined goal, no matter how long term, and NASA must be held accountable for progress toward its achievement. The overall goal should not change with every U.S. president or, worse yet, with every new NASA administrator. At the moment, the national goal appears to be manned exploration of the solar system, which I take to mean landings on neighboring planets.

At the risk of drawing fire from many quarters, I confess that I cannot find any sensible reason to go back to the moon. We proved we can do it. Considering the high cost of such undertakings, we ought at least to have a very good reason to do it again. The vague argument of this being a "necessary step on the way" was hollow for the space station and I think it will prove invalid for the moon as well. As to the scientific reasons for going to the moon, in my opinion they would not stand up in a competitive forum. Furthermore, supporting a moon colony—even the smallest of colonies—in that sterile environment would consume a great deal (if not most) of our resources with no appreciable gain, which is what happened with the space station.

I have completely different feelings about a manned mission to Mars. Certainly Mars exploration would be interesting in its own right. I also find the doomsday scenarios described by Martin Rees compelling enough that the idea of a repository for the seeds of human civilization is attractive.[9] It would still be attractive even if we do not exterminate one another in the near future. To begin with, given the current state of our technology, Mars is the only place we can dream of colonizing. Even so, it would clearly be an incredibly complex undertaking. Still there is nothing wrong in trying to bring a dream to reality. We could start studying the possibility of a Mars landing from many different points of view—development of flight vehicles, mission profiles, utilization of in situ resources, Mars habitats, long-term storage of biological specimens, and radiation protection, to name a few areas. To the extent that such a program could be funded incrementally and executed in a progression of clear, simple steps, it might receive a reasonable degree of national support. I only regret I will not be able to see it.

The Nobel Prize

In October 2002, shortly after finishing my research, I received a completely unexpected call. When the telephone rang at six o'clock in the morning, Mirella and I jumped up, worried that something had happened to our daughters. On the phone were several people whom I could not readily recognize, speaking to me one after another in such a way that, half asleep as I was, it was difficult for me to follow. I did, however, get the gist of the message, which was that I had been awarded the Nobel Prize in Physics for 2002 for x-ray astronomy, together with Raymond Davis Jr. and Masatoshi Koshiba for their work on neutrinos. I reassured my wife that everything was all right, and that either I was the victim of an elaborate joke or I had become a Nobel laureate.

I have often been asked whether I expected this recognition. The work for which I was being recognized was done prior to 1980. Even at that time, some overenthusiastic friends were assuring me, unasked, that they had learned through the grapevine that I would soon receive this honor. I knew that the discoveries in x-ray astronomy deserved some recognition, so the expectation was not entirely implausible.

It was my great fortune that I was able to ignore rumors and expectations. I had known and worked closely with two scientists, Giuseppe (Beppo) Occhialini and Robert W. Thompson, who, in my estimation, deserved the Nobel Prize but never received it. Beppo did not seem bothered by this outcome and continued his life's work with his usual enthusiasm. Thompson, on the other hand, was devastated by the lack of recognition, and that put an end to his scientific work, and practically to his life. I was sure that I wanted to follow the example of Occhialini.

I became convinced that this award could not be given to all the people who deserved it, the only consolation being that it was seldom given to those who did not merit it. Whether one did or did not receive it depended on many circumstances, including politics and the timeliness of the research. In time I came to believe that the opportunity for me had passed, and I certainly was not about to waste my life pining for it. So the award was indeed a surprise.

That morning from 6:30 on, Mirella and I were besieged by phone calls and reporters showing up at our door. Letters of congratulation came from all over the world, and I tried to answer as many as I could. Because AUI headquarters was staffed with only a few people, it was difficult to cope with it all, even with some temporary help.

My children Anna Lee and Guia and my grandchildren Alexandra and Colburn loved it, and the entire family worried about the appropriate wardrobe necessary to participate in all of the activities that surround the award ceremony. I was happy to be able to invite Herbert Gursky, Ethan Schreier, and Harvey Tananbaum, the people most closely involved in my x-ray work, and their wives to take part with us in the ceremonies in Stockholm in early December (Figures 20.6 and 20.7). I was also pleased and honored to meet so many other Nobel winners in different disciplines.

Beyond the pomp and circumstance, I was really touched by the atmosphere of great warmth that seemed to pervade Sweden at that time of the

FIG. 20.6. *The author and his wife, Mirella, in Stockholm, 2002.*

year. The fact that it was centered on a celebration of reason, the arts, and the sciences made it extraordinary. As I was the youngest of the Nobel laureates in physics for the year, my elders took advantage of me and asked that I give the thank-you speech at the December 10 banquet, attended by 1,200 guests. I had little time to prepare and decided to give it from the heart. It still accurately represents my feelings at the time:

Your Majesties, Your Royal Highnesses, Ladies and Gentlemen:

A dinner speech is harder for me to give than a lecture. Therefore I will simply share with you some of the feelings and thoughts I have had about this Nobel week.

At first, upon learning that we have become Nobel laureates there is a feeling of personal pride which we share with family and friends. Then the realization hits us that the work for which we are honored is the result of the cooperative effort of many, many people over the years. Finally we feel a sense of continuity with the quest, initiated thousands of years ago, for an understanding of the cosmos in which we live. While enormous strides have been made in the last decades, some of the most fundamental questions still have no answers, and so the quest goes on.

But other feelings and thoughts have come to the fore since reaching Stockholm. The warmth and kindness we have found here is expressed sometime with solemnity and sometimes with whimsy and humor. We all feel we are participating in a great festival of lights. This week we are celebrating not only the light of reason, but also the light of poetry and art, the light of human solidarity and finally the lights of Santa Lucia.

It is a splendid reaffirmation of the human spirit and of our capacity to give meaning and warmth to life.

We are very grateful to you for the privilege of participating in these events. Thank you.

I was told that life changes after you receive the Nobel Prize. In fact I have received invitations to give speeches everywhere in the world and to participate in all kinds of initiatives. I decided early on that one should choose what one wants to do, and that one does not have to accept all invitations nor sign all of the many appeals for this or that cause that are circulated. I also did not want to become a professional Nobel Prize winner, a specimen that I have often encountered, who is supposed to know everything about everything. I decided that I would agree to speak only about subjects on which I felt com-

FIG. 20.7. *Ethan Schreier, the author, Herbert Gursky, and Harvey Tananbaum (back row) with the author's grandchildren, Alexandra and Colburn, at the Nobel ceremony, Stockholm, 2002.*

373

petent because of firsthand experience, and in settings where I could do some good.

I do enjoy participating in initiatives aimed at young people. I found the Lindau Nobel Lectures (held in Lindau, Germany) for five hundred students from all over the world stimulating, and I have enjoyed meeting with some of my colleagues there in 2004 and 2005. I accepted an invitation to lecture in Bangkok, Thailand, in January 2005 as part of a program hosted by the International Peace Foundation, where I hope my discussions on the management of large projects may have been as useful as the scientific talks.

I retired from AUI at the end of 2004, although I still consulted for them from time to time until 2006, and at present I hold the honored position of university professor at the Johns Hopkins University. In this capacity I come into contact with younger colleagues and share in their successes and in the intoxicating pleasure of intellectual combat with peers. I even hope that one day I will see some of the work I left unaccomplished brought to fruition by them and will help them bring this about. I would like to pass to them my quest for excellence, and the joy of pursuing great science in an honest and open way.

There is one more aspect of scientific research and its diffusion to the general public that I mention in closing. I have done my very best to share with the general public the fruits of discoveries in astronomy. In the United States, Hubble is a household word, and its pictures and descriptions are loved by one and all. I think ESO has been doing an excellent job of promoting educational activities throughout Europe. Yet I wonder whether these efforts are just providing entertainment. I often feel that we intellectuals have essentially failed in our job of creating a more enlightened society. While our society relies increasingly on technology, based on science and reason, the effects on society of teaching rational thinking are not at all obvious, and a wave of fundamentalist revival has swept the entire world. There is a chilling description in *Fateless* by Imre Kertesz, who won the Nobel Prize in Literature in 2002, of a tree planted by Goethe more than a hundred years ago near Weimar, which in the 1940s found itself in the middle of Buchenwald. The contrast between the brutality of the surroundings and the enlightened tradition that inspired the planting of the tree could not be sharper. Although I am doubtful about how successfully one individual can counter these deep and pervasive currents, I have pledged myself to do what I can still do in my life to spread the light of reason.

Acronyms and Abbreviations

AFCRL	Air Force Cambridge Research Laboratories
AGN	active galactic nucleus
ALMA	Atacama Large Millimeter and Submillimeter Array
AMBER	Astronomical Multiple-Beam Recombiner
AMOS	Advanced Mechanical and Optical Systems
ANS	Astronomische Nederlandse Satelliet
AOSO	Advanced Orbiting Solar Observatory
APL	Applied Physics Laboratory
AS&E	American Science and Engineering
AUI	Associated Universities, Inc.
AURA	Association of Universities for Research in Astronomy
AXAF	Advanced X-Ray Astrophysics Facility
CCD	charge-coupled device
CERN	European Organization for Nuclear Research
CFA	Center for Astrophysics (Harvard-Smithsonian)
CNRS	Centre National de la Recherché Scientifique
COSTAR	Corrective Optics Space Telescope Axial Replacement
DOD	Department of Defense
DOE	Department of Energy

ECF	European Coordinating Facility (ESO)
EGG	evaporating gaseous globule
ESA	European Space Agency
ESO	European Southern Observatory
EVLA	Expanded Very Large Array
FGS	fine-guidance sensor
FOC	faint-object camera
FORS	Focal Reducer Low-Dispersion Spectrograph
FOS	faint-object spectrograph
FPCS	focal plane crystal spectrometer
GHRS	Goddard high-resolution spectrometer
HEAO	High Energy Astronomical Observatory
HRI	high-resolution imager
HSP	high-speed photometer
ICBM	intercontinental ballistic missile
IPC	imaging proportional counter
IRAF	Image Reduction and Analysis Facility
IRAM	Institut de Radioastronomie Millimetrique
ISAAC	Infrared Spectrometer and Array Camera
JPL	Jet Propulsion Laboratory
LMXRB	low-mass x-ray binary
LOXT	Large Orbiting X-Ray Telescope
LRL	Lawrence Radiation Laboratory
LSA	Large Southern Array
MACAO	Multi-Application Curvature Adaptive Optics
MIDAS	Munich Image Data Analysis System
MIDI	Mid-Infrared Interferometer
MIT	Massachusetts Institute of Technology
MMA	Millimeter and Submillimeter Array
NAOJ	National Astronomical Observatory of Japan
NAS	National Academy of Sciences
NRAO	National Radio Astronomical Observatory
NRL	Naval Research Laboratory
NTT	New Technology Telescope
OAO	Orbiting Astronomical Observatory
OMB	Office of Management and Budget
OSO	Orbiting Solar Observatory
OWL	Overwhelmingly Large Telescope
PERT	Program Evaluation and Reporting Technique

POGO	Polar Orbiting Geophysical Observatory
PRIMA	Phase-Referenced Imaging and Micro-arcsecond Astrometry
ROSAT	Roentgen Satellite
SAS	Small Astronomy Satellites
SOFI	Son of ISAAC
SSS	solid state spectrometer
SUSI	Superb Seeing Imager
STScI	Space Telescope Science Institute
TDRSS	Tracking and Data Relay System
TIROS	Television and Infrared Observational Satellite
URA	University Research Association
UVES	Ultraviolet High-Resolution Spectrograph
VINCI	VLT Interferometer Commissioning Instrument
VLA	Very Large Array
VLBA	Very Long Baseline Array
VLT	Very Large Telescope
VLTI	Very Large Telescope Interferometer
WBS	Work Breakdown Structure
WF/PC	wide field / planetary camera
XMM	X-Ray Multi-Mirror
XRB	x-ray background

Notes

ONE: My Italian Roots

1. L. La Rosa and E. La Rocca, "Oscar Chisini." Wikipedia entry in the section contributed by the School of Mathematics and Statistics, University of St. Andrews, Scotland.

2. R. Giacconi, A. Lovati, A. Mura, and C. Succi. 1956. "Osservazioni preliminary sullo sviluppo della cascata nucleare in piombo prodotta dalla radiazione cosmica a 3500 metri." *Il Supplemento del Nuovo Cimento* series X, 4:892–895.

3. E. Fiorini, R. Giacconi, and C. Succi. 1957. "Una nuova camera di Wilson di grandi dimensioni." *Il Nuovo Cimento* series X, 6:355.

4. R. W. Thompson, A. V. Buskirk, L. R. Etter, C. J. Karmak, and R. H. Rediker. 1953. "The disintegration of the V^0 particle." *Physical Review* 90:329.

TWO: New World

1. R. W. Thompson, J. R. Burwell, and R. W. Huggett. 1956. "The θ^0 meson." *Il Supplemento del Nuovo Cimento* series X, 4:286–318.

2. Thompson et al., 1956.

3. Thompson et al., 1956.

4. R. Giacconi, W. Blum, and G. T. Reynolds. 1959. "Detection of high-energy mu mesons by an air Cerenkov counter." *Il Nuovo Cimento* series X, 11:102–107.

5. R. Giacconi, H. Gursky, and A. Hendel. 1959. "A search for a mass 500 m_e particle." Annual Meeting of the American Physical Society, Session X A9, Washington, D.C. American Physical Society vol. 4-289.

THREE: Introducing X-Ray Astronomy

1. National Academy of Sciences. 1977. *The National Academy of Sciences: First hundred years.* Washington, D.C.: National Academy of Sciences, p. 555.

2. L. Goldberg. 1959. *Summary of the report from the Committee on Optical and Radio Astronomy.* Washington, D.C.: National Academy of Sciences, Space and Science Board, p. 5.

3. R. F. Hirsh. 1979. "Science, technology and public policy: The case of x-ray astronomy, 1959 to 1978." PhD thesis, University of Wisconsin, Madison.

4. A. H. Compton. 1965. "X-rays as a branch of optics." Nobel Prize lecture, December 12, 1927. Amsterdam: Elsevier.

5. S. L. Mandel'shtam and A. I. Efremov. 1958. "Research on shortwave solar ultraviolet radiation." In *Russian literature of satellites.* Moscow: Academy of Sciences of the USSR, pp. 47–65. Translation by International Physical Index, New York.

6. R. Giacconi and B. B. Rossi. 1960. "A 'telescope' for soft x ray astronomy." *Journal of Geophysical Research* 65:773.

7. R. Giacconi, G. W. Clark, and B. B. Rossi. 1960. "A brief review of experimental and theoretical progress in x ray astronomy." Document AS&E-TN-49. Cambridge, Mass.: American Science and Engineering.

8. R. Giacconi. 1960. "An instrument for gathering x-ray flux." In *Proceedings of the Conference on X-Ray Astronomy at the Smithsonian Astrophysical Observatory,* ed. A. I. Berman. Cambridge, Mass.: Smithsonian Astrophysical Observatory, p. 51.

FOUR: The First Celestial X-Ray Source

1. H. Wolter. 1952. "Spiegelsysteme streifenden Einfalls als abbildende Optiken für Röntgenstrahlen." *Annalen der Physik* 445(1): 94–114.

2. R. Giacconi, H. Gursky, F. Paolini, and B. B. Rossi. 1962. "Evidence for x rays from sources outside the solar system." *Physical Review Letters* 9:442.

3. H. Gursky, R. Giacconi, F. Paolini, and B. B. Rossi. 1963. "Further evidence for the existence of galactic x-rays." *Physical Review Letters* 11:530–535.

4. S. Bowyer, E. T. Byram, T. A. Chubb, and H. Friedman. 1964. "X-ray sources in the galaxy." *Nature* 201:1307.

5. R. F. Hirsh. 1979. "Science, technology and public policy: The case of x-ray astronomy, 1959 to 1978." PhD thesis, University of Wisconsin, Madison, p. 96.

6. P. C. Fisher and A. J. Meyerott. 1964. "Stellar x-ray emission." *Astrophysical Journal* 139:123.

7. S. Bowyer. 1964. "An alternative interpretation of the paper 'Stellar x-ray emission' by Fisher, P. C., and Meyerott, A. J." *Astrophysical Journal* 140:820b.

8. P. C. Fisher and A. J. Meyerott. 1964. "A response to S. Bowyer comments." *Astrophysical Journal* 140:821f.

FIVE: Plans and Progress in X-Ray Astronomy

1. R. Giacconi and H. Gursky. 1963. "An experimental program of extra-solar x-ray astronomy." September. Proposal by AS&E to NASA. Document AS&E-449.

2. S. Bowyer, E. T. Byram, T. A. Chubb, and H. Friedman. 1964. "Lunar occultation of x-ray emission from the Crab Nebula." *Science* 146:912–917.

3. R. Giacconi. 1965. "A program of high angular resolution studies of celestial x-ray sources using sounding rockets." Proposal by AS&E to NASA. Document AS&E-1070, p. iv.

4. R. Giacconi, H. Gursky, and J. Waters. 1965. "Spectral data from the cosmic x-ray sources in Scorpius and near the galactic center." *Nature* 207:572–575.

5. M. Oda, G. Clark, G. Garmire, M. Wada, R. Giacconi, H. Gursky, and J. Waters. 1965. "Angular sizes of the x-ray sources in Scorpio and Sagittarius." *Nature* 205:554.

6. H. Gursky, R. Giacconi, P. Gorenstein, J. Waters, M. Oda, H. Bradt, G. Garmire, and B. Sreekantan. 1966. "Measurement of the angular size of the x-ray source Sco X-1." *Astrophysical Journal* 146:310.

7. A. R. Sandage, P. Osmer, R. Giacconi, P. Gorenstein, H. Gursky, J. R. Waters, H. Brant, G. Garmire, B. Sreekantan, M. Oda, K. Osawa, and J. Jugaku. 1966. "On the identification of Sco X-1." *Astrophysical Journal* 146:316.

8. G. R. Burbidge. 1967. "Theoretical ideas concerning x-ray sources." In *Radio astronomy and galactic systems. Proceedings of the IAU Symposium 31,* ed. G. Burbidge. New York: Academic Press, pp. 465–466.

SIX: The First Orbiting X-Ray Observatory

1. R. Giacconi, P. Gorenstein, H. Gursky, E. M. Kellogg, and H. Tananbaum. 1964. "An x-ray explorer to survey galactic and extragalactic sources." April. Proposal by AS&E to NASA. Document AS&E-2149.

2. M. Buckley. 2002. "An x-ray vision: APL-built satellite collects Nobel Prize-winning data." *APL News,* Winter, p. 8.

3. E. Mathieson and P. W. Sanford. 1963. "Pulse shape discrimination in proportional counters." In *Proceedings of the International Symposium on Nuclear Electronics,* Paris, p. 65.

4. P. Gorenstein and S. Mickiewicz. 1968. "Reduction of cosmic background in an x-ray proportional counter through risetime discrimination." *Review of Scientific Instruments* 39:815.

5. R. Giacconi, P. Gorenstein, H. Gursky, E. M. Kellogg, and H. Tananbaum. 1968. "The X-Ray Explorer experiment." Document AS&E 2149. Cambridge, Mass.: American Science and Engineering, p. 19.

6. R. Giacconi, E. M. Kellogg, P. Gorenstein, H. Gursky, and H. Tananbaum. 1971. "An x-ray scan of the galactic plane from Uhuru." *Astrophysical Journal Letters* 165:L27–L35.

7. H. Tananbaum, E. M. Kellogg, H. Gursky, S. Murray, E. Schreier, and R. Giacconi. 1971. "Measurement of the locations of the x-ray sources Cygnus X-1 and Cyg X-2 from Uhuru." *Astrophysical Journal* 165:L37.

8. H. Gursky, E. Kellogg, C. Leong, H. Tananbaum, and R. Giacconi. 1971. "Detection of x-rays from the Seyfert galaxies NGC 1275 and NGC 4151 from the Uhuru satellite." *Astrophysical Journal* 165:L43–L48.

9. E. M. Kellogg, H. Gursky, C. Leong, E. Schreier, H. Tananbaum, and R. Giacconi. 1971. "X-ray observation of the Virgo Cluster, NGC 5128, and 3C273 from the Uhuru satellite." *Astrophysical Journal* 165:L49.

SEVEN: Breakthrough

1. W. H. Lewin, G. R. Richer, and J. E. McClintok. 1971. "X-rays from a new variable source GX 1+4." *Astrophysical Journal* 169:L117.

2. E. T. Byram, T. A. Chubb, and H. Friedman. 1966. "Cosmic x-ray sources, galactic and extragalactic." *Science* 152:66–71. J. W. Overbeck and H. Tananbaum. 1968. "Time variations in Scorpius X-1 and Cygnus XR-1." *Astrophysical Journal* 153:899.

3. M. Oda, P. Gorenstein, H. Gursky, E. M. Kellogg, E. Schreier, H. Tananbaum, and R. Giacconi. 1971. "X-ray pulsations from Cygnus X-1 observed from Uhuru." *Astrophysical Journal* 166:L1.

4. D. J. Cooke and K. A. Pounds. 1971. "X-ray astronomy—Some southern hemisphere sources." *Nature* 229:144.

5. R. Giacconi, H. Gursky, E. M. Kellogg, E. Schreier, and H. Tananbaum. 1971. "Discovery of periodic x-ray pulsations in Cen X-3 from Uhuru." *Astrophysical Journal* 167:L67.

6. E. Schreier, R. Levinson, H. Gursky, E. M. Kellogg, H. Tananbaum, and R. Giacconi. 1972. "Evidence for the binary nature of Centaurus X-3 from Uhuru x-ray observations." *Astrophysical Journal* 172:L79. H. Tananbaum, H. Gursky, E. M. Kellogg, R. Levinson, E. Schreier, and R. Giacconi. 1972. "Discovery of a periodic pulsating binary source in Hercules from Uhuru." *Astrophysical Journal* 174:L134.

7. Schreier et al., 1972.

8. Tananbaum et al., 1972.

9. R. Rothschild, E. A. Bolt S. S. Holy, and P. J. Serlemitzos. 1974. "Millisecond temporal structure in Cygnus X-1." *Astrophysical Journal* 189:L13.

10. S. Rappaport, W. Zaumen, and R. Doxsey. 1971. "On the location of Cygnus X-1." *Astrophysical Journal* 168:L17.

11. L. Braes and G. Miley. 1971. "Detection of radio emission from Cygnus X-1." *Nature* 232:246. R. M. Hjellming and C. M. Wade. 1971. "Radio emission from x-ray sources." *Astrophysical Journal* 168:L21.

12. B. Webster and P. Murdin. 1972. "Cygnus X-1: A spectroscopic binary with a heavy companion." *Nature* 235:37. C. Bolton. 1972. "Identification of Cygnus X-1 with HDE 226868." *Nature* 235:271.

13. C. Rhoades and R. Ruffini. 1974. "Maximum mass of a neutron star." *Physical Review Letters* 32:324.

14. E. M. Kellogg, S. Murray, R. Giacconi, H. Tananbaum, and H. Gursky. 1973. "Clusters of galaxies with a wide range of x-ray luminosities." *Astrophysical Journal* 185:L13–L16.

15. H. Gursky, A. Solinger, E. M. Kellogg, S. Murray, H. Tananbaum, R. Giacconi, and A. Cavaliere. 1972. "X-ray emission from rich clusters of galaxies." *Astrophysical Journal* 173:L99.

16. T. Matilsky, H. Gursky, E. M. Kellogg, H. Tananbaum, S. Murray, and R. Giacconi. 1973. "The number intensity distribution of X-ray sources observed by Uhuru." *Astrophysical Journal* 181:753.

EIGHT: Constructing X-Ray Telescopes

1. L. G. Parrat and C. F. Hempstead. 1954. "Anomalous dispersion and scattering of x-rays." *Physical Review* 94:1593.

2. A. H. Compton and S. K. Allison. 1963. "Dispersion theory applied to x-rays." In *X-rays in theory and experiment.* New York: D. Van Nostrand, pp. 263–311.

3. R. Giacconi, W. P. Reidy, G. S. Vaiana, L. P. Van Speybroeck, and T. F. Zehnpfennig. 1969. "Grazing incidence telescopes for x-ray astronomy." *Space Science Reviews* 9(1):3–57.

4. L. G. Parrat. 1954. "Surface studies of solids by total reflection of x-rays." *Physical Review* 95:359.

5. R. Giacconi and B. B. Rossi. 1960. "A 'telescope' for soft x-ray astronomy." *Journal of Geophysical Research* 65(2):773.

6. H. Wolter. 1952. "Verallgemeinerte Schwarzschildsche Spiegelsysteme strifender Reflexion als Optiken für Röntgenstrahlen." *Annalen der Physik* 445(4):286–295.

7. R. Giacconi, N. F. Harmon, R. F. Lacey, and Z. Szilagy. 1964. "An x-ray telescope." December. Report by American Science and Engineering submitted to NASA. NASA CR-41.

8. P. A. Sturrock. 1955. *Static and dynamic electron optics.* Cambridge: Cambridge University Press.

9. R. Giacconi, W. P. Reidy, T. F. Zehnpfennig, J. C. Lindsay, and W. S. Muney. 1965. "Solar x-ray image obtained using grazing incidence optics." *Astrophysical Journal* 142:1274.

10. Giacconi et al., 1965.

11. F. R. Paolini, R. Giacconi, O. Manly, W. P. Reidy, G. S. Vaiana, and T. F. Zehnpfennig. 1968. "Preliminary results from the AS&E spectroheliograph on OSO-IV." *Astronomical Journal* 73:72.

12. G. S. Vaiana, W. P. Reidy, T. F. Zehnpfennig, and R. Giacconi. 1968. "X-ray structures of the sun during the importance 1 N flare of 8 June 1968." *Science* 161:564–567.

13. G. S. Vaiana and R. Rosner. 1978. "Recent advances in coronal physics." In *Annual Review of Astronomy and Astrophysics.* Palo Alto, Calif.: Annual Reviews, pp. 393–428.

14. National Academy of Sciences. 1966. *Space research directions for the future.* Report of a study by the Space Science Board at Woods Hole, H. Friedman, chair. NAS publication 1403. Washington, D.C.: National Academy of Sciences.

15. National Aeronautics and Space Administration. 1969. *A long-range program in space astronomy.* Position paper of the Astronomy Mission Board. NASA SP-213. Washington, D.C.: Astronomy Mission Board.

16. Consortium scientists of HEAO-3. 1969. "Preliminary study: Telescopes and scientific subsystems for a High Energy Astronomy Observatory." August. Preliminary report prepared for NASA.

NINE: Plans for Space and Realities on the Ground

1. G. H. Hale. 1928. "The possibilities of large telescopes." *Harper's Magazine,* April, pp. 639–646.

2. American Science and Engineering. AS&E proposal for LOXT, volumes I, II, and III. R. Giacconi Archives, Smithsonian Institution, Washington, D.C.

3. R. Giacconi, P. Gorenstein, S. S. Murray, E. Schreier, F. Steward, H. Tananbaum, W. H. Tucker, and L. Van Speybroeck. 1981. "The Einstein Observatory and future x-ray telescopes." In *Telescopes for the 1980s,* ed. G. Burbidge and A. Hewitt. Annual Reviews Monograph. Palo Alto, Calif.: Annual Reviews, pp. 195–278. R. Giacconi, G. Branduardi, U. Briel, A. Epstein, D. Fabricant, E. Feigelson, W. Forman, P. Gorenstein, J. Grindlay, H. Gursky, F. R. Harnden, J. P. Henry, C. Jones, E. Kellogg, D. Kock, S. S. Murray, E. Schreier, F. Seward, H. Tananbaum, K. Topka, L. Van Speybroeck, S. S. Holt, R. H. Becker, E. A. Boldt, P. J. Serlemitzoa, G. Clark, C. Canazares, T. Markert, R. Novick, D. Helfand, and K. Long. 1979. "The Einstein HEAO-2 x-ray observatory." *Astrophysical Journal* 230:540–550.

TEN: The Einstein Results

1. M. Elvis, ed. 1990. *Imaging x-ray astronomy, a decade of Einstein achievements.* Cambridge: Cambridge University Press.

2. J. Linsky. 1990. "Einstein and stellar sources." In Elvis, 1990, p. 39. G. S. Vaiana. 1990. "X-ray emission from stars: A sharper and deeper view of our galaxy." In Elvis, 1990, p. 61.

3. F. Seward. 1990, "Einstein observations of supernova remnants." In Elvis, 1990, p. 319.

4. H. Tananbaum, Y. Avni, G. Branduardi, M. Elvis, G. Fabbiano, E. Feigelson, R. Giacconi, J. P. Henry, J. P. Pye, A. Soltan, and G. Zamorani. 1979. "X-ray studies of quasars with the Einstein observatory." *Astrophysical Journal* 234:L7–L13.

5. E. Schreier, E. Feigelson, J. Delvaille, R. Giacconi, J. Grindlay, and D. A. Schwartz. 1979. "Einstein observations of the x-ray structure of Cen A—Evidence for the radio-lobe energy source." *Astrophysical Journal* 234:L39–L43.

6. S. Holt. 1990. "Spectra of nonthermal sources." In Elvis, 1990, p. 137.

7. H. Gursky, E. M. Kellogg, S. Murray, C. Leong, H. Tananbaum, and R. Giacconi. 1971. "A strong x-ray source in the Coma Cluster observed by Uhuru." *Astrophysical Journal Letters* 169:L81. E. M. Kellogg, H. Gursky, H. Tananbaum, R. Giacconi, and K. Pounds. 1972. "The extended x-ray source at M 87." *Astrophysical Journal Letters* 174:L65.

8. E. M. Kellogg. 1974. "Extragalactic x-ray sources." In *X-ray astronomy*, ed. R. Giacconi and H. Gursky. Dordrecht and Boston: D. Reidel, pp. 321–358.

9. R. J. Mitchell, J. L. Culhane, P. J. Davidson, and J. C. Ives. 1976. "ARIEL 5 observations of the x-ray spectrum of the Perseus Cluster." *Monthly Notices of the Royal Society* 176:1–29. P. J. Serlemitzos, B. W. Smith, E. A. Boldt, S. S. Holt, and J. A. Swank. 1977. "X-ray radiation from clusters of galaxies." *Astrophysical Journal Letters* 211:L63.

10. J. P. Henry, G. Branduardi, U. Briel, D. Fabricant, E. Feigelson, S. Murray, A. Soltan, and H. Tananbaum. 1979. "Detection of x-ray emission from distant clusters of galaxies." *Astrophysical Journal Letters* 238:L15.

11. W. Forman and C. Jones. 1990. "Hot gas in clusters of galaxies." In *Clusters of galaxies*, ed. W. R. Oegerle, M. J. Fitchett, and L. Danly. STScI Symposium Series. Cambridge: Cambridge University Press, pp. 257–277.

12. R. Giacconi and R. Burg. "Cluster research with x-ray observations." In *Clusters of galaxies*, ed. W. R. Oegerle, M. J. Fitchett, and L. Danly. STScI Symposium Series. Cambridge: Cambridge University Press, pp. 377–395.

13. G. S. Vaiana and S. Sciortino. 1986. "X-ray activity." *Advances in Space Research* 6:99.

ELEVEN: Transitions

1. B. B. Rossi and B. M. Belli. 1974. *Astronomia in raggi X: lezioni tenute nel febbraio e marzo 1972.* Bibliographic code: 1974QB472.R67. Rome: Accademia nazionale dei Lincei.

2. S. Freud. "Letter to Fliess, Vienna, February 1, 1900, IX, Berggasse 19." In *The complete letters of Sigmund Freud to Wilhelm Fliess, 1887–1904*, ed. and trans. J. M. Masson. Cambridge, Mass.: Belknap Press of Harvard University Press, pp. 397–398.

3. R. Giacconi and H. Tananbaum. 1976. "Study of the 1.2-meter x-ray telescope national space observatory." April. Smithsonian Astrophysical Observatory Proposal to NASA. P605-4-76.

4. Space Science Board of the National Academy of Sciences. 1976. "Institutional arrangements for the Space Telescope." Report from the Space Science Board meeting, July 19–30, Woods Hole, Mass.

TWELVE: The Hubble Space Telescope
and the Space Telescope Science Institute

1. Space Science Board of the National Academy of Sciences. 1976. "Institutional arrangements for the Space Telescope." Report from the Space Science Board meeting, July 19–30, Woods Hole, Mass.

2. R. W. Smith. 1989. *The Space Telescope.* Cambridge: Cambridge University Press.

3. Smith, 1989, p. 343.

4. John Teem, 1981 telephone conversation with John Naugle, as related to the author.

5. Smith, 1989.

6. C. R. O'Dell. 1981. "The Space Telescope." In *Telescopes for the 1980s,* ed. G. Burbidge and A. Hewitt. Annual Reviews Monograph. Palo Alto, Calif.: Annual Reviews, pp. 129–193.

7. Smith, 1989, Appendix 3.

THIRTEEN: Paradigm Shifts

1. R. W. Smith. 1989. *The Space Telescope.* Cambridge: Cambridge University Press, pp. 346–347.

2. M. Capaccioli, E. V. Held, H. Lorenz. G. M. Richter, and R. Ziener. 1988. "Application of an adaptive filter to surface photometry of galaxies. I—The method tested on NGC 3379." *Astronomische Nachrichten* 309(2):69–80.

3. R. L. White, M. Postman, and M. G. Lattanzi. 1992. "Compression of the guide star digitized Schmidt plates." In *Proceedings of the Conference on "Digitised Optical Sky Surveys,"* ed. H. T. MacGillivray and E. B. Thomson. Dordrecht: Kluwer Academic, p. 167.

4. T. L. Friedman. 2005. *The world is flat: A brief history of the twenty-first century.* New York: Farrar, Straus and Giroux.

FOURTEEN: The Space Telescope Science Institute

1. R. Giacconi. 1992. "Annual report to the AURA Board of Directors." March. Baltimore: Space Telescope Science Institute.

2. C. J. Burrows, J. A. Holtzman, S. M. Faber, P. Y. Bely, H. Hasan, C. R. Linds, and D. Schroeder. 1991. "The imaging performance of the Hubble Space Telescope." *Astrophysical Journal Letters* 369(2):L21–L25.

3. R. A. Brown and H. C. Ford, eds. 1991. *A strategy for recovery.* Baltimore: Space Telescope Science Institute.

4. R. A. Brown and R. Giacconi. 1987. "New directions for space astronomy." *Science* 238: 617–619.

FIFTEEN: Science at the Space Telescope Science Institute

1. C. Burrows, R. Burg, and R. Giacconi. 1992. "Optimal grazing incidence optics and its application to wide field imaging." *Astrophysical Journal Letters* 392:760.

2. U. Grothkopf, B. Leibundgut, D. Macchetto, J. P. Madrid, and C. Leitherer. 2005. "Comparison of science metrics among observatories." ESO *Messenger,* March, pp. 45–49.

SIXTEEN: The European Southern Observatory

1. H. van der Laan. 1992. "The idea of the European Southern Observatory." ESO *Messenger,* December, p. 3.

2. R. N. Wilson. 2003. "The history and development of the ESO active optics system." ESO *Messenger,* September, pp. 2–9.

3. ESO. 1999. "ESO-wide review." Collected presentations, February 1–4, 1999, books one and two. R. Giacconi Archives, Smithsonian Institution, Washington, D.C.

4. ESO, 1999, 9:30, February 2, book one and 14:25, February 3, book two.

5. M. R. Rosa. 1999. "Predictive calibration and forward data analysis of spectroscopic data." In *Optical and infrared spectroscopy of circumstellar matter.* Astronomical Society of the Pacific (ASP) Conference Series 188, ed. E. W. Guenther, B. Stecklum, and S. Klose. San Francisco: Astronomical Society of the Pacific, p. 351.

6. ESO, 1999.

SEVENTEEN: Building the Very Large Telescope

1. P. Dierickx. 1996. "All VLT primary mirror blanks delivered." ESO *Messenger,* December, p. 9.

2. E. Ettlinger, R. Giordano, and M. Schneermann. 1999. "Performance of the VLT mirror coating unit." ESO *Messenger,* September, p. 4.

3. G. Monnet and R. Bacon. 2003. "VLT/VLTI instrumentation lessons learned." ESO *Messenger,* September, p. 10.

4. O. von der Luhe, D. Bonaccini, F. Derie, B. Koehler, S. Levegne, E. Manil, A. Michel, and M. Verola. 1997. "A new plan for VLTI." ESO *Messenger,* March, p. 8.

5. B. Koehler. 1994. "Hunting the bad vibes at Paranal." ESO *Messenger,* June, p. 4.

6. B. Koehler, F. Koch, and L. Riveria. 1995. "Impact of the micro seismic activity at Paranal." ESO *Messenger,* March, p. 4.

7. J. Glancey. 2003. "Stars in their eyes." *Guardian,* June 9. http://arts.guardian.co.uk/critic/feature/0,,973537,00.html.

EIGHTEEN: The Role of ESO in Major European Astronomy Programs

1. ESO. 1999. "ESO-wide review." Collected presentations, February 1–4, 1999, books one and two. R. Giacconi Archives, Smithsonian Institution, Washington, D.C.

2. R. Giacconi. 1998. "Cost of VLT." ESO *Messenger,* September, p. 8.

3. R. Giacconi. 1998. "Role of ESO in European astronomy." ESO *Messenger,* March, pp. 1–8.

4. "Executive summary" (ESO Cou 582), "Organization and work assignment 1997–2003" (ESO Cou 583), "ESO VLTI new plan" (ESO Cou 584), "Plan for restructuring La Silla around self-management teams" (ESO Cou 585), "VLT schedule documentation" (ESO Cou 586), and "Summary long-range plan work breakdown structure for 1996–2003" (ESO Cou 587).

5. Giacconi, 1998, "Role."

6. G. Gilmore. 2002. "ESO and the UK: Why does the UK need more astronomy?" ESO *Messenger,* September, pp. 4–6.

7. R. Gilmozzi and P. Dierickx. 2000. "OWL concept study." ESO *Messenger,* June, pp. 1–10.

NINETEEN: Radio Astronomy on the Radar

1. D. S. Heeschen. 1981. "The Very Large Array." In *Telescopes for the 1980s,* ed. G. Burbidge and A. Hewitt. Annual Reviews Monograph. Palo Alto, Calif.: Annual Reviews, pp. 1–62.

TWENTY: First Loves and Last Words

1. R. A. Brown and R. Giacconi. 1987. "New directions for space astronomy." *Science* 238: 617–619.

2. H. Tananbaum and M. Weisskopf. 2001. "A general description and current status of the Chandra X-Ray Observatory." In *Astronomical Society of the Pacific Conference Proceedings,* vol. 251, ed. H. Inue and H. Krineda, pp. 4–10.

3. R. P. Kraft, W. Forman, C. Jones, A. T. Kenter, S. S. Murray, T. L. Aldcroft, M. S. Elvis, I. N. Evans, G. Fabbiano, T. Isobe, D. Jerius, M. Karovska, D.-W. Kim, A. H. Prestwich, F. A. Primini, D. A. Schwartz, E. J. Schreier, and A. A. Vikhlinin. 2000. "Chandra high resolution image of Centaurus A." *Astrophysical Journal* 531:L9–L12. R. Giacconi and the CDFS Team. 2001. "First results from the x-ray and optical survey of the Chandra Deep Field South." *Astrophysical Journal* 551:624–634. P. Rosati and the CDFS Team. 2002. "The Chandra Deep Field South: The 1 million second exposure." *Astrophysical Journal* 566:667–674.

4. P. Tozzi and the CDFS Team. 2001. "New results of x-ray and optical survey of the Chandra Deep Field South: The 300,000 seconds exposure." *Astrophysical Journal* 562:42.

5. R. Giacconi and the CDFS Team. 2002. "The Chandra Deep Field South One Million Seconds Catalog." *Astrophysical Journal Supplement* 139:369–410.

6. E. J. Schreier, A. M. Koekemoer, N. A. Grogin, R. Giacconi, R. Gilli, L. Kewley, C. Norman, G. Hasinger, P. Rosati, A. Marconi, M. Salvati, and P. Tozzi. 2001. "Hubble Space Telescope imaging in the Chandra Deep Field South: Multiple active galactic nucleus populations." *Astrophysical Journal* 560:127–138.

7. W. Zheng. 2004. "Photometric redshift of x-ray sources in the Chandra Deep Field South." *Astrophysical Journal Supplement* 155:73–87.

8. M. Paolillo, E. J. Schreier, R. Giacconi, A. M. Koekemoer, and N. A. Grogin. 2004. "Prevalence of x-ray variability in the Chandra Deep Field South." *Astrophysical Journal* 611:93–106.

9. M. Rees. 2003. *Our final hour: A scientist's warning: How terror, error, and environmental disaster threaten humankind's future in this century—On Earth and beyond.* New York: Basic Books.

Name Index

Abbe, Ernst, 45
Abell, George, 265
Academy of Sciences (Soviet Union), 38
Accademia Nazionale dei Lincei, 190–191
Advanced Mechanical and Optical Systems (AMOS), 317
Air Force Cambridge Research Laboratories (AFCRL), 43, 50, 55, 56, 59, 63
Alenia Spazio, 268
Alessandri, Arturo, 293
Alikhanian, A. I., 30
Allègre, Claude, 329, 330
Allen, Lew, 252
Allen, Ron, 246
Aller, Lawrence, 34
Allied Research Associates, 33
Allison, S. K., 115
Allison, Samuel, 35
Almanez Castilla, Jorssy, 296
American Academy of Arts and Sciences, 191
American Association for the Advancement of Science, 257
American Association of Variable Stars Observers, 238
American Astronomical Society, 190, 263

American Physical Society, 31
American Science and Engineering (AS&E), 32, 33, 35, 41, 43, 50, 51, 54–59, 60–64, 70–76, 78, 79, 92, 94, 112, 122, 126, 127, 129, 140, 141, 143, 146–147, 152–154, 156, 180–185
AMOS. *See* Advanced Mechanical and Optical Systems
Andersen, Johannes, 283, 286
Andersen, Torben, 309–310
Anderson, Carl, 24
Angel, Roger, 255, 308
Annis, Martin, 32, 33, 34, 40, 54, 182
Anouilh, Jean, 1
Ansaldo Energia, 306–307, 308
APL. *See* Applied Physics Laboratory
Apollonius of Perga, 138
Appenzeller, Imo, 312
Applied Physics Laboratory (APL), 80, 84, 86
Artaza, Mario, 293
AS&E. *See* American Science and Engineering
Association of Universities for Research in Astronomy (AURA), 204, 215–217, 221–224, 269, 274, 281, 328, 353
Associated Universities, Inc. (AUI), 196, 205, 339–341, 374

Subject Index

DECONVOLUTION, 270

Credits

Figure 1.1: figure 3 from R. Giacconi, A. Lovati, A. Mura, and C. Succi. 1956. "Osservazioni preliminary sullo sviluppo della cascata nucleare in piombo prodotta dalla radiazione cosmica a 3500 metri." *Il Supplemento del Nuovo Cimento* series X, 4:892–895. **Figure 1.2:** figure 2 from E. Fiorini, R. Giacconi, and C. Succi. 1957. "Una nuova camera di Wilson di grandi dimensioni." *Il Nuovo Cimento* series X, 6:355. **Figure 1.3:** figure 1 from R. W. Thompson, J. R. Burwell, and R. W. Huggett. 1956. "The θ^0 meson." *Il Supplemento del Nuovo Cimento* series X, 4:286–318.

Figure 2.1: figure 9 from R. W. Thompson, J. R. Burwell, and R. W. Huggett. 1956. "The θ^0 meson." *Il Supplemento del Nuovo Cimento* series X, 4:286–318. **Figure 2.2:** figure 1 from R. Giacconi, W. Blum, and G. T. Reynolds. 1959. "Detection of high-energy mu mesons by an air Cerenkov counter." *Il Nuovo Cimento* series X, 11:102–107.

Figure 3.1: figure 2 from R. Giacconi and B. B. Rossi. 1960. "A 'telescope' for soft x ray astronomy." *Journal of Geophysical Research* 65:773.

Figure 4.1: author's collection. **Figure 4.2:** figure 1.1 from R. Giacconi. 1974. "Introduction." In *X-ray astronomy,* ed. R. Giacconi and H. Gursky. Dordrecht: D. Reidel. **Figure 4.3:** figure 2.2 from H. Gursky and D. Schwartz, "Observational techniques," in Giacconi and Gursky 1974. **Figure 4.4:** R. Giacconi, H. Gursky, F. Paolini, and B. B. Rossi. 1962. "Evidence for x rays from sources outside the solar system." *Physical Review Letters* 9:442. **Figure 4.5:** figure 2 from S. Bowyer, E. T. Byram, T. A. Chubb, and H. Friedman. 1965. "Observational results of x-ray astronomy." *Annales d'Astrophysique* 28(4):791–803.

Figures 5.1 and 5.2: figures 2 and 1 from R. Giacconi and H. Gursky. September 1963. "An experimental program of extra-solar x-ray astronomy." Proposal by AS&E to NASA. Document AS&E-449. **Figures 5.3 and 5.4:** figures 6 and 7 from S. Bowyer, E. T. Byram, T. A. Chubb, and H. Friedman. 1965. "Observational results of x-ray astronomy." *Annales d'Astrophysique* 28(4):791–803. **Figure 5.5:** M. Oda, G. Clark, G. Garmire, M. Wada, R. Giacconi, H. Gursky, and J. Waters. 1965. "Angular sizes of the x-ray sources in Scorpio and Sagittarius." *Nature* 205:554.

Figures 5.6 and 5.7: H. Gursky, R. Giacconi, P. Gorenstein, J. Waters, M. Oda, H. Bradt, G. Garmire, and B. Sreekantan. 1966. "Measurement of the angular size of the x-ray source Sco X-1." *Astrophysical Journal* 146:310. **Figure 5.8:** A. R. Sandage, P. Osmer, R. Giacconi, P. Gorenstein, H. Gursky, J. R. Waters, H. Brant, G. Garmire, B. Sreekantan, M. Oda, K. Osawa, and J. Jugaku. 1966. "On the identification of Sco X-1." *Astrophysical Journal* 146:316.

Figure 6.1: figure 2 from R. Giacconi, P. Gorenstein, H. Gursky, E. M. Kellogg, and H. Tananbaum. April 1964. "An x-ray explorer to survey galactic and extragalactic sources." Proposal by AS&E to NASA. Document AS&E-2149. **Figures 6.2, 6.3, and 6.4:** pp. 73, 74, and 66 of W. Tucker and R. Giacconi, *The x-ray universe.* Cambridge, Mass.: Harvard University Press, courtesy Agenzia Spaziale Italiana (6.2 and 6.3) and American Science and Engineering (Robert Plourde) (6.4). **Figure 6.5:** figure 4 from R. Giacconi, E. M. Kellogg, P. Gorenstein, H. Gursky, and H. Tananbaum. 1971. "An x-ray scan of the galactic plane from Uhuru." *Astrophysical Journal Letters* 165:L27–L35.

Figure 7.1: W. Forman, C. Jones, L. Cominski, P. Julien, S. Murray, G. Peters, H. Tananbaum, and R. Giacconi. 1978. "The fourth Uhuru catalog," *Astrophysical Journal Supplement* 38:357–412. **Figure 7.2:** figure 3 from M. Oda, P. Gorenstein, H. Gursky, E. M. Kellogg, E. Schreier, H. Tananbaum, and R. Giacconi. 1971. "X-ray pulsations from Cygnus X-1 observed from Uhuru." *Astrophysical Journal* 166:L1. **Figures 7.3 and 7.4:** E. Schreier, R. Levinson, H. Gursky, E. M. Kellog, H. Tananbaum, and R. Giacconi. 1972. "Evidence for the binary nature of Centaurus X-3 from Uhuru x-ray observations." *Astrophysical Journal* 172:L79. **Figures 7.5, 7.7, and 7.8:** figures 7, 8, and 9 from R. Giacconi. 2002. *The dawn of x-ray astronomy.* Stockholm: Les Prix Nobel. **Figure 7.6:** American Science and Engineering original in author's collection. **Figure 7.9:** R. Rothschild, E. A. Bolt, S. S. Holt, and P. J. Serlemitzos. 1974. "Millisecond temporal structure in Cyg X-1." *Astrophysical Journal* 189:L13. **Figure 7.10:** E. M. Kellogg. 1974. "Extragalactic x-ray sources." In *X-ray astronomy,* ed. R. Giacconi and H. Gursky. Dordrecht: D. Reidel. **Figure 7.11:** D. Schwartz and H. Gursky, "The cosmic x-ray background," in Giacconi and Gursky 1974. **Figure 7.12:** T. Matilsky, H. Gursky, E. M. Kellogg, H. Tananbaum, S. Murray, and R. Giacconi. 1973. "The number intensity distribution of X-ray sources observed by Uhuru." *Astrophysical Journal* 181:753. **Plate 1:** figure 10 from Giacconi 2002, illustration by Robert Plourde, courtesy American Science and Engineering. **Plate 2:** figure 12 from Giacconi 2002, illustration by Lois Cohen, courtesy TRW.

Figures 8.1, 8.2, 8.4, 8.6, and 8.8: figures 7, 8, 14/15, 16a, 16b from R. Giacconi, W. P. Reidy, G. S. Vaiana, L. P. Van Speybroeck, and T. F. Zehnpfennig. 1969. "Grazing incidence telescopes for x-ray astronomy." *Space Science Reviews* 9(1):3–57. **Figure 8.3:** figure 7 from R. Giacconi, N. F. Harmon, R. F. Lacey, and Z. Szilagy. December 1964. "An x-ray telescope. Report by American Science and Engineering submitted to NASA." (NASA CR-41). **Figure 8.5:** figure 6 from R. Giacconi, W. P. Reidy, T. F. Zehnpfennig, J. C. Lindsay, and W. S. Muney. 1965. "Solar x-ray image obtained using grazing incidence optics." *Astrophysical Journal* 142:1274. **Figure 8.7:** cover image from G. S. Vaiana, W. P. Reidy, T. F. Zehnpfennig, and R. Giacconi. 1968. "X-ray structures of the sun during the importance 1 N flare of 8 June 1968." *Science* 161:564–567. **Figure 8.9:** frontispiece and figure XXII from F. B. Cavalieri. 1632. *Lo specchio ustorio, ovvero trattato delle sezioni coniche* [*The burning mirror, or treatise on conic sections*]. Bologna: H. H. del Dozza, from the collection of the Library of Congress. **Plate 3:** courtesy NASA/MSFC. **Plate 4:** figure 15 from R. Giacconi. 2002. *The dawn of x-ray astronomy.* Stockholm: Les Prix Nobel, courtesy L. Golub.

Figure 9.1: figure 1.5-1 from American Science and Engineering. AS&E proposal for LOXT, volumes I, II, and III. R. Giacconi Archives, Smithsonian Institution, Washington, D.C. **Figures 9.2, 9.5, and 9.7:** figures 4, 13, and 7 from R. Giacconi, P. Gorenstein, S. S. Murray, E. Schreier, F. Steward, H. Tananbaum, W. H. Tucker, and L. Van Speybroeck. 1981. "The Einstein Observatory and future x-ray telescopes." In *Telescopes for the 1980s,* ed. G. Burbidge and A. Hewitt. Annual Reviews Monograph. Palo Alto, Calif.: Annual Reviews, pp. 195–278. **Figures 9.3 and**

9.4: courtesy NASA/Marshall Space Flight Center. **Figure 9.6:** R. Giacconi, G. Branduardi, U. Briel, A. Epstein, D. Fabricant, E. Feigelson, W. Forman, P. Gorenstein, J. Grindlay, H. Gursky, F. R. Harnden, J. P. Henry, C. Jones, E. Kellogg, D. Kock, S. S. Murray, E. Schreier, F. Seward, H. Tananbaum, K. Topka, L. Van Speybroeck, S. S. Holt, R. H. Becker, E. A. Boldt, P. J. Serlemit-zoa, G. Clark, C. Canazares, T. Markert, R. Novick, D. Helfand, and K. Long. 1979. "The Einstein HEAO-2 x-ray observatory." *Astrophysical Journal* 230:540–550.

Figure 10.1: figure 1 from R. H. Becker, B. W. Smith, N. E. White, R. F. Mushotzsky, S. S. Holt, E. A. Boldt, and P. J. Serlemitsos. 1980. Letter to the editor. *Astrophysical Journal* 235:L5–L8. **Figures 10.2, 10.7, and 10.8:** figures 8 (p. 23), 1 (p. 310), and 1 (p. 324) from M. Elvis, ed. 1990. *Imaging x-ray astronomy, a decade of Einstein achievements.* Cambridge: Cambridge University Press. **Figure 10.3:** courtesy NASA and Center for Astrophysics. **Figure 10.4:** figure 7 from I. M. Gioia, T. Maccacaro, R. E. Schild, J. T. Stocke, J. W. Liebert, I. J. Danziger, D. Kinth, and J. Lub. 1984. "The medium sensitivity survey—A new sample of X-ray sources with optical iden-tifications." *Astrophysical Journal* 283:495. **Figure 10.5:** courtesy W. Forman and C. Jones, Center for Astrophysics. **Figure 10.6:** author's collection. **Table 10.1:** adapted from table 2 of W. Forman and C. Jones. 1990. "Hot gas in clusters of galaxies." In *Clusters of galaxies,* ed. W. R. Oegerle, M. J. Fitchett, and L. Danly. Cambridge: Cambridge University Press. **Plates 5 and 6:** author's collection, courtesy NASA and Center for Astrophysics.

Plate 7: courtesy J. Trümper.

All figures for Chapter 12 are courtesy NASA/AURA/STScI. **Figure 12.1:** Space Telescope Science Institute. 2005. *Yearbook.* Baltimore: Space Telescope Science Institute. **Figures 12.2, 12.3, 12.4, 12.5, 12.6, 12.7, and 12.8:** pp. 4, 6, 8, 9, 10, 3, and 11 from Space Telescope Science In-stitute. 1986. *Hubble Space Telescope handbook for amateur astronomers.* Baltimore: Space Tele-scope Science Institute. **Plate 8:** http://hubblesite.org/gallery/spacecraft/22/lg_web.jpg, courtesy NASA/MSFC.

Figure 14.1: redrawn from figure 1.1 from R. Giacconi. March 1992. "Annual report to the AURA Board of Directors." Baltimore: Space Telescope Science Institute. **Figure 14.2:** C. J. Bur-rows, J. A. Holtzman, S. M. Faber, P. Y. Bely, H. Hasan, C. R. Linds, and D. Schroeder. 1991. "The imaging performance of the Hubble Space Telescope." *Astrophysical Journal Letters* 369(2):L21–25.

All plates for Chapter 15 are courtesy NASA/AURA/STScI. **Plate 9:** http://hubblesite.org/gallery/album/the_universe_collection/pr2004007a/. **Plate 10:** http://imgsrc.hubblesite.org/hu/db/1992/27/images/b/formats/large_web.jpg. **Plate 11:** http://hubblesite.org/gallery/album/nebula_collection/pr/1995044a/. **Plate 12:** http://hubblesite.org/gallery/album/star_collection/pr199905c. **Plate 13:** http://sci.esa.int/science-e/www/object/index.cfm?fobjectid=30592&fareaid=31, courtesy NASA-ESA.

All figures and plates for Chapter 16 are courtesy ESO. **Figure 16.1:** R. Giacconi, "Executive Summary of ESO Long Range Plan 1999–2006." In European Southern Observatory. 1999. "ESO-wide review." Collected presentations, February 1–4, 1999, books one and two. R. Giacconi Archives, Smithsonian Institution, Washington, D.C. **Figure 16.2:** figure on p. 5 from R. N. Wilson. 2003. "The history and development of the ESO active optics system." ESO *Messenger,* September, pp. 2–9. **Figure 6.3:** J. Beletic, "Optical detectors." In European Southern Observa-tory. 1999. **Figure 16.4:** author's collection. **Plate 14:** www.ls.eso.org/lasilla/user-info/images/LaSilla.jpg. **Plate 15:** ESO PR photo 08/94-2.

All figures and plates for Chapter 17 are courtesy ESO. **Figures 17.1 and 17.2:** www.eso.org/projects/vlt/images/vltfv6/gif and www.eso.org/outreach/utfl/whitebook/images/wb-fig-0301.gif. **Figure 17.3:** M. Tarenghi. 1999. "Seeing—Paranal Observatory." In European Southern Observatory. 1999. "ESO-wide review." Collected presentations, February 1–4, 1999, book two. R. Giacconi Archives, Smithsonian Institution, Washington, D.C. **Figure 17.4:** figure 2 from S. Stanghellini. 1996. "The M1 cell–M3 tower of the VLT: Design overview and manu-facturing process." ESO *Messenger,* September, pp. 2–5. **Figures 17.5A and 17.5B:** figures 1 and

6 from E. Ettlinger, P. Giordano, and M. Schneermann. 1999. "Performance of the VLT mirror coating unit." ESO *Messenger,* September, pp. 4–8. **Figure 17.6:** http://eso.org/outreach/press-rel/pr-1995/phot-37-95.gif. **Figure 17.7:** T. Andersen. 1996. "VLT Systems Engineering Group moving ahead." ESO *Messenger,* September, pp. 6–8. **Figure 17.8:** http://eso.org/outreach/press-rel/pr-12-98.html. **Figure 17.9:** figure 2 from N. Devillard, Y. Jung, and J.-G. Cuby. 1999. "ISAAC pipeline data reduction." ESO *Messenger,* March, pp. 5–8. **Figure 17.10:** figure 1 from A. Glindemann, P. Ballester, B. Bauvir, E. Gugueño, M. Cantzler, S. Correia, et al. 2001. "First fringes with ANTU and MELIPAL." ESO *Messenger,* December, pp. 1–2. **Plate 16:** figure 2 from P. Dierickx. 1996. "All VLT primary mirror blanks delivered." ESO *Messenger,* December, p. 9. **Plate 17 (left):** R. Giacconi. 1995. "ESO 1993 to 2000+." ESO *Messenger,* December, pp. 18–19. **Plate 17 (right):** R. Mueller, H. Hoeness, J. Espiard, J. Paseri, and P. Dierickx. 1993. "The 8.2-m primary mirrors of the VLT." ESO *Messenger,* September, p. 6. **Plates 18 and 19:** figures 2 and 4, T. Andersen. 1996. "VLT Systems Engineering Group moving ahead." ESO *Messenger,* September, pp. 6–8. **Plate 20:** ESO PR photo 47d/98. **Plate 21:** bottom figure on p. 23 from ESO. 1999. " 'First Light' for VLT high-resolution spectrograph UVES." ESO *Messenger,* September, pp. 22–23, courtesy S. d'Odorico. **Plate 22:** figure on p. 55 from S. White. 1998. "Weighing young galaxies—An occasional observer goes to Paranal." ESO *Messenger,* December, pp. 54–55. **Plate 23:** www.eso.org/outreach/press-rel/pr-2000/phot-05c-02-preview.jpg. **Plate 24:** www.eso.org/outreach/press-rel/pr-2000/phot-05e-02-preview.jpg, courtesy M. Tarenghi.

All figures for Chapter 18 are courtesy ESO. **Figure 18.1:** M. Tarenghi, "Paranal Observatory," In European Southern Observatory. 1999. "ESO-wide review." Collected presentations, February 1–4, 1999, book two. R. Giacconi Archives, Smithsonian Institution, Washington, D.C. **Figure 18.2:** figures 1 and 2, R. Giacconi. 1998. "The cost of the VLT." ESO *Messenger,* September, pp. 8–10. **Figure 18.3:** www.eso.org/projects/owl/images/High-resolution/OWLE_1200.jpg, courtesy R. Gilmozzi.

All figures and plates for Chapter 19 are courtesy NSF/AUI/NRAO, except plate 28. **Figure 19.1:** presentation by Marian W. Pospieszalsky, May 16, 2005. **Plate 25:** www.gb.nrao.edu/gallery/gbt/main/telescope/Previews/telescopes-1866.jpg, courtesy William C. Keel. **Plate 26:** www.nrao.edu/imagegallery/php/level3.php?Id=455, courtesy photographer Kelly Gatlin, digital composite by Patricia Smiley. **Plate 27:** www.nrao.edu/imagegallery/php/level3.php?Id=549, Earth image courtesy SEAWiFS Project NASA/Goddard and ORBIMAGE. **Plate 28:** cover art, ESO *Messenger,* June 2003, p. 1, image by Herbert Zodet.

Figures 20.1 and 20.5: figures 13 and 10 from R. Giacconi and the CDFS Team. 2002. "The Chandra Deep Field South One Million Seconds Catalog." *Astrophysical Journal Supplement* 139:369–410. **Figure 20.2:** E. J. Schreier, A. M. Koekemoer, N. A. Grogin, R. Giacconi, R. Gilli, L. Kewley, C. Norman, G. Hasinger, P. Rosati, A. Marconi, M. Salvati, and P. Tozzi. 2001. "Hubble Space Telescope imaging in the Chandra Deep Field South—Multiple active galactic nucleus populations." *Astrophysical Journal* 560:127–138, courtesy E. Schreier. **Figure 20.3:** P. Rosati and the CDFS Team. 2002. "The Chandra Deep Field South the 1 million second exposure." *Astrophysical Journal* 566:667–674, courtesy P. Rosati. **Figure 20.4:** figure 8 from W. Zheng. 2004. "Photometric redshift of x-ray sources in the Chandra Deep Field South." *Astrophysical Journal Supplement* 155:73–87, courtesy W. Zheng. **Figures 20.6 and 20.7:** author's collection. **Plate 29:** http://chandra.harvard.edu/resources/illustrations/chandra_earth.html, courtesy NASA / MSFC / Chandra X-Ray Observatory. **Plate 30:** http://chandra.harvard.edu/photo/0052/index.html, courtesy NASA / Chandra X-Ray Observatory. **Plate 31:** http://chandra.harvard.edu/photo/2001/0157blue/index.html, courtesy NASA / Chandra X-Ray Observatory and Kraft et al. **Plate 32:** figure 20 from R. Giacconi. 2002. *The dawn of x-ray astronomy.* Stockholm: Les Prix Nobel, courtesy G. Hasinger. **Plate 33:** R. Giacconi and the CDFS Team. 2002. "The Chandra Deep Field South One Million Seconds Catalog." *Astrophysical Journal Supplement* 139:369–410. **Plate 34:** author's collection.